WASTEWATER TREATMENT BY A NATURAL WETLAND:
THE NAKIVUBO SWAMP, UGANDA
PROCESSES AND IMPLICATIONS

Wastewater Treatment by a Natural Wetland: The Nakivubo Swamp, Uganda
Processes and Implications

DISSERTATION
Submitted in fulfillment of the requirements of
the Board of Deans of Wageningen Agricultural University
and the Academic Board of the International Institute for Infrastructural,
Hydraulic and Environmental Engineering for the Degree of DOCTOR
to be defended in public
on Thursday, 20 May 1999 at 10:00 h in Delft, the Netherlands

by
FRANK KANSIIME
born in Bushenyi District, Uganda

and
MAIMUNA NALUBEGA
born in Kampala, Uganda

A.A. BALKEMA / ROTTERDAM / BROOKFIELD / 1999

This dissertation has been approved by the promoters:

Dr G.J. Alaerts	Professor in Sanitary Engineering (IHE)
Dr P. Denny	Professor in Wetland and Aquatic Ecology (IHE)
Dr L. Lijklema	Professor in Aquatic Ecology and Water Quality Management (LUW)

Co-promoter:

Dr J.J.A. van Bruggen	Senior lecturer at the Department of Environmental Science and Water Resources Management (IHE)

Published by
A.A. Balkema, P.O. Box 1675, 3000 BR Rotterdam, Netherlands
Fax: +31.10.4135947; E-mail: balkema@balkema.nl; Internet site: http://www.balkema.nl

A.A. Balkema Publishers, Old Post Road, Brookfield, VT 05036-9704, USA
Fax: 802.276.3837; E-mail: info@ashgate.com

ISBN 90 5410 420 1

Printed in the Netherlands

Contents

Acknowledgment

First we are very grateful to the Dutch Government who through the International Institute for Infrastructural Hydraulic and Environmental Engineering (IHE) and Wageningen Agricultural University (WAU), in The Netherlands, financed this study, under the Ecotechnology Project. This project also involved Makerere University and Fisheries Research Institute (FIRI), both in Uganda.

We are particularly indebted to our promoters: Professors Alaerts, G. J., Denny, P., and Lijklema, L. and Co-promoters: Dr. Bugenyi, F.W.B. and Dr. van Bruggen, J. J. A., for their constructive discussions and advices both during the Research work and write up of the thesis. We are very grateful to Aalderink, R.H. for his kind assistance and contribution throughout this research.

Additional thanks go to Azza, N. G. T.; Berhane, M.; Kaarsemaker, S.; Omoit, S.; Oosthoek, J.; Otoi, S. P.; Sloot, J. and Van der Veen, M., who participated in the data collection.

Our sincere thanks also go to Dr. de Ruyter, E and Dr. Tukahirwa, E. for coordinating the project activities in the Netherlands and Uganda respectively. We would also like to thank Schakenraad, R., Melis, F. and Personnel in their section for administrative arrangements.

Thanks also go to Dr. Kiggundu, Department of Civil Engineering Makerere University for allowing us to use the Public Health Engineering Laboratory in his Department. At this juncture, we also thank members of the Laboratory who in one way or another made our work a success.

We extend further thanks to Vermaat, J. Hootsmans, M., Jacobs, A.A.J., (for statistical advice), Lubberding, H., Kelderman, P. and de Laat, P.J.M., who read some chapters of this thesis and made constructive suggestions and Dr. Hooijmans, T., for her efforts during the first part of the study. We are also grateful to the Computer Group and Laboratory staff at IHE for their technical and moral support.

We also acknowledge and thank our field assistants, who made sure that both we and our supervisors never did our research while 'swimming' under the swamp mat. To the residents of Bukasa Village and the wardens of Luzira Prison who never denied us access to the Nakivubo Swamp during data collection, we say thank you.

Lastly, our heartfelt thanks go to our parents who have always prayed for our success.

Abstract

An investigation to assess the capacity of the Nakivubo swamp, Kampala-Uganda (which has been receiving partially treated sewage from the city for more than 30 years now), to remove nutrients and pathogens was carried out. The aim of the study was to evaluate the potential of this swamp to remove nutrients and pathogens from wastewater in a sustainable way, with emphasis on describing and quantifying their pathways, transformations and budgets.

From field studies, water balance terms of channel discharges, rainfall, subsurface flows, evapotranspiration and seiches were measured or calculated from existing hydrometeorological data to form a water balance. Nutrients (N and P) and faecal coliforms (FC) transformations in the swamp were studied from four transects cut across the swamp. Vertical and longitudinal profiles of nutrients and pathogens were also constructed. Laboratory simulations were carried out to estimate nutrient fluxes into the plant and sediment compartments and to estimate the removal mechanisms of FC from the water column.

In this study differences in the morphological, hydraulic, physico-chemical, floristic and overall wastewater treatment performance between areas covered by the two major vegetation types *Cyperus papyrus* L. and *Miscanthidium violaceum* Robyns (about 80% and 20% of the study area, respectively) were elucidated. Papyrus is emergent at the swamp edges where the water level is more affected by the seasons (rainfall). It floats towards the centre and closer to the lake. The loose rhizomatous raft over which papyrus floats allows for fairly free fall-through of plant debris and decomposing matter onto the sediment via the water column resulting in high suspended solids content in the underlying water. This possibly slows, and sometimes restricts water flow in some areas. Due to the lower flows closer to the edges, a thick (up to 60 cm) layer of peaty material is also formed. The loose mat facilitates vertical mixing between the interstitial mat water and the water beneath the mat during the rise and fall of water/mat levels. This lead to a less steep gradient of nutrients over the vertical profile and facilitates nutrient uptake from the water column by papyrus vegetation.

In comparison, *Miscanthidium* vegetation is restricted to the middle of the swamp and is characterised by a thick (0.9 to 1.6 m) mat with highly interlaced roots, but low bulk density (60 - 300 kg/m^3, surface to bottom). The thick mat helps the retention of falling plant debris on to its surface, where low rate decomposition and further mat accretion take place. The combination of material retention onto the mat surface and high water flows beneath results into a clearer water column and a very thin peat layer (maximum 10 cm) of poorly decomposed plant material. Further, the mat structure prevents free vertical and lateral mixing of the mat water with the water column beneath. This leads to reduced interactions of the plants with wastewater in these zones, and therefore less nutrient abstraction by plants from the wastewater in these zones.

The average waste water discharge in the swamp was estimated at 103,575 m^3/d. Water flow is highly channelised and hydraulic retention times in the swamp during the rainy periods may be as low as 18 hours. Seepage is negligible. Water quality variations within the swamp showed that wastewater is not evenly transported to all parts of the swamp as it flows through.

The nutrient load into the swamp was 770 $gN/m^2/yr$ and 66 $gP/m^2/yr$. Different nutrient uptake rates and plant tissue contents (N=1.3%, P=0.21% for papyrus and N=0.64% and P=0.15% for the *Miscanthidium* vegetated zones) plus the above structural differences in flows and retention times are partly responsible for the disparate purification efficiencies between the vegetation zones. In the papyrus vegetated zones, the average purification efficiencies were 67% N, and TP and 99.3% FC while in the *Miscanthidium* vegetated zones, it was lower at 55% N, 33% TP and 89.3% FC. The lower flows (about 20%) that went through the papyrus vegetated zones enabled higher retention times for these zones. The major mechanisms of nutrient removal in papyrus vegetated zones were identified to be plant uptake for the nutrients and attachment onto particulates followed by sedimentation, for FC and P. Predation and natural die-off of FC may be high especially in the root zones where micro-aerobic zones exist (mostly in papyrus zones).

The thick mat of *Miscanthidium* limits the number of live roots that can reach the water column to get nutrients from there. Since the bulk (80% near the lake) of the wastewater goes through this zone, then it means that the overall (swamp-wide) nutrient and pathogen removal efficiency from the wastewater is low (56% N, 40% TP and 91% FC).

Very low levels of oxygen were observed in the Nakivubo swamp (and very infrequently) due to the high oxygen demand exerted by decomposing organic matter in the swamp. Mostly, either hypoxic or anoxic conditions existed in most compartments of the swamp limiting nitrification although most physical and chemical variables were the range that would favour the survival of nitrifying bacteria. In the *Miscanthidium* mat, the low pH also possibly limited the viability and the activity of the nitrifiers in this zone.

The sharp decline in the concentration of pollutants from the swamp interface to the open waters of the Inner Murchison Bay can be explained by mixing and dilution in the lake. Combined effects of solar radiation, temperature, pH, biocides and the grazing protozoa may also be responsible for the lower FC numbers.

To protect the swamp and use it sustainably, efforts should not only concentrate on halting reclamation but also reducing the loads of effluents/pollutants being discharged into the swamp. Distribution of water over the large expanse of the upper and lower Nakivubo swamps in addition to creating a supplementary buffer system in the form of a forest wetland in the upper are suggested as the best sustainable management options. This should be supplemented with a proper wastewater collection and treatment to at least secondary level within the city.

Chapter 1

Introduction

Kansiime, F. and Nalubega, M.

1.1 Global wastewater problems

Across countries, water is being put to various uses: drinking, personal hygiene, fisheries, navigation, industrial production, recreational activities, etc. On the other hand, water has been considered, since ancient times, as the most suitable medium to cleanse oneself and dispose of wastes. The wastes subsequently become distributed throughout the water body presenting a risk to downstream water users. This is the more problematic if the water body is at the same time used as a source of drinking water.

In industrialised countries, faecal contamination of water supply sources caused serious health problems, such as typhoid and cholera in large cities until in the mid 1800's. By that time, cities in Europe and North America began constructing sewer networks to route domestic wastes further away downstream of water intakes (UNEP, 1995). This led to a significant reduction of water-borne diseases. In developing countries, the situation is still in a state of despair, as in addition to eutrophication of surface water and their oxygen depletion, water-borne diseases are occurring with alarming frequencies, claiming millions of lives every year (Bartram and Ballance, 1996).

Pollutants in wastewater include: (i) organic substances which can be degraded by bacteria using dissolved oxygen in the water and rendering it unfit for human and animal use. The anoxic conditions created by depleting the dissolved oxygen in water may result in the death of higher forms of aquatic life like fish; (ii) excessive concentration of nutrients (N and P) which may stimulate growth of plants and notably algae; when in excess, these nutrients result in eutrophication of water bodies; (iii) pathogenic (disease causing) microorganisms from human and animal excreta and human wastewater; and (iv) heavy metals and organic micro-pollutants which may be toxic to plants, animals and humans.

The combination of rapid population growth, industrialisation and associated urbanisation resulted in increased wastewater generation. This necessitates treatment and disposal of the generated wastewater in a safe way in order to minimise environmental pollution and prevent transmission of water-borne diseases. Treatment is necessary to remove oxygen depleting substances, the major nutrients (nitrogen and phosphorus) as well as pathogenic bacteria.

1.2 Wastewater problems in developing countries

The links between water quality and disease have been described in much detail (Feachem *et al.*, 1983). The existence of these links is now generally recognized: so has the importance of the provision of safe drinking water to the population. This recognition is largely due to the impact of the United Nations International Drinking Water Supply and Sanitation Decade (IDWSSD), 1981 - 1990. During this period, there were accelerated and concerted efforts by the governments of the world=s developing countries to expand water supply and sanitation services to deprived populations. This led to considerable progress on the implementation side as shown in Table 1.

Table 1. Water supply and sanitation coverage in developing countries showing portions of population (in percentages) with access to adequate facilities [1]

	Year		
	1980	1990	1996
Urban water supply	75 (66)	82 (67)	84 (64)
Rural water supply	34 (22)	51 (35)	60 (37)
Urban sanitation	60 (54)	60 (65)	65 (55)
Rural sanitation	31 (20)	34 (23)	35 (24)

[1] values between brackets are for Africa. Source: WHO- (1992, 1996).

The percentage of people in developing countries provided with water supply and sanitation facilities increased. However, growth in sanitation was much slower than that for water supply. This, as suggested by Cairncross (1992), is because any form of sanitation is relatively more expensive than its water supply equivalent. The main water supply and sanitation facilities remain concentrated in urban areas, though most people live in rural areas.

Although the water supplies are constantly being upgraded, urban sanitation is not well developed, even in the largest cities in developing countries. Many cities and towns have no sewerage at all. One of the major problems is the very large investment required to construct it. Programmes for improving the water supply to overpopulated cities have gone ahead while the communities continue to depend on traditional means of waste disposal. This policy of continuing to provide for ever-increasing demand for water without making an improvement in the sanitation system increases the risk of major epidemics and reduces the general quality of urban life (Pacey, 1978).

Since independence in the 1960's most governments in the developing countries (and Africa in particular), were the main providers of improved water supply and sanitation facilities, especially in rural areas. They adopted the socialist stance whereby the state provided the basic services. The rationale behind this was that the communities were poor and could not afford to develop water and sanitation facilities. With frequent rises in inflation in most countries and reduction in western support, the water sector management has weakened (Wood, 1994). Other factors which have hindered the sector's progress include investments focused on high cost technology and absence of a development strategy to maintain the few technologies that have already been installed (Luong *et al.*, 1994). This was the case during the United Nations

IDWSSD whereby governments could provide up to 64% of the input costs, though emphasis was put on construction; maintenance of new systems was not prepared whereas the older ones were neglected, posing a potential danger to environment and public health. The figure of 40% of the systems not working has often been reported for Africa (Wood, 1994). Currently the funding of the water sectors by the central governments in Africa is less than 10% of investment costs, the rest coming from donor agencies.

Though most people in Africa live in rural areas (80% of the pop.), water supply and sanitation facilities have been concentrated in urban areas (Wood, 1994). Even organisations like the World Bank, which are supporting water supply and sanitation projects in developing countries, are concentrating on cities and major towns (Howard-Humphrey and Partners, 1980, Haskoning and M-Consult, 1989; Gauff and Parkman, 1989). This has resulted in the concentration of most of the water and sanitation programmes in urban areas. This creates an imbalance of water supply and sanitation between urban and rural communities.

Poor sanitation conditions in developing countries like Africa are related to the lack of a linkage between sanitation and public health. To improve public health, sanitation is as important as provision of drinking water. Even the UN IDWSSD put most emphasis on supplying drinking water and as consequence public health did not improve significantly (Alaerts *et al.*, 1993). This resulted in limited public health benefits from investments made in water supply schemes.

Access to sufficient safe water and adequate sanitation facilities do not ensure an automatic improvement in public health. Health benefits of clean water supply and adequate sanitation come only through proper functioning and use of these facilities. This often also requires improved hygiene behaviour which can only be achieved through communication, like health education. For instance, despite the installation of a new safe communal water source, people may still return to their sometimes closer but polluted water sources instead of paying the low fees for water (Neidrum, 1994). This is usually the case when there is no community participation during the installation of the new water supply or sanitation facility and people feel that they are not part of the programme and are being bypassed in the decision making.

1.3 Wastewater treatment in developing countries

In most urban and semi-urban areas of developing countries, wastewater treatment is through conventional methods. These require considerable input of electro-mechanical installations, energy, chemicals and skilled labour resulting in high investment, operation and maintenance costs. In the event that municipal authorities cannot meet these costs owing to financial and other restrictions, as is often the case in developing countries, the treatment plants deteriorate. This often leads to an eventual breakdown some years after connection.

Most of the conventional wastewater treatment systems are primarily directed to satisfy the needs of cities and the wealthy class rather than the smaller rural units and people who live at subsistence level or who have little economic resources. The high per capita costs of sewage collection and treatment put the high-technology treatment systems beyond the reach of most, if not all, rural population. In such circumstances other solutions must be sought. This calls for technologies which can be adapted to given social, cultural, political, religious and functional conditions of a particular situation and are based on low investment, organisational and operational costs. In addition, this technology should benefit as many people as possible, be adaptable to the changing needs of the community and should not conflict with the local ecology. These characteristics play a significant role in the acceptability of a given technology to the community (Marks, 1993).

On site sanitation, and waste stabilisation ponds seem to be the preferred methods (mimicking natural processes) of wastewater treatment in many tropical countries (Mara *et al.*, 1992) especially for towns, institutions and hospitals. These systems have been identified as the most appropriate treatment option not only because of the low energy requirements but also due to the sustainability of their operation and maintenance. Despite the advantages, these systems are often neglected, not maintained or rehabilitated and hence may be inefficient. Furthermore, large areas of land are required and they might not be suitable for small communities like individual households and farms.

The great majority of houses in Africa especially in peri-urban areas or slums, which are not connected to sewerage systems, have on-site sewage disposal, mostly pit latrines or septic tanks. Pit latrines are used by 80% of these households (Muller *et al.*, 1994). The residents of these areas have usually migrated from the rural to the urban areas hoping for jobs and better services. However, the cities and towns lack the technical and financial capacity to set up the infrastructure to cope with this influx. In addition, the migrants' problems are compounded by the fact that they lack the necessary skills to secure good jobs. The effect is that they are eventually marginalised to the slums and informal settlements which are substandard, unplanned and lack the necessary infrastructure for basic services.

A technology for wastewater treatment which is not well exploited in developing countries is the use of wetlands (natural or constructed), though in these countries natural wetlands have long been used for wastewater disposal (Chale, 1987; Taylor, 1991; Osborne and Totome, 1994). The use of wetlands for wastewater treatment is well established in Europe and in the United States. These systems are already being installed in South Africa (Wood, 1990), and the technology is picking up in other African countries like Kenya and Uganda (Denny, 1997). Wetlands are potentially cost effective systems for purifying wastewater because in comparison with conventional treatment systems, they have comparatively low investment, operational and maintenance costs as well as low skilled labour and energy requirements. However, like stabilisation ponds, more land surface is required and this may be expensive where wetlands are not already existing. Additional disadvantages are that wetlands can harbour pests and act as breeding grounds for mosquitoes. In addition to wastewater treatment wetlands have other uses like biomass utilisation and acting as a habitat of fauna and flora.

Although much research has been devoted to understand the physical, biological, and chemical processes in constructed wetlands (Hammer, 1989; Cooper and Findlater, 1990; Bavor and Mitchell, 1994), the use of wetlands for nutrient removal remains controversial, despite the large international research effort on the topic. Natural wetlands are characterized by extensive variability in the type, system and functioning, which makes it difficult to extrapolate from one experience, and to predict their response to wastewater application and to translate results from one geographic region to another (Hammer and Knight, 1994; Kadlec and Knight, 1996). Most of the information so far on the use wetlands for wastewater treatment is on temperate systems and little information is available on tropical systems.

1.4 Wastewater problems in Uganda

The status of water management in Uganda is closely related to the water supply level and economic wealth of a given area. Whereas about 51% of the urban population is supplied with tap water and 11.5% gets well water, only 3.4% of the rural population has piped water, with 6.3% getting their water from unprotected wells (SOE, 1994). This water supply coverage means that only a small percentage of the population can use water-borne sanitation. The bulk of Ugandans, especially in peri-urban and rural areas, employ on-site sanitation, mostly in the form of pit latrines.

The National Water and Sewerage Corporation (NWSC), established in 1972, is responsible for developing and operating urban water supply and sanitation systems. It is presently responsible for nine towns, namely Kampala, Jinja, Entebbe, Masaka, Mbarara, Mbale, Tororo, Gulu and Lira. Even then, only about 5 - 10% of these urban areas are covered by sewerage systems. Only Kampala (the capital city) and Masaka town are using conventional treatment systems (SOE, 1994). Other towns (Entebbe, Jinja, Mbale, Mbarara and Tororo) are using waste stabilisation ponds and these systems are recognized by the government (NWSC) for being the most appropriate water treatment technology.

1.5 Wetlands and wastewater treatment in Uganda

Wetlands cover about 10 - 13% of Uganda=s total land surface area and provide a wide variety of biophysical and social economic functions (Ministry of Natural Resources, 1995). Wetlands are the main sources of drinking water for most rural communities in Uganda. However, with urbanisation and industrialisation proceeding in towns, wetlands are increasingly becoming recipients of wastewater (SOE, 1994). A number of the wastewater treatment plants discharge their effluent into wetlands, e.g., Kampala, Jinja and Masaka.

1.6 Wetlands and water quality of Lake Victoria

Lake Victoria, shared by Uganda, Kenya and Tanzania is the largest lake in Africa with a surface area of 68,800 km^2 (Crul, 1995). The lake shoreline is long and convoluted and has small bays and wetlands along the coasts. Many of the wetlands bordering the shores of the lake are being encroached upon and are used as disposal sites for different wastes (Kampala, Jinja and Entebbe, Kisumu and Mwanza). A substantial portion of the urban and industrial growth has taken place along the shores of Lake Victoria and this trend is expected to continue.

The lake is the main fresh water resource for the people living in the vicinity and along the river Nile, who depend on it for fish and drinking water among other uses. The biology and chemistry of the lake have changed profoundly through the last decades as evidenced by the reduction of biodiversity, increase in chlorophyll concentrations and nutrient levels (Ochumba and Kibaara, 1989; Hecky 1993; Mugidde, 1993). The wetlands around the lake form an important buffering ecotone which stores flood water temporarily, filters and traps nutrients and sediments from the catchment area.

The latest ecological change in Lake Victoria is the infestation of the lake by the water hyacinth (*Eichhornia crassipes*) which is spreading at a very fast rate. The presence of the water hyacinth may cause further deterioration of the water quality, especially along the coasts, reduction in biodiversity and in fish catches and will affect navigation. Hypotheses postulated to explain these changes include increased nutrient inputs, changes in the lakes' trophic structure and climate (Hecky, 1993) and degradation of wetlands (Bugenyi, 1993). These changes may be due partly to human activities in the lake basin, which have polluted the lake. In Mwanza (Tanzania), the gold rush puts additional stress on the environment, with discharges of heavy metals such as mercury into the lake. Kisumu town in Kenya discharges its wastewater into Nyanza Gulf, Lake Victoria (Crul, 1995). In Uganda Kampala, Jinja and Entebbe areas discharge domestic sewage, industrial wastewater and other effluents into the ecotones bordering Lake Victoria. The rapid population growth in the riparian states Kenya, Tanzania and Uganda will continue to exert pressure on natural resources and environmental water hyacinth may cause further deterioration of the water quality, especially along the coasts, reduction in biodiversity and in fish catches and will affect navigation. Hypotheses postulated to explain these changes include increased nutrient inputs, changes in the lakes' trophic structure and climate (Hecky, 1993) and degradation of wetlands (Bugenyi, 1993). These changes may be due partly to human activities in the lake basin, which have polluted the lake. In Mwanza (Tanzania), the gold rush puts additional stress on the environment, with discharges of heavy metals such as mercury into the lake. Kisumu town in Kenya discharges its wastewater into Nyanza Gulf, Lake Victoria (Crul, 1995). In Uganda Kampala, Jinja and Entebbe areas discharge domestic sewage, industrial wastewater and other effluents into the ecotones bordering Lake Victoria. The rapid population growth in the riparian states Kenya, Tanzania and Uganda will continue to exert pressure on natural resources and environmental conditions of Lake Victoria. Changed land use, with intensified agriculture and livestock husbandry, expanding industry etc. will contribute to the ongoing deterioration of the Lake's water quality and its changed ecological conditions.

1.7 Problem statement and significance of the study

The Nakivubo swamp, which separates the city of Kampala and the Inner Murchison Bay of Lake Victoria, has been receiving secondary treated effluent from the Bugolobi sewage treatment works and run-off from Nakivubo channel for more than 30 years now. Also, raw sewage from Luzira prisons is discharged into the Nakivubo swamp. As a result, information on water quality in the swamp and Inner Murchison Bay is needed for the planning of the drinking water supply of Kampala and for the control of eutrophication in the Bay and Lake Victoria in general. It is currently assumed that the Nakivubo swamp retains the nutrients and pathogens carried with the wastewater, but there is no quantification of this function. Such information is necessary for the efficient planning of long-term sustainable use of the Nakivubo swamp. The Kampala city water supply was expanded in 1991, with the intake continuing to be at Gaba in the Inner Murchison Bay which is downstream from both diffuse and point source sewage discharges from Kampala. There is a possibility that nutrients (N and P) reach the lake and cause eutrophication. The eutrophication of the Inner Murchison Bay, could result in the clogging of the filters at Gaba water works by algae and the wider nutrient enrichment in Lake Victoria thus 'feeding' *Eichhornia*. Pathogens might be transported to the water works posing a threat to public health.

Previous studies related to the Nakivubo swamp, using the 'in-out' (black box) approach concentrated on the wastewater purification capability of Nakivubo swamp (Kizito, 1986; Taylor, 1991) and the possible impact of the discharge of pollutants on the operations of the concerns include eutrophication of Lake Victoria, of which the growth of *Eichhornia* along the edge of the Inner Murchison Bay is an indicator. There is a potential of pathogens to reach the water intake of the city of Kampala at Gaba, about 4 km south of the swamp, under favourable conditions. Because of the uncertainty regarding the functioning of the Nakivubo swamp and the lack of information on the processes taking place therein, this study was carried out to address some of the unanswered questions.

1.8 Aims of this thesis

Prior to this study, the Nakivubo swamp was assumed to be removing pollutants from wastewater discharged into it, thus protecting the water supply for Kampala located 4 km south east at Gaba water works. However, this function is not quantified. Therefore this study was aimed at (1) evaluating the potential of the existing natural wetland downstream of Kampala to remove nutrients and pathogens from wastewater discharged into it in a sustainable way; (2) describing the pathways, conversions, sinks and budgets of nitrogen, phosphorus and pathogens; and (3) suggesting options for sustainable management of the wetland, including the potential of the swamp to be used in a more optimal way for wastewater treatment while maintaining ecological and biodiversity quality. Heavy metals and micro pollutants nor oils, detergents etc. were looked at in this study as the priority was given to oxygen concentration, nutrients and pathogens.

1.9 Scope of the thesis

This thesis is organised such that it highlights the fundamental ecological and hydrological principles and processes determining the functioning of the Nakivubo swamp, a natural tropical floating swamp, as a treatment system for wastewater.

Chapter 2 gives a description of Nakivubo swamp with specific reference to geomorphological, hydrological and floristic functions of the swamp. The coexistence of floating papyrus and *Miscanthidium* vegetation is emphasised in view of the role of the swamp to treat wastewater.

Chapter 3 gives a full account of the key hydraulic and hydrological flow characteristics and water balance of the Nakivubo swamp. The effects of seiches due the interactions between the lake and the swamp on the water balance and its influence on the functioning of the whole swamp are described.

The wastewater flow paths and the water quality of Nakivubo swamp and the emerging difference between the functioning of papyrus and *Miscanthidium* floating vegetation mats are presented in Chapter 4.

The functioning of the Nakivubo swamp with respect to wastewater treatment is generally thought to be related to the aquatic macrophytes therein. Chapter 5 describes the role of dominant aquatic macrophytes (papyrus and *Miscanthidium*) of the swamp in nutrient removal. The influence of the floating mats of these macrophytes on the functioning of the swamp is also described.

The role of Nakivubo swamp in retaining pathogens and the possible contamination of the water intake at Gaba water works is described in Chapter 6. The possible transport of pathogens to the water works is discussed.

Chapter 7 outlines the role of the nitrification process in the transformation of nitrogen in the wastewater discharged into the swamp and its eventual loss from the swamp through denitrification. Microbial plant associations and the role of vegetation were given special attention. The role of lake seiches and aquatic macrophytes in supplying oxygen into the swamp is discussed.

Unlike nitrogen, phosphorus does not have a gaseous form. This means that it can only be transformed from one form into another and retained within the swamp. Apart from nutrient removal through biomass harvest, sediments act as major sinks of nutrients, especially phosphorus. This is described in chapter 8.

For a better understanding the retention of nutrients and pathogens in the Nakivubo swamp, a mass balance approach was used (Chapter 9).

Nakivubo swamp in addition to acting as recipient for wastewater, has other anthropogenic uses. These include agriculture, biomass utilisation and protection of biodiversity, all which are presented in Chapter 10.

Chapter 11 provides a general discussion of results in other chapters and considers management issues of the Nakivubo swamp in relation to the sustainable use of the swamp for wastewater treatment and other anthropogenic uses.

2.0 References

Alaerts, G. J., Veenstra, S., Bentvelsen, M. and van Duijl, L. A. (1993). Feasibility of anaerobic sewage treatment in Sanitation strategies in developing countries. *Wat. Sci. Tech. 27:* 179 - 186.

Bartram, J. and Ballance, R. (1996). *Water quality monitoring: a practical guide to the design and implementation of fresh water quality studies and monitoring programmes*. UNEP, WHO Publication, E&FN Spon, Chapman and Hall, London.

Bavor, H. J. and Mitchell, D. S . (eds.) (1994). Wetland systems in water pollution control. *Wat. Sci. Tech. 29*: (4)

Bugenyi, F.B.W. (1993). Some considerations of the functioning of tropical riparian ecotones. *Hydrobilogia*, *25*: 33 - 38.

Cairncross, S. (1992). *Sanitation and water supply: practical lessons from the decade*. The International Bank for Reconstruction and development / The world Bank, Washington DC.

Chale, F. M. M. (1987). Plant biomass and nutrient levels of a tropical macrophyte (*Cyperus papyrus* L.) receiving domestic wastewater. *Hydrobiol. Bull.* 21*:* 167 - 170.

Cooper P. F. and Findlater, B. C. (1990). *Constructed wetlands for water pollution control.* Pergamon Press, Oxford.

Crul, C. M. R. (1995). *Limnology and Hydrobiology of Lake Victoria. Monographs of the African Great Lakes*. UNESCO/IHP - IV Project M- 5.1.

Denny, P. (1989). *Wetlands*. In: UNEP. Report IX; Strategic planning in Uganda.

Denny, P. (1997). Implementation of constructed wetlands in developing countries. *Wat. Sci. Tech. 35*: 27 - 34.

Faechem, R. G. Bradley, D. J. Garelick, H and Mara D. D. (1983). *Sanitation and disease: health aspects of excreta and wastewater management. Studies in water supply and sanitation 3*. The World bank, Washington, D. C.

Gauff/Parkman, (1983). *Water quality surveys. Gaba, Entebbe and Jinja. Update of the WHO/ Resource studies report C/D. Lake Investigations studies report `B'. Water quality*. Report prepared for the Ministry of Water and Mineral Development and National Water and Sewerage Corporation; Kampala - Uganda.

Gauff, H. P. (1988). *Seven town water supply and sewerage expansion project. Water quality studies*. Report prepared for the Ministry of Water and Mineral Development, and National water and Sewerage Corporation; Kampala - Uganda.

Hammer, D. A. (1989). *Constructed wetlands for wastewater treatment*. Lewis Publishers, Inc. Michigan.

Hammer, D. A. and Knight, R. L. (1994). Designing constructed wetlands for nitrogen removal. *Wat. Sci. Tech. 29*: (4) 15 - 27.

Haskoning and M-Konsult (1989). *Study of solid management and pollution caused by sewerage in Dar es Salaam.* World Bank report.

Hecky. R. E. (1993). The eutrophication of Lake Victoria. *Int. Ver. Theor. Angew. Limnol Verh. 25:* 39-48.

Howard Humphrey and Partners (1980). *Dar es Salaam sewerage and sanitation study, master plan*. World Bank report.

Kadlec, R.H. and Knight, R. L. (1996). *Treatment wetlands*. Lewis, Publishers, Boca, Raton.

Kizito, S. (1986). *The evaluation of pollution levels in Nakivubo Channel, Kampala- Uganda.* M.Sc. Thesis, Makerere University, Uganda.

Luong, T. V., Njau, F. and Kahesa, A. Y. (1994). Towards self management: water and sanitation. *In:* Pickford *et al.*, (eds.). *Water, sanitation, environment and development. Selected papers of the 19th WEDC conference*. Accra, Ghana, 1993. pp 9 - 11.

Marks, R. F. 1993). Appropriate sanitation options for Southern Africa. *Wat. Sci. Tech. 27:* 1-10.

Mara, D. D., Alabaster, G. P., Pearson, H. W. and Mills, S.W. (1992). *Waste stabilisation Ponds: a design manual for eastern Africa.* Lagoon Technology International Leeds, England.

Ministry of Natural Resources (1995). National Policy for the Conservation and Management Wetland Resources. Kampala, Uganda.

Mugidde, R. (1993). The increase in phytoplankton productivity and biomass in Lake Victoria (Uganda). *Ver. Intern. Verein. Limnol. 25:* 846-849.

Muller, M., Kirango, J. and Rijnsburger, J. (1994). An alternative pit latrine emptying system. In: Pickford *et al.*, (eds.). *Water, sanitation, environment and development.* Selected papers of the 19th WEDC conference, Accra, Ghana, 1993. pp 101 - 103.

Neidrum, S. (1994). The need for hygiene education. *In*: Pickford *et al.*, (eds.). *Water, sanitation, environment and development.* Selected papers of the 19th WEDC conference, Accra, Ghana, 1993. pp 52 - 54.

Ochumba, P. B. O. and Kibaara, D. I. (1989). Observations on blue green algal blooms in the open waters of Lake Victoria, Kenya. *Afr. J. Ecol. 27:* 23-34.

Osborne, P. L. and Totome, R .G. (1994). Long-term impacts of sewage disposal on a tropical wetland. *Wat. Sci. Tech. 29*: (4) 111 - 117.

Pacey, A. (1978). *Sanitation in developing countries.* John Wiley & sons Ltd, Chichester.

SOE (1994). *State of Environment for Uganda.* Ministry of Natural Resources, Kampala - Uganda.

Taylor, A. R. D. (1991). *Status of Nakivubo and Luzira swamps in Kampala.* Report prepared for the Ministry of Environment Protection, National Wetlands Conservation and Management programme, Kampala- Uganda.

UNEP, (1995). Water Quality of World River Basins. UNEP Environment Library No 14. Nairobi.

WHO/Resources Group London (1970). *Master plans for Water Supply and Sewerage for greater Kampala and Jinja Areas (1970 - 2000). Special Studies report ``B'' Water quality. Special studies Reports ``C'' and ``D'' Lake investigations.*

WHO (1992). *The drinking water supply and sanitation decade. Review of the mid-decade progress (as of December 1990)*. CWS unit, Division of Environmental Health, WHO, Geneva. WHO (1996). *Water Supply and Sanitation Sector Monitoring report (as of December 1994)*. WHO, Water supply and sanitation collaborative council and UNICEF, Geneva.

Wood, A. (1990). The application of artificial wetlands in South Africa. *In:* Cooper P. F. and Findlater, B. C. (eds). *Constructed wetlands for water pollution control.* Pergamon Press, Oxford. pp 235 - 244.

Wood, M. (1994). Privatisation of rural water supply. *In:* Pickford *et al.*, (eds.). *Water, sanitation, environment and development. Selected papers of the 19th WEDC conference*. Accra, Ghana, 1993. pp 76 - 78.

Chapter 2

Description of the Nakivubo swamp

Nalubega, M. and Kansiime, F.

2.1 Introduction

The Nakivubo swamp in Uganda is one of the many shallow drowned valley swamps that occupy the northern fringes of Lake Victoria which is the second largest freshwater lake in the world: 69,400 km^2. It is predominantly a sedge (papyrus) and grass (*Miscanthidium*), rheotropic (fed by surface water system), lacustrine swamp, on the outskirts of Kampala (Fig. 2.1). Although Nakivubo and other swamps in the Lake Victoria basin were originally valley swamps, tectonic activity and lake level rises flooded the valleys (Welcomme, 1966) which resulted in these swamps becoming valley swamps colonised by wetland macrophytes.

The factors that normally control the types and distribution of swamps include water depth and its seasonal variability and the nutrient status of the water (Howell *et al.*, 1988). These, in turn, depend on the geomorphology and the local or regional climate of the area (Patten, 1990). Climate plays an important role for a wetland established in a depression but may also be important for adjusting the size of lacustrine swamps through the seasonal wet and dry regimes of the edges. Human interventions like waste water discharges and drainage can also affect the wetland's hydrobiology.

Ellery *et al* (1990) suggested that during the establishment of a wetland, an initial colonising site or nucleus such as organic sudd or debris bank is required before stolon or rhizomatous growth of aquatic macrophytes can take place to expand the mat. When water recedes from flooded valleys, papyrus plants spread across the residual mud around the deeper waters of the lake edge as reported for lake Naivasha, Kenya (Gaudet, 1977). From there, papyrus pioneers out into the water by means of floating rhizomes, leaving behind a substratum of organic matter. This, together with silt washed from the land, might form a suitable basis for the growth of other vegetation types like *Miscanthidium violaceum* Robyns, *Phragmites mauritianus* Kunth and *Typha latifolia* L. The process would be accelerated if at the same time the water level continues to fall. Once papyrus has established itself, it can float over the surface of water by the help of rhizomatous rafts.

This chapter gives an overview of the geomorphological, hydrological and floristic features of the Nakivubo swamp area and their effects on the waste water treatment and other functions of this wetland. Further details of the hydrology and vegetation of the Nakivubo swamp area are discussed in Chapters 3 and 5, respectively.

2.2 Location of the Nakivubo swamp

Nakivubo swamp is situated between latitudes 00°17' and 00°19' N and longitudes 32°37' and 32°39' E, at an altitude of 1135 m above mean sea level. The swamp (a tropical perennial natural wetland) lies about 5 km south-east of Uganda's capital city, Kampala (Fig. 2.1, inset B), and connects the city to the Inner Murchison Bay and Lake Victoria.

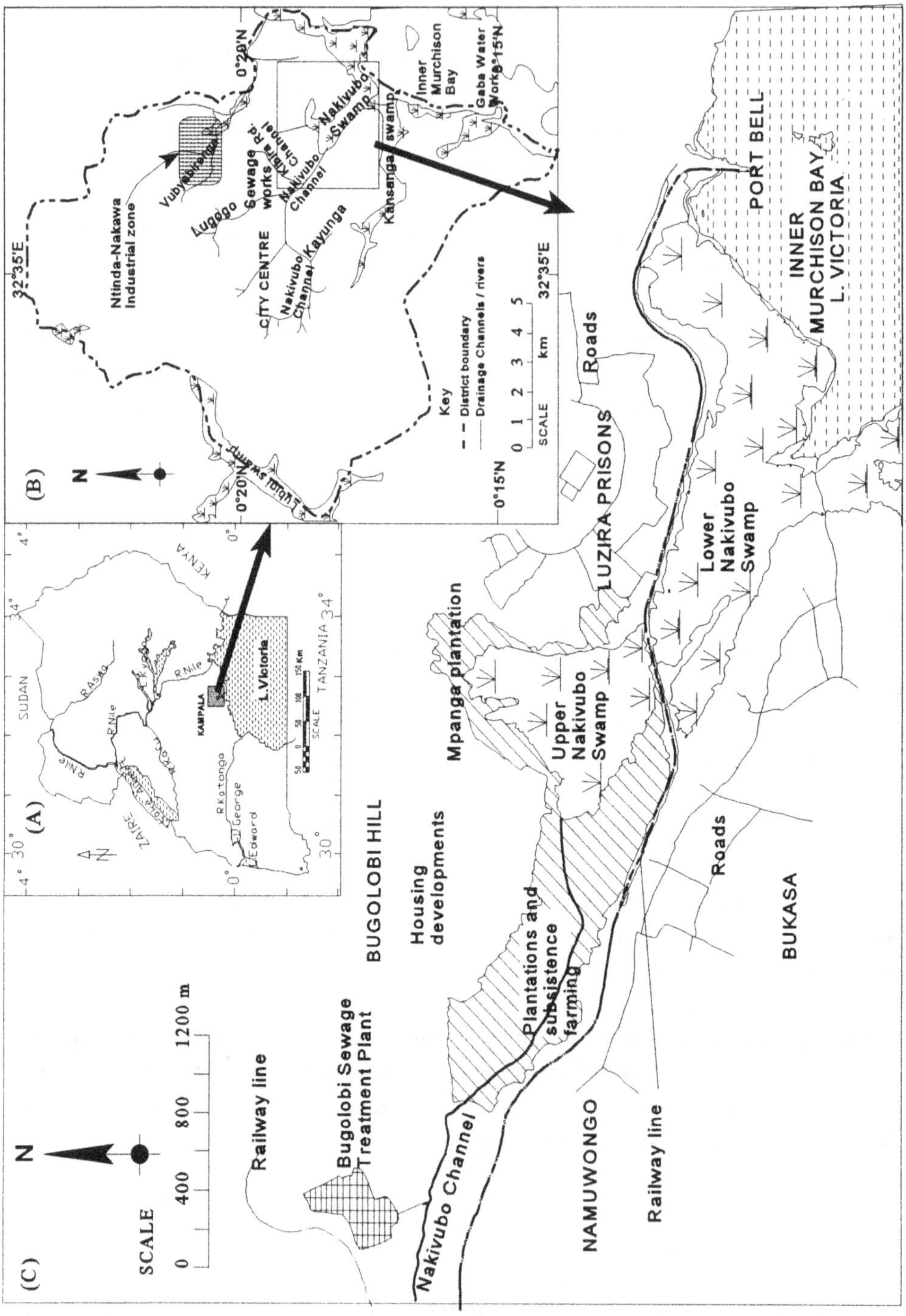

Fig. 2.1. Map showing location and extent of the Nakivubo channel and swamp. The railway line traverses the upper and lower swamps. Inset A is the map of Uganda showing the location of the capital city Kampala on the northern shores of Lake Victoria. Inset B shows the extent of Kampala city, with the Nakivubo swamp, Murchison Bay and the location of the water treatment plants at Gaba.

The 2.5 km^2 swamp drains a total area of about 50 km^2 from mainly the central zone of Kampala. The Kampala - Port Bell railway line crosses over the swamp demarcating the upper and lower swamps (Fig. 2.2).

2.3 Characteristics of the catchment area for the Nakivubo swamp

2.3.1 Central Kampala City

Kampala City is a hill and valley complex (originally termed the city of seven hills), with a substantial swamp system connecting to Lake Victoria via the Kansanga, Nakivubo and Wankolokolo swamps. It receives its water supply (presently about 100,000 m^3/d augmented from about 55,000 m^3/d in 1991) from Gaba Water Works located in the Inner Murchison Bay of Lake Victoria (Fig. 2.1).

Although 51% of Kampala's one million people use water-borne sanitation, only 11 % is connected to the sewerage system (MFEP, 1994). The remaining 40% of the population that uses water-borne sanitation employs on-site sanitation systems notably the septic tank. The sewered population discharges its waste water via the Bugolobi sewage treatment works (capacity is about 16,000 m^3/d but it presently treats only about 3,000 m^3/d of wastewater). The effluent from the treatment works, the rest of the generated wastewater (industrial and domestic) as well as subsurface flows and stormwater discharges are all drained via the Nakivubo channel, through the Nakivubo swamp and back into the Inner Murchison bay, about 4 km upstream of the water abstraction point.

Other swamps, like the Wankolokolo swamp (Fig. 2.4), may become increasingly important as the Ntinda-Nakawa industrial zone (Fig 2.1, inset B) expands.

2.3.2 The Inner Murchison Bay

The Inner Murchison Bay (Fig. 2.2) is a comparatively small water body (area: 16.8 km^2, length: 5.6 km, width: 3.5 km and mean depth: 6.7 m) of limited volume (113 x 10^6 m^3). It is in a semi-enclosed basin off the main part of Lake Victoria. The total catchment area of the bay is about 279 km^2 of which 52 km^2 is fringing swamps (Gauff/Parkman, 1988).

The city, swamp and bay water cycle

The Inner Murchison Bay is both the source of Kampala's water supply and the recipient of its waste water. Enhanced water supply and development of industries in the catchment area of the bay, and continued discharge of sewage into the swamp are therefore principal potential threats to the water quality of the bay, its ecological functions and to other values of the swamp. Thompson (1985) noted evidence of floristic changes in species distribution into the Nakivubo swamp as a consequence of nutrient enrichment from wastewater.

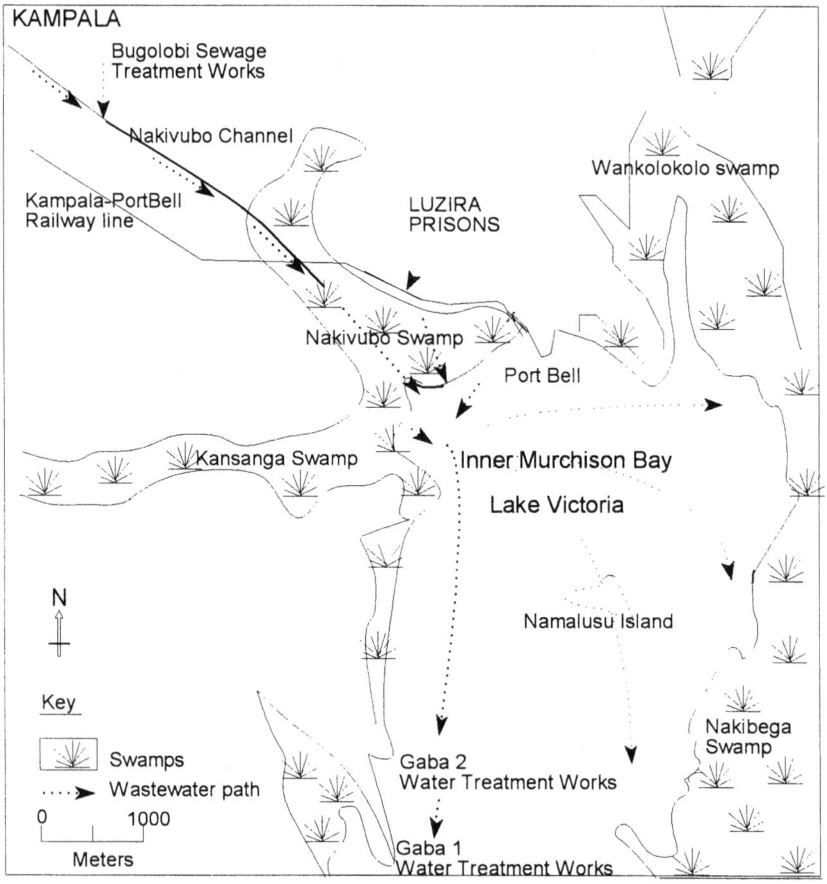

Fig. 2.2. Map showing the relationship between the wastewater paths, Nakivubo Swamp and the Gaba Water Works in the Inner Murchison Bay of Lake Victoria.

It is speculated that the water in the bay is cleansed from the polluting effects of the wastewater discharged by such mechanisms as the tertiary treatment potentials of the fringing swamps, water interchange between the main lake and the bay, and self-purification in the bay through, e.g. reoxygenation. Three studies carried out between 1969 and 1988 to assess the changes in the water quality of the bay because of the increasing discharges showed excellent raw water quality for water supply (Gauff/Parkman, 1988). Although this study generally found similarly good water quality in the bay close to Gaba Water Works (Chapters 4 and 6), the fact that high loads of the pollutants reached the bay was unrefutable. Gaudet (1976) and Denny (1989) expressed concern about the purification ability of the swamp and recommended a comprehensive study. It was deemed imperative that the capacity of the swamp to receive increased pollutant loads be closely assessed in this study, since the bay-swamp-city interaction is fundamental for the health and well being of Kampala's population.

2.3.3 Meteorology of the swamp area

The Nakivubo swamp is within the equatorial belt, and has a moist sub-humid climate. It receives a bi-seasonal rainfall in the periods of March to May and September to November. The rainfall is linked to the Inter Tropical Convergence Zone (ITCZ), the altitude, local topography and the lake. Short duration tropical thunderstorms are particularly common around Lake Victoria and Kampala. The latter is reported to receive more thunder storms than any other capital city in the world (Thompson and Hamilton, 1983). This rainfall frequency and reliability favour the formation of peatlands and swamps. The presence of a large adjacent water body also ensures a both reliable and fairly stable hydrological regime (always humid, annual water level variations about 0.5 m). This is a requirement for papyrus, the dominant wetland macrophyte in the Nakivubo swamp.

Climatic data were obtained from the Hydrometeorological Department in Kampala for two stations within a distance of 3 to 8 km of the swamp (Makerere University and Kibira road). An analysis of the moving average on monthly rainfall data showed no major difference in the quantities and distribution of rainfall for the area. Fig. 2.3 (a) and (b) show the mean daily climatic parameters distributions for a 2 year period (1995-1996) compared with the monthly mean daily rainfall for 1962 - 1994 for the Kibira road station. Graph in b) shows no significant difference between the two rainfall periods (p= 0.386).

There is very little annual variation in air temperature and limited variation in irradiation,

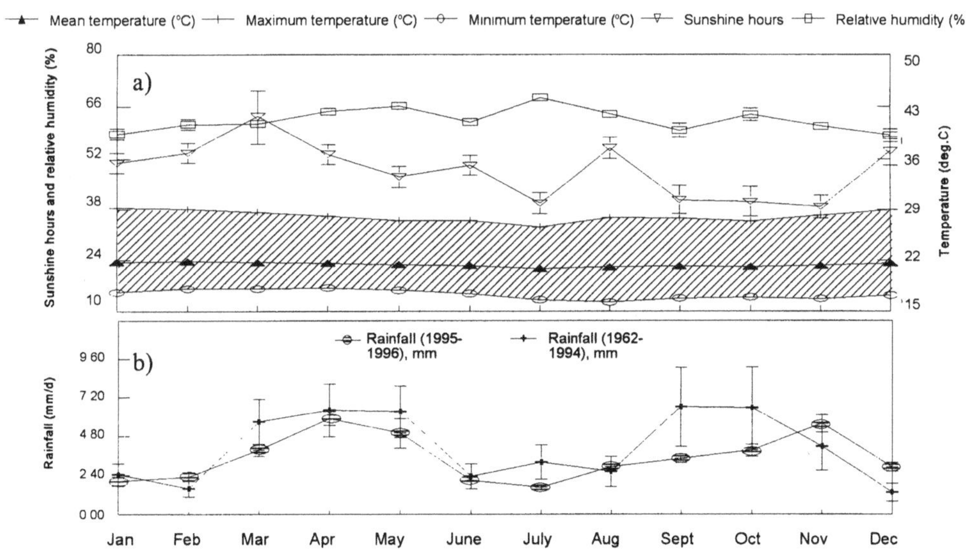

Fig. 2.3. Mean daily climatic data for the Nakivubo swamp area. (a) Mean temperature, maximum and minimum temperature, percentage sunshine hours and relative humidity for (1995 - 1996). (b) Rainfall data over 2 different seasons (r =0.69, p =0.386).

hence probably little seasonal variations in most biological processes within the aquatic environment will occur. The average annual precipitation for the area amounts to about 1500 mm. Mean temperature is 22.7 ± 0.1°C. Average potential evapotranspiration from the swamp calculated on the basis of the Penman-Monteith equation was 4.8±0.2 mm/d (Chapter 3). This means that the wetland crop evapotranspiration is about 120% of the average annual rainfall over the area, an indication of the high reliance of the system on surface water and other water sources.

The SE trade winds are most prevalent and they are strongest in May to July. Wind speeds higher than 60 km/h are not sustainable for periods beyond 2 minutes and the average wind run is only about 50 km/d. Diurnal on and offshore winds lead to gradual displacement of surface water northwards. These affect the hydro biological processes within the swamp by modifying the hydraulic retention time and through transport of materials (notably oxygen and suspended matter) across the swamp-lake interface. The annual lake level fluctuation is 0.4-0.7 m and varies with the annual rainfall pattern (Chapter 3).

2.3.4 Geology of the swamp area

The soils of the Nakivubo swamp area are alluvial and lacustrine sands, silts, and clays overlying granite gneisses (Fig. 2.4). Indeed the gneisses overlay most of the Lake Victoria basin north of Kagera River, the main tributary of Lake Victoria located in the SW (White, 1983). Within the swamp, the alluvial soils range from semi-liquid organic material in the very upper layers of the emergent vegetation zones, through reddish ferruginous (high contents of dissolved iron in run-off water) loams to clays. Resistivity studies (Chapter 3) identified the underlying soils to consist of up to 30 m thick of impervious clays, implying a thick barrier to free mixing between ground water and swamp water.

Like in other drowned valley swamps on the shores of lake Victoria in Uganda (eg. Kawaga: Gaudet, 1976; and Namiro: Lind and Visser, 1962), peat formation is very poor, possibly due to the effects of the flood regime whereby materials are exported from the swamp into the lake, the near neutral pH which is favourable for microbial activities and the all-year-round warm temperatures which increase decomposition rates and oxygen inputs from turbulence and seiches. This contrasts with the ombrotrophic valley swamps of Kigezi in south-western Uganda, for example.

2.3.5 Drainage of Nakivubo Swamp

The major surface water drain into Nakivubo swamp is the Nakivubo Channel (Fig.2.5). In addition, Port Bell and Luzira waste water channels and a number of minor culverts discharge their water (some seasonally) into the lower Nakivubo swamp. The catchment area into the lower Nakivubo swamp is about 1.1 km^2 from the Luzira watershed, 2.5 km^2 for the Bukasa watershed

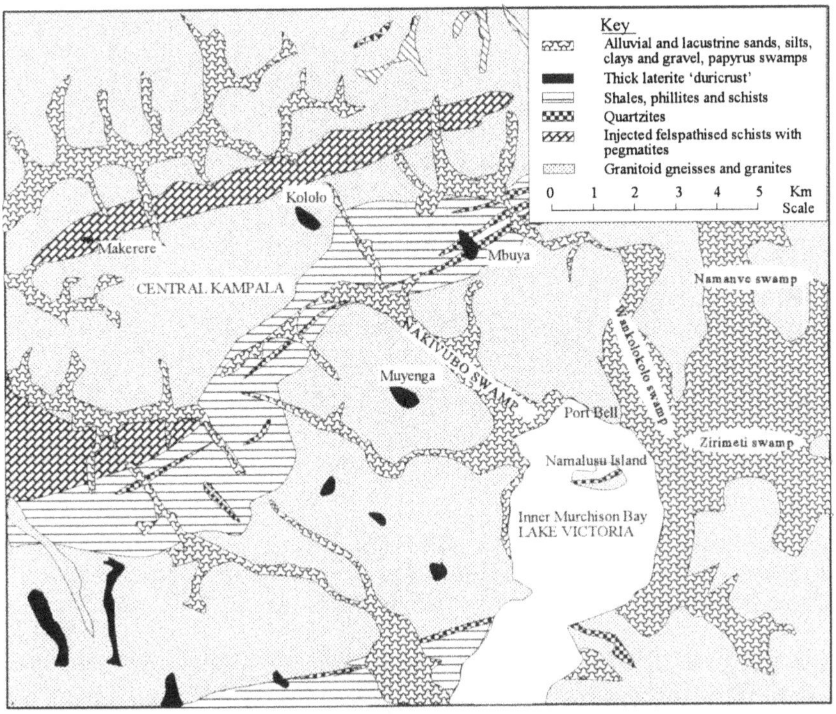

Fig. 2.4. Geological map of the Nakivubo swamp area, modified from the 1:100,000 geological map of Kampala (Adapted from Geological Survey of Uganda, 1957).

and about 50 km^2 from the city centre via the Nakivubo Channel and the upper Nakivubo swamp. These, with the exception of Luzira Prisons' effluent and the Nakivubo channel, also carry rainfall and contribute different amounts of water into the swamp (Chapter 3).

Other important surface water source for the swamp are the high frequency seiches (average about 11 cycles per day, Chapter 3) from the lake which lead to series of building-up and lowering of water levels. These result into the flushing of swamp water, solids and nutrients into the bay.

The Nakivubo Channel

The Nakivubo channel was constructed prior to 1958 (as seen from the topographical maps) and was designed to carry storm water as fast as possible from the city of Kampala into the lake. During its about 12.3 km journey, the Nakivubo channel is joined by a number of constructed and natural channels including Nakulabye, Kitante and Lugogo. It also receives wastewater from industries like Mukwano (soap and oil), the city abattoir, Kampala meat packers, as well as wash-offs from oil depots before the confluence with the Bugolobi sewage works effluent at a point about 2.5 km from the start of the upper swamp. The 'channel length' in the upper swamp is about 1.1 km and it is 1.2 km in the lower swamp.

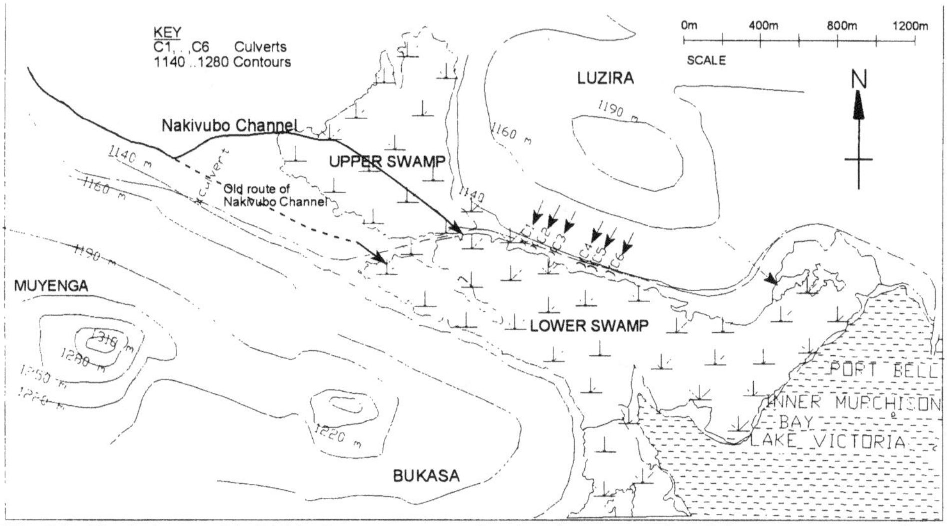

Fig. 2.5. Physical map of the Nakivubo swamp area showing the topography of the swamp surroundings and the drainage channels / culverts into the area. The old route of the Nakivubo channel is also shown. The arrows marked C1 - C6 refer to drainage culverts from Luzira.

A comparison of the 1958 topographic maps for Kampala (1:5000) and the aerial photographs of February 1992 from the Mapping and Surveys Department in Entebbe, and a field survey of the area, revealed that the original path of the Nakivubo channel was altered during the wars of the 1980s at a point upstream of the upper swamp to its present path (Fig. 2.5). This could have implications as far as the distribution of nutrient content in the sediments is concerned.

Due to the relatively small water discharge quantities involved (9,000 to 30,000 m^3/d) in the dry weather flow, it is often possible to find the quality of the water in the channel upstream of the channel's confluence with the sewage outfall to be more polluted than the treated sewage (data from the National Water and Sewerage Corporation).This indicates the raw sewage discharged from the city into the channel. On average, the channel discharges up to about 103,000 m^3/d (Chapter 3) of a mixture of untreated waste water from the city, treated effluent of Bugolobi sewage treatment works plus large quantities of storm water and subsurface flows from the catchment area.

2.4 Dimensions, topography and morphology of the Nakivubo swamp

2.4.1 Recent history and extent of the swamp

The 1958 map of Kampala city shows that the Nakivubo swamp originally extended to and partly surrounded the Bugolobi Hill (Fig. 2.1), but part of the area was later planted with trees (Mpanga plantation) and developed for housing and subsistence agriculture. The floods that occurred in Uganda in the early 1960s led to a water level increase in Lake Victoria of about 2.5 m and must have altered the swamp morphology and size: probably as a landward expansion and shift of the swamp's vegetation cover. The water level has receded but remained at least 1 m above the 1961 level.

In 1991, Taylor estimated the swamp area to be about 2.8 km^2. Presently however, the swamp area has actually dwindled owing to human encroachment. Through drainage, the swamp has been reduced from its original 4.8 km to less than 3 km length. This study estimates the swamp area under indigenous swamp vegetation at only about 1.9 km^2. The rooted edges of the swamp, and to some extent the floating zones in the main wastewater flow paths, are increasingly drained for subsistence agriculture. Table 2.1 gives an overview of the swamp vegetation characteristics and a comparison between findings of our study and the report of Taylor (1991). This study shows that the area covered by floating swamps is bigger than earlier predicted. Further human encroachment is on the increase.

2.4.2 Topography and morphology of the lower Nakivubo swamp

The Nakivubo swamp is located at an altitude of about 1135 m.a.s.l (1958 map of Kampala from the Department of Mapping and Surveys in Entebbe adjusted upwards by 1 m to take into account the effect of the 1962 floods). It is surrounded by three ridges: Bukasa (1250 m.a.s.l) and Luzira (1190 m.a.s.l) on the western and eastern ends, respectively (Fig. 2.5), and the Bugolobi hill in the northern end. Muyenga hill is also in the proximity of the swamp.

Table 2.1 Nakivubo swamp vegetation characteristics and impacts of human encroachment

Characteristic	Taylor (1991)	This study (1996)
Total swamp area under natural vegetation (m^2)	2.2×10^6	1.9×10^6
Total length of lower swamp (m)	1200	1200
Length of disturbed but rooted swamp (m)	500	350
Length of undisturbed but rooted swamp (m)	400	250
Length of floating swamp (m)	300	600
Cultivated area of lower swamp (km^2)	$<0.05 \times 10^6$	0.225×10^6

Swamp morphology

There is a distinct relationship between swamp vegetation, hydrological conditions and morphology. In the lower Nakivubo swamp, the rooted (emergent) vegetation of the swamp edges gradually transforms into a floating vegetation zone as described under 2.5. The undisturbed portions of the swamp are bordered by a rooted papyrus vegetation that is inundated by either surface water runoff or rainfall, and sometimes seiches from the lake.

It is interesting to note the transition from a rooted emergent vegetation at the edges of the swamp into a free-floating regime. A typical cross-section of the swamp showing the relative vertical vegetation, mat, water and sediment distributions along the longitudinal section from the swamp inlet to the lake is shown in Fig. 2.6. The figure shows increasing water depth lakeward, high detrital matter beneath the edge papyrus and variable mat thicknesses (less than 0.6 m for papyrus and over 1 m for *Miscanthidium*). The situation across a typical transect traversing the *Miscanthidium* vegetation zone would be similar only that the western edge would be quite similar to the eastern edge, with emergent vegetation.

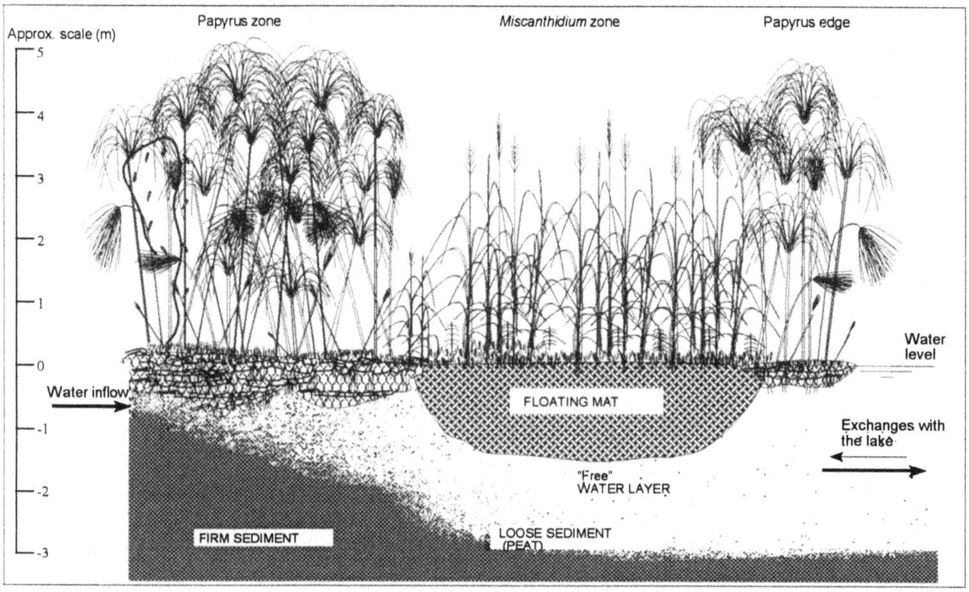

Fig. 2.6. A typical longitudinal section of the lower Nakivubo swamps showing the dominant vegetation types and the related morphological features . Not to horizontal scale.

2.5 Vegetation of the Nakivubo swamp

The Nakivubo swamp vegetation is co-dominated by two aquatic macrophytes namely *Cyperus papyrus* L. and *Miscanthidium violaceum* Robyns[1]. Papyrus covers a vast expanse of the swamp from the landward edge up to the shores of the Inner Murchison Bay (Fig.2.7). At the swamp edges, papyrus is rooted in the sediment whereas in the middle parts, and more so towards the lake, the plants are rooted in a floating mat. On the other hand, *Miscanthidium* is only restricted to the middle part of the swamp and is held in a thick mat floating over a column of water 1.5-2 m deep (Fig. 2.6). The dominant vegetation types are associated with

1 Plant species were identified based on observations of characteristics and using the expertise of staff from the Department of Botany, Makerere University.

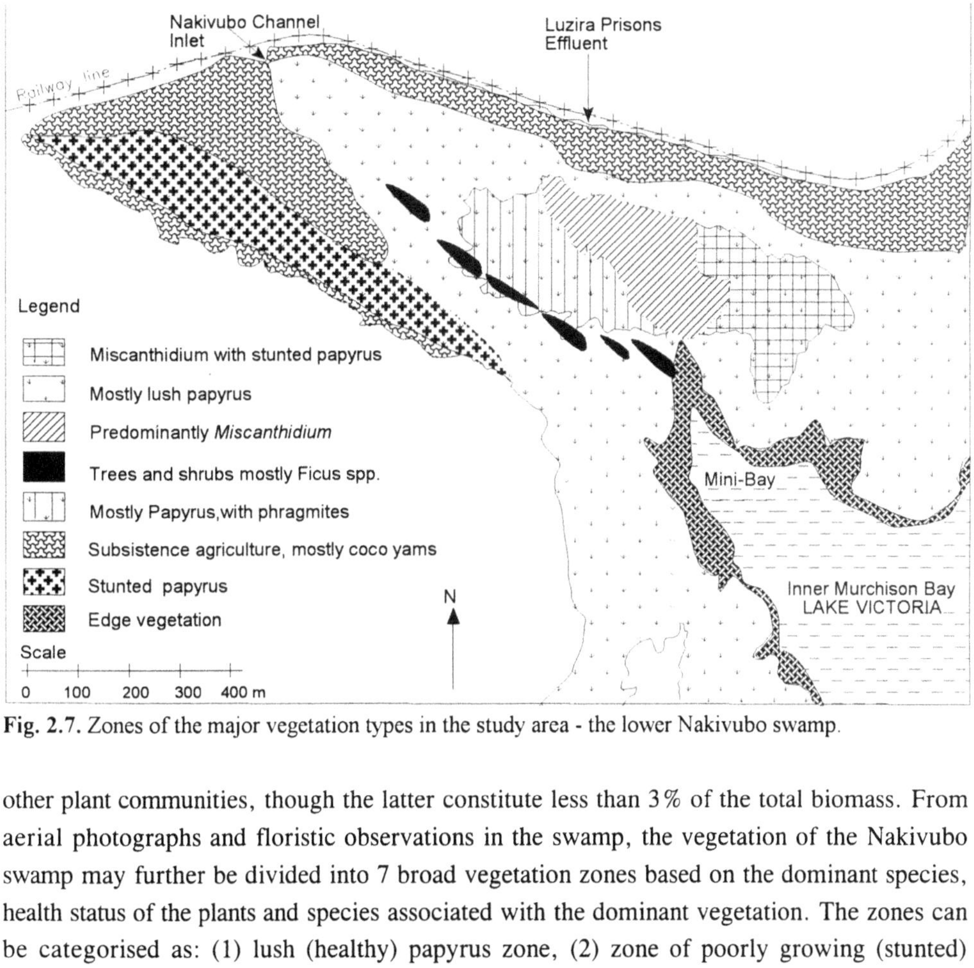

Fig. 2.7. Zones of the major vegetation types in the study area - the lower Nakivubo swamp.

other plant communities, though the latter constitute less than 3% of the total biomass. From aerial photographs and floristic observations in the swamp, the vegetation of the Nakivubo swamp may further be divided into 7 broad vegetation zones based on the dominant species, health status of the plants and species associated with the dominant vegetation. The zones can be categorised as: (1) lush (healthy) papyrus zone, (2) zone of poorly growing (stunted) papyrus, (3) *Miscanthidium* zone, (4) zone of a mixture of papyrus and *Phragmites,* (5) tree zone, (6) surface floating (obligate acropleustophytic) lake-swamp interface vegetation zone and (7) zone of vegetation bordering the swamp.

2.5.1 Healthy (lush) papyrus

A large portion of Nakivubo swamp is covered by healthy growing papyrus especially in areas where the waste water flows. In this zone, the plants are lush as revealed by thick and tall culms with dark green umbels. The average height of plants in this region is 3 - 4 m and they may reach 5 m high. Here, papyrus mostly grows as a monoculture and forms a dense shade which completely outcompetes the undergrowth leaving the mat surface bare. In some places, dense umbels of papyrus prop up each other forcing the culms to bend and lean. This allows light penetration which favours creepers and scramblers to grow. These opportunistic plants also flourish in areas where papyrus has been recently harvested or burnt.

Of the climbers associated with papyrus, *Ipomoea rubens* Choisy is commonly encountered. In some cases, the cover of *Ipomoea* is very dense and may climb up to and intertwine with culms and umbels forcing them to bend. Where there are pools of waste water in the papyrus zone, *Enhydra flactuans* tends to flourish. This plant grows very fast and appears to prefer nutrient-rich areas. It is common at the swamp inflows of both the Luzira drain and the Nakivubo channel. It exhibits a fast growth rate and decays easily contributing to peat formation. Where the free water layer is close to the mat surface (mostly due to a loose mat structure), other competitive species like *Phragmites* species and *Miscanthidium* species are absent.

2.5.2 Stunted papyrus zones

The western edge of the Nakivubo swamp is dominated by poorly growing (stunted) papyrus characterised by yellowish-green umbels. The culms are thin and short with an average height of 2 - 3 m. This part of the swamp is not under the direct influence of waste water and is only inundated during rainy seasons. Climbers and scramblers dominated by *Ipomoea rubens* are common. The stunted papyrus stretches from T1 to T2 (Transect 1 to Transect 2) (Fig. 2.8). At the landward edge of this part of swamp communities of *Hyparrhenia rufa* and *Echinochloa pyramidalis* are also present. The papyrus in this region flourishes during rainy seasons when the mat surface becomes flooded with water.

2.5.3 Papyrus - *Phragmites* zone

In the middle of the swamp, a zone dominated by papyrus interspersed with *Phragmites mauritianus* is encountered. The two vegetation types coexist on a thick floating mat (1.2 m) overlying water of about 1.5 m. Climbers and scramblers dominated by *Ipomoea* are common features of this zone. Since *Phragmites* has been reported to be intolerant to flooding (Thompson and Hamilton, 1983) its existence in the middle of the swamp may be due to the support provided by the thick mat. In this region, *Phragmites* stems are on average 3 m tall. In Lake Bunyonyi (Uganda), Denny (1973) recorded that the octaploid *Phragmites australis* was rooted to a depth of 4.2 m.

Interesting to note is that in the neighbourhood of this zone, where there are open pools of water or where water is close the mat surface, papyrus dominates and grows as a monoculture. This suggests that flooding and regular water availability favours the growth and dominance of papyrus over *Phragmites* and other macrophytes intolerant to flooding. Other isolations of *Phragmites* are restricted to the swamp edges on the Luzira side of the swamp.

2.5.4 *Miscanthidium* zone

Miscanthidium is a tough perennial grass with erect stalks of 2 - 4 m high and an inflorescence of 25-40 cm long . The grass is found in the middle of Nakivubo swamp, floating on a mat of about 1 - 1.6 m thick. *Miscanthidium* grows in dense healthy tussocks whilst associated plants, including papyrus, are dwarfed and ailing.

The species diversity was much less in *Miscanthidium* areas than under papyrus. A unique feature of the *Miscanthidium* zone was the existence of small open patches occupied mostly by *Cyperacaea*. The strongest presence on these patches was that of *Cyperus nitidus*. Associated plants, papyrus inclusive, were yellow and stunted in appearance. The presence of stunted papyrus, suggests that this plant may be on the verge of being out competed by *Miscanthidium.* The poor survival of papyrus in this region may be attributed to the thick root mat of *Miscanthidium* in this region which restricts nutrients and water supply to papyrus, one of the factors which appear to be controlling the dominance of papyrus in the Nakivubo swamp.

One of the striking features in the Nakivubo swamp is that *Miscanthidium* does not grow attached to a bottom substratum in open pools of water but only in a thick soggy floating mat. This contrasts observations by some authors who indicated that *Miscanthidium* does not have the ability to float and tends to be restricted to shallow flooding areas (Lind and Visser, 1962; Thompson and Hamilton, 1983). *Miscanthidium* in the Nakivubo swamp is rooted in a firm floating mat, and unless one accesses the water beneath, say, by auguring through the mat, it might be easily concluded that the plants are rooted in the bottom sediment. In fact, the local population thought that the *Miscanthidium* zone was a rocky island on which it was possible to erect a house.

A probable reason for its present location in the middle of the swamp could be the lake water level rise after the heavy rains of the early 1960s. These rains increased the water level in Lake Victoria by some 2.5 m. Before 1960, the Nakivubo swamp could have been a marsh in which *Miscanthidium* could have been more extensive and rooted to the bottom substratum. In his work in the Namanve swamp close to the Nakivubo swamp (Fig. 2.4), Eggeling (1935) reported the recent start of burning of *Miscanthidium* in the Luzira (Nakivubo) swamp and the accompanying change of species. The mat could have consisted of over 0.5 m of the spongy *Miscanthidium* litter forming a weak bond with the underlying clay substratum. The sudden rise in water level may have detached this loosely bound vegetation and mat from the bottom. Subsequent deposition of litter onto its mat surface and the low rate of decomposition of the lignin therein (Visser, 1964) are responsible the present mat thickness. The compact, thick but low density mat of *Miscanthidium* (its survival strategy) may have protected it from take over by papyrus. The papyrus behind *Miscanthidium* in the hydrosere of this swamp may be attributed to continuous water inflow from the Nakivubo channel.

Another characteristic feature of *Miscanthidium* in the Nakivubo swamp is that it is not located at the dry edges of the swamp, suggesting that this plant prefers soggy areas although it can not withstand flooding. It has never been reported to break away from the swamp edges into the open water of the lake even in areas like Entebbe where it grows close to the shores of Lake Victoria, because of its firm and thick mat which keeps the plants firmly held and resistant to water currents.

The plants growing in association with *Miscanthidium* and papyrus in the Nakivubo swamp are listed in Table 2.2. A species comparison of the papyrus vs. *Miscanthidium* dominated zones shows that less than 10% of the species are common to both, an indication that the swamp conditions are indeed different for each zone.

2.5.5 Surface-floating vegetation zone (acropleustophytic)

The swamp-lake interface corresponding to the outlet of the Nakivubo Channel into the Inner Murchison bay ("mini-bay") is bordered by a community of floating plants. Common species in this zone include *Cyperus mundtii* (Nees) Kunth., *Enhydra fluctuans* Lour., *Eichhornia crassipes* (Mart.) Sloms-Laub. and *Vossia cuspidata* (Roxb.) Griff (not an acropleustophyte). Over open water, *E. crassipes* and *C. mundtii* co-dominate whilst at the edges of the zone where a kind of mat covered the water *E. fluctuans* dominated, growing in association with *Ipomoea rubens* and *Impatiens irvingii*. Lake-ward from the papyrus edge towards *Eichhornia,* stands of *Vossia cuspidata* are prominent. The association between *Vossia* and *Eichhornia* interspersed with other plants, forms a firm floating mat which is strong enough to support the weight of a mature person. Occasionally floating papyrus and *Vossia* are detached from the main beds and drift down into the Inner Murchison Bay, with *Eichhornia*.A similar habitat occurs in the sudd vegetation of the Upper Nile (Denny, 1985).

The flanking *Vossia* and the floating plant debris in this part of the bay provide a footing for papyrus to quickly expand lake ward. This is probably because on their own, small islets of papyrus are unable to grow across the open waters since the rhizomes would not have intertwined and developed adequately to support the plant weight over open waters. As a result this part of the bay is narrowing and the plant communities are slowly being dominated by papyrus.

2.5.6 Tree zone

In the middle of the swamp between T1A and T4 a strip of dense vegetation of shrubs and trees is well established. The shrubs and trees found here were *Alchornea cordifolia* (Shum et Thonn). Muell. Arg., *Triumfetta macrophylla* K.Schum, *Ficus trichopoda* Bak. *Maesa lanceolata* Forsic, *Syzygium cordatum* Sond and *Briddelia bridelifolia* (Pax) Fidde. *Ficus* spp. are dominant in this area. These are occasionally harvested by the local population for fuel.

Table 2.2 Plants growing in association with *C. papyrus* and *M. violaceum* in the Nakivubo Swamp

PAPYRUS AREAS		*MISCANTHIDIUM* AREAS	
Plant	**Family**	**Plant**	**Family**
Climbers, creepers and scramblers			
Ipomoea rubens Choisy	Convolvulaceae	*Cissampelos mucronata* (A. Rich)	Merispermaceae
Zehneria capillaceae (Schum) C. Jeffrey	Cucubitaceae	*Stephania abyssinica*(Dill. & Rich.) Walp.	Merispermaceae
Stephania abyssinica (Dill. & Rich.) Walp	Merispermaceae		
Vigna luteola var *luteola* (Jacq.) Benth.	Thabaceae		
Cissus oliveri (Engl.) Gilg.	Vitaceae		
Cyphostema adenocaule (A.Rich) Desc.	Vitaceae		
Other small herbs			
Achyranthes aspera L.	Asteraceae	*Lygodium micropyllum* (Cav.) R. Br	Schizaeaceae
Agerantum conyzoides L.	Asteraceae	*Otomeria elatior* (DC) Verdc.	Rubiaceae
Crussocephalum pincridifolium	Amaranthacea		
Impatiens irvingii Hook. F.	Balsaminaceae		
Dissotis rotundifolia (Sm.) Triana	Melastomastaceae		
Fern			
Thelypteris striata (Sch.) Ching	Thelypteridaceae	*Thelypteris striata* (Sch.) Ching	Thelypteridaceae
Grasses			
Leersia hexandra SW.	Poaceae (Graminaea)	*Panicum chionache* Mez.	Poaceae
Sedges			
		Cyperus nitidus Lam.	Cyperaceae
		Cyperus rotundus L.	Cyperaceae
		Cyperus denadatus L.f.	Cyperaceae
		Fimbristylis subaphylla Boeck.	Cyperaceae
Mosses		*Sphagnum* cuspidatum Ehrh	

2.5.7 Swamp-edge Vegetation

Both the western (Bukasa) and the eastern (Luzira) edges of the swamp have wetland plants which proliferate in drier areas. The common species are, *Hyparrhenia rufa*, *Imperata cylindrica*, *Typha domingensis* **Pers.**, *Phragmites mauritianus* and *Echnoccloa sp*. These plant species occupy semi-dry areas at the edges of the swamp which are usually inundated only during the rainy season. Apart from *Phragmites*, these species were not observed anywhere inside the swamp.

2.6 Outline of functions of the swamp

Global wetland values are classified differently by different authors (Mercer, 1990; Williams, 1990; Denny, 1993a; Mitsch and Gosselink, 1993). Wetland biodiversity, ecosystem and related global values are threatened by increased pollutant loads in the waste water that is discharged into them, by draining and polder development, by pollution transported from the open water and by other forms of human encroachment, eg., subsistence farming.

Despite its small size, the Nakivubo swamp is an important wetland on both local and national scales. Table 2.3 is an adaptation of the US fish and wildlife values classification showing the perceived value of the Nakivubo swamp.

A fundamental function of the Nakivubo swamp is the regulation and buffering of nutrient inputs and filtration of anthropogenic pollutant material, thus playing a role in controlling pollution of the bay. For over 30 years, the swamp is the overall recipient of partially treated waste water from the city of Kampala as well as of storm water. Other sources of raw waste water into the ecotone include a prison and factories (including a brewery) in the north east end of the swamp.

The swamp's existence may be threatened due to human encroachment through burning, unsustainable cutting and increasing subsistence farming. Subsistence farming is on the increase considering that prior to 1993, only a small portion of the lower swamp (very close to the railway line) was under cultivation and now about 22.5 ha is cultivated (Fig. 2.7 and Table 2.1). Further discussion on the socio-economic aspects of the Nakivubo swamp are presented in Chapter 10.

2.7 Transects and sampling sites

To study the processes within the swamp and the transformations in water quality as the waste water flows through, access transects were cut through the swamp. The main study was

Table 2.3 Values of the Nakivubo swamp: ★ is fairly perceivable, ★★★ is very perceivable and — is not perceived

Wetland value or function	Perception level for the Nakivubo swamp
Environmental quality:	
• water quality maintenance(pollution filters, sediment removal, and chemical nutrient absorption	★★★
• aquatic productivity	★
• microclimate regulation	★
• world climate functions (ozone layer)	—
Fish and wildlife values - habitat	
• habitats for fish production (near the lake)	★★★
• water fowl and other birds	★
• sitatunga and other wildlife	★
Socio-economic values - consumptive use and non-consumptive use	
• water supply	★
• timber and other natural products (fuel - peat, reeds; housing materials - reeds, soils; fishing)	★★★
• livestock grazing	—
• recreation (bird watching)	—
• aesthetics and spiritual considerations	—
• education and scientific research	★★★
• agricultural production	★★★
• hunting	★
Hydrological functions	
• flood control	★★★
• ground water recharge/discharge	—
• shoreline anchorage and dissipation of erosive forces	—
• wave damage protection	—
• erosion control	★

conducted in the lower swamp because (i) it was easier to quantify a smaller region, (ii) the major surface water flows were more readily accessible for measurement, (iii) the effluent of this part of the swamp had direct influence on the water quality in the bay, and (iv) the diverse vegetation zones in this swamp made the research more intriguing.

Five sampling transects of about 1.5 m wide were cut across the width of the swamp from Bukasa side towards the Luzira side (Fig. 2.8), at different periods of the research. Transects 1 and 3 were cut first to study the in-swamp processes. Water samples were also taken from the bay. When the bay became too infested with water hyacinth, making it difficult to use a boat, T4 was prepared to enable water collection from as close to the swamp outflow as possible. T2 was also cut to interpolate between the rather distant T1 and T3. T1A was prepared much later, mainly to confirm some results of the topographical survey. The transects are not exactly parallel because of the dense vegetation which made it difficult to see much further ahead during the cutting.

As noted by Carter (1954) and Denny (1985), it may be possible to walk through a floating papyrus vegetation by stepping quickly from one rhizomatous clump to another. This will not be possible if one has to spend some time at a site to either collect samples or make observations. Papyrus mats wiggle and swing over a distance of as much as 10 m as one walks over. Any reasonable exertion of weight on the floating mat leads to sagging, with a pool of water rapidly collecting over one's feet. In fact, cases of drowning in papyrus swamps have been reported, especially in Western Uganda.

To access T1, where the floating raft could not support the research team, a channel was cut along the transect at a point beyond 200 m from the western edge. From this channel, sampling points were accessed in a canoe at points off the channel. For other transects, access was possible after bridging the cut paths with papyrus culms cut and tied into bundles. There was no problem in the *Miscanthidium* zones due to the thick and tightly interwoven mat.

Sediment, plant and water samples were taken from points along the transects at selected distances (basically multiples of 25 m). Marking poles and later bamboo posts were stationed at a number of points to 'permanently' mark the measuring points.

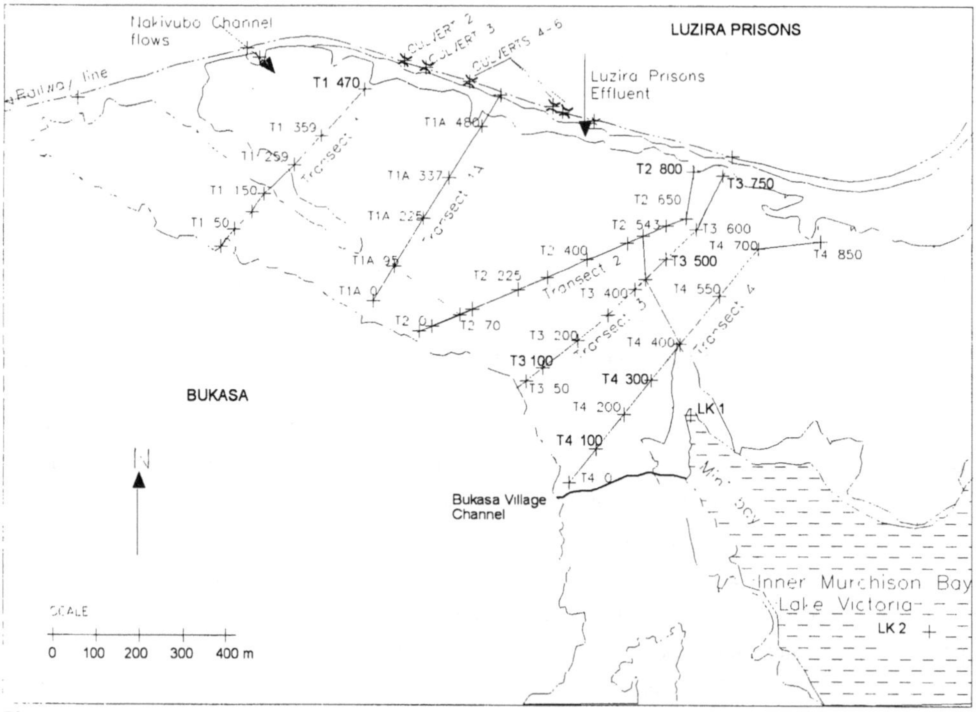

Fig. 2.8. Location of sampling points of the lower swamp used in this study. T1 50 refers to the transect name (Transect 1) and the corresponding distance in metres from the western (Bukasa) edge of the swamp (50).

Randomness of selection of measuring points

The sampling points used in the measurement of physical-chemical parameters were tested for complete spatial randomness (CRS) using the computer software program ILWIS (Integrated Land and Water Information Systems). The results (Fig. 2.9) show that the points were randomly distributed and could therefore be used for interpolation within the accuracy of the distance between the points.

Determination of effective swamp areas

Estimates of the different areas of the swamp covered by different vegetation types or under the influence of wastewater (notably in the major flow paths versus the minor flow paths) were made on the basis of visual observations in the swamp, aerial photographs of 1993 from the Department of Mapping and Surveys in Entebbe and water quality and depth profile measurements. The coverages were included in the digital map of the area and read off using the computer program AUTOCAD (Version 12).

A summary of these areas is as follows:

Total area of the swamp (upper and lower swamps) .... 2,343,000 m^2.
Total area of lower swamp .... 1,686,000 m^2.
Total study area .... 1,150,000 m^2.
Area covered by Miscanthidium/ mixed vegetation .... 235,000 m^2.
Study area covered by papyrus .... 915,000 m^2.

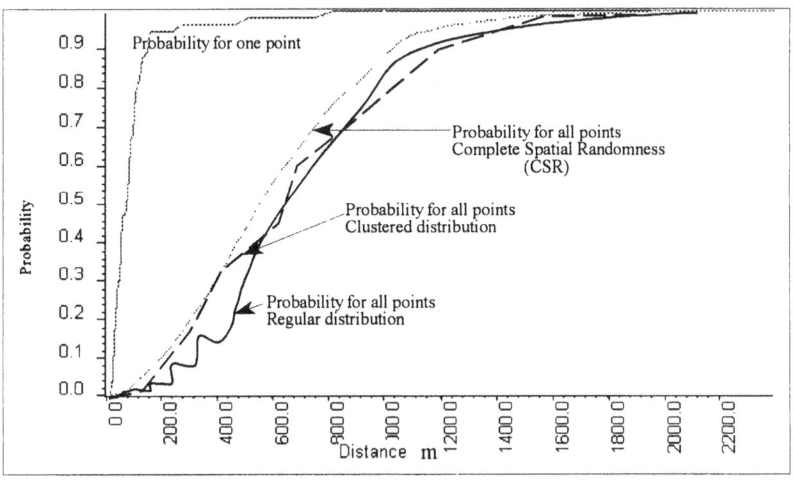

Fig. 2.9. Pattern analysis for randomness of distribution of data points used in the measurement of physico-chemical water quality parameters in the lower Nakivubo swamp. The vertical axis is the probability that another point exists within a corresponding distance from any given point. Probability of one point is that probability that within a given distance, at least one other point exists. 'Probability for all points' is the average probability that within a given distance, 1 or 2 or ... or (n-1) points exist. The generated curves showed complete spatial randomness (compared to clustered or regular distribution patterns).

2.8 Conclusion

The Nakivubo swamp is a dynamic system that has had a fair share of human threat, notably waste water discharges and encroachment. Besides human factors, long term and daily variations of meteorological conditions and lake levels have affected its morphology and hydrobiology. Whereas natural phenomena cannot be easily controlled, it is important that man's role in affecting the extent and functions of this system be checked if the population of Kampala is to continue abstracting good quality raw water from the bay, as well as enjoy the other values of this wetland. In this respect, it is important to quantify the impact of continued and increased waste water discharge into the swamp (vice versa) and the capacity of the swamp to treat waste water in view of increased encroachment. This is the purpose of this study and will form the content of the following chapters.

2.9 References

Carter, G. S. (1954). The papyrus swamps of Uganda. W. Heffer and Sons Ltd. Cambridge England.

Denny, P. (1973). Lakes of south-western Uganda. II. Vegetation studies of Lake Bunyonyi. Freshwat. Biol.(3):123-135.

Denny P., Ed. (1985). The ecology and management of African wetland vegetation - A botanical account of African swamps and swallow water bodies. Dr.W.Junk Publishers, Netherlands.

Denny, P. (1989). Wetlands. In: Strategic Resources Planning in Uganda. UNEP report IX.

Denny, P. (1993a). Wetland use and conservation: Introduction. *In*: Whigham, D. F., Dykyjová, D. F. and Hejný, S. (eds.), *Wetlands of the world, I. Inventory, Ecology and Management*. Kluwer Academic Publishers, Dordrecht. pp 111-114.

Denny. P. (1993b). Wetlands of Africa. *In*: Whigham, D. F., Dykyjová, D. F. and Hejný, S. (eds.), *Wetlands of the world, I. Inventory, Ecology and Management.*, Kluwer Academic Publishers, Dordrecht. pp 1- 46

Eggeling, W.J. (1935). The vegetation of Namanve swamp, Uganda. *J. of Ecol.* 23:422-435.

Ellery, K., Eller, W. N., Rogers, K.H. and Walker, B. H. (1990). Formation, colonization and fate of floating sudds in the Maunachira river system of Okavango Delta, Botswana. *Aquat. Bot. 38:* 315 - 329.

Gaudet, J. J. (1976). Nutrient relationships in the detritus of a tropical swamp. *Arch. Hydrobiol.* Vol. 78, pp 213-239.

Gaudet, J .J. (1977). Natural Drawdown on lake Naivasha, Kenya, and the formation of papyrus swamps. *Aquat. Bot. 3:* 1- 47.

Gauff/Parkman. (1988). Kampala Water supply expansion Project. Final design report. I. water Supply Studies. Report prepared for the Ministry of Water and Mineral Development and National Water and Sewerage Corporation; Kampala - Uganda.

Howell, P., Lock, M., Cobb, S. (1988). The Jonglei canal: impact and opportunity. Cambridge University Press, Cambridge.

Lind, E. M. and Visser, S.A. (1962) A study of a swamp at the north end of Lake Victoria. *J. Ecol. 50:* 599-613.

Mercer, C. D. (1990). Recreation and wetlands: impacts, conflict and policy issues. In: *Wetlands: A threatened landscape*. Williams, M. (Ed.). Basil Blackwell, Oxford. pp 267-295.

MFEP (1994). *The 1991 population and housing census (national summary).* Statistics Department, Uganda.

Mitsch J.W. and Gosselink G.J. (1993). *Wetlands.* 2nd. Ed. van Nostrand Reinhold, New York.

Patten, B.C. (1990). *Wetlands and shallow continental water bodies, Vol. I. Natural and Human relations.* SPB Academic Publishing, The Hague.

Taylor, A.R. D. (1991). Status of Nakivubo and Luzira swamps in Kampala. Report prepared for the Ministry of Environment Protection, National Wetlands Conservation and Management Programme, Kampala- Uganda.

Thompson, K. (1985). Emergent plants of permanent and seasonally-flooded wetlands. In: Denny, P. ed.: The ecology and management of African wetland vegetation. Dr. W. Junk Publishers, Dordrecht.pp 43-108.

Thompson, K. and Hamilton, A. C. (1983). Peatlands and swamps of the African continent. In: Gore, A. J. P. (ed.), *Mires: Swamp, Bog, Fen and Moor.* Elsevier, Ecosystems of the world series, 4B, (Regional studies), Amsterdam. pp 331 - 373.

Visser, S.A. (1964). A study in the decomposition of *Cyperus papyrus* in the swamps of Uganda, in natural peat deposits, as well as in the presence of various additives. *East African Agricultural and Forestry Journal*, 29 (3): 268-287.

Welcomme, R.L. (1966). The biological and ecological effects of climatic changes on some fishes in Lake Victoria basin. PhD Thesis. University of East Africa (Makerere), Kampala. 87p.

White, F. (1993). The vegetation of Africa. UNESCO, Paris.

Williams, M. (1990). Wetlands: A threatened landscape. Basil Blackwell, Oxford.

WMO (1974). Hydrometeorological survey of the catchments of Lake Victoria, Kyoga and Albert. WMO report, Geneva, Switzerland.

Chapter 3

Water Balance and Hydrodynamics of the Nakivubo Swamp

Nalubega, M.

3.1 Introduction

3.1.1 The hydrology of wetland systems

The importance of hydrology in maintaining the structure and functioning of a wetland has been emphasized by various researchers (*i.a.*, Hammer and Kadlec, 1986; Mitsch and Gosselink, 1986; Brown and Stark, 1989; Kadlec and Knight, 1996). While Mitsch and Gosselink (1986) speculated that hydrology is the single most important determinant for the establishment and maintenance of specific types of wetlands and wetland species, Howell *et al* (1988) specified water depth, its seasonality and nutrient status as the controlling factors for the distribution of wetland types.

Hydrologic parameters, namely rainfall, ground water characteristics, surface water flow and evapotranspiration (which control the amount, frequency and flow rate/dynamics of water) in wetlands are known to affect abiotic factors such as salinity, soil anaerobicity and nutrient availability. These, in turn, determine the flora and fauna that develop in the wetland. *Cyperus papyrus* L., for example, grows in permanently inundated sites and survives on nutrients in the water over which it usually floats. *Typha latifolia* on the other hand, develops at sites remote from the main deep water body, in shallow water of usually low nutrient concentration, where it is firmly rooted in the substrate (Howell *et al*, 1988). At the same time, the biotic factors in turn affect the wetland's hydrology and biochemistry.

High water flow rates and wind effects (in case of shallow water bodies) create intense energy dissipation and dynamism in the system. The resulting erosion and re-suspension lead to faster flux of materials and hence an enhanced nutrient recycling rate. Conversely, slower renewal rates due to lower energy dissipation in the system lead to accumulation of materials. This changes the basin geometry and the wetland may begin to mature and diversify (Mitsch and Gosselink, 1986).

The hydrology thus affects nutrient flows, vegetation ecology and microbial ecology. These, as discussed in Chapters 4 to 8, influence the capacity of the wetland to remove pollutants from waste water. This capacity in particular depends on (i) the amount of waste water flowing into the wetland, (ii) system geometry which affects the retention time and distribution of waste water, and (iii) the interaction processes between the water, wetland sediments, the suspended solids, plants and microbial populations. In natural wetlands, hydrology is also known to directly influence temporal changes in depth and topography of the wetland bottom (Kadlec and Knight, 1996).

In order to assess the performance of a treatment wetland, a determination of the hydrological characteristics and the basin geometry is necessary. A consideration of all terms for inflows and outflows gives a simple water balance or budget of the wetland. The terms include inflows (rainfall, seepage, surface runoff and discharges) and outflows (evapotranspiration, infiltration, effluent). Exchanges with the lake may either be input or output.

The terms of the water balance for a wetland such as the Nakivubo swamp are represented in Equation 3.1 and Fig. 3.1.

$$\frac{dV}{dt} = Q_{into} + R - ET + Q_g + Q_l + Q_s, \text{ (all terms in e.g., m}^3\text{/d)} \quad (3.1)$$

where dV/dt = rate of change of water volume (storage) in the wetland, R = precipitation multiplied by the surface area, ET is evapotranspiration multiplied by the surface area, Q_{into} = inflows from adjacent rivers and waste streams, Q_s = basin sub-surface flows, Q_g = ground water seepage and infiltration, Q_l = inflows and outflows from and into the lake. The importance of each of these terms depends on the wetland type and regional factors like climate, geology and physiography of the area (Walton *et al*, 1996).

Precise determination of inflows versus outflows is useful for the evaluation of the overall retention time in the wetlands. In a natural wetland like the Nakivubo swamp, with a complex morphology and a flow regime complicated by the presence of seiches from Lake Victoria, there is also a need to characterize the internal water flows. Without this, local transport and pollutant removal processes cannot be described and quantified beyond values that are average

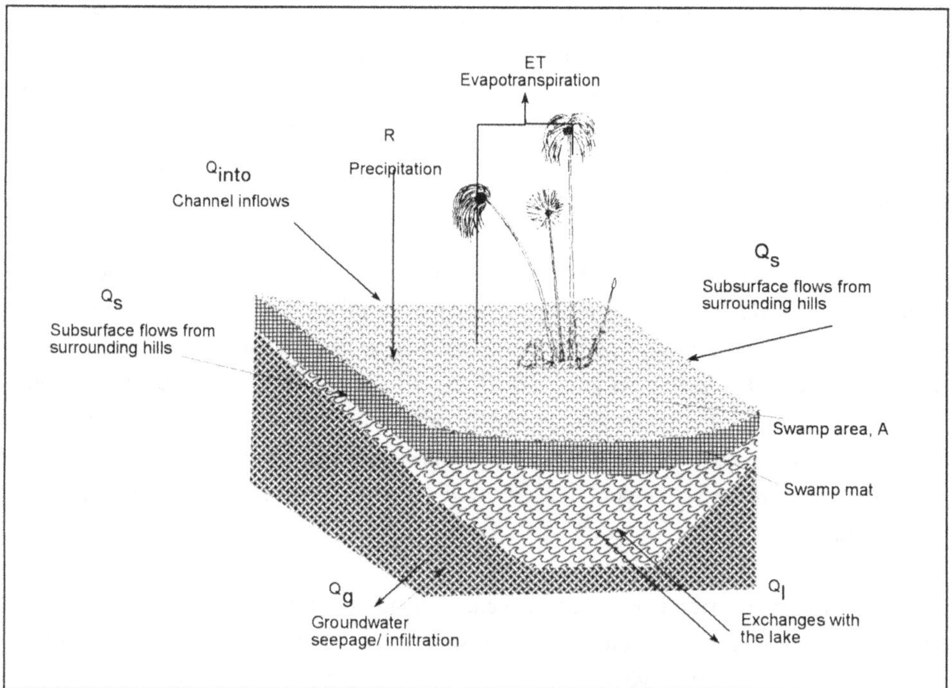

Fig. 3.1. Schematic presentation of water sources and losses in Nakivubo swamp.

for the whole body. Given the scale and internal hydraulic heterogeneity of the wetland, this requires an extensive measuring and monitoring network both within and across the boundaries of the system.

Accurate description of local water flow in a wetland is difficult (van Heut, 1991). The available data are usually inadequate to yield accurate results for all circumstances (e.g. under dry and rainy seasons; Boyt *et al*, 1977). Mathematical models based on a number of assumptions and simplifications are, therefore, valuable tools to supplement field data and test working hypotheses in estimating the role of the different processes (Mitsch and Reeder, 1991).

Wetlands typically have channelized flows and short-circuiting (Gaudet, 1980; Kadlec and Knight, 1996) and areas with stagnant water which cause a retention time distribution. Short circuiting causes short retention and contact time between in-flowing water and the bulk of the wetland system, and possibly low levels of nutrient exchange and purification. The use of artificial tracers to study flow dynamics, even in constructed wetlands, has been found laborious (Pilgrim *et al*, 1992), and their dispersivity was found to remain quite unpredictable due to the heterogeneity and temporal variability in these systems. Natural tracers such as electrical conductivity (EC) can act as, and are reported to be simple tools that can provide a semi-quantitative description of the wetland transport processes (Gaudet, 1979; Mitsch and Reeder, 1991). Van Heut (1991) for example, found that in Lake Tjeuke in the Netherlands, EC was strongly correlated to the conservative and inert Cl^- ($r = 0.98$, $p = 0.001$), which is a suitable tracer. In a freshwater wetland receiving wastewater flow, conductivity of the diluted wastewater is usually higher than the background values, making it a possible tracer.

3.1.2 Study objectives

The Nakivubo swamp in Kampala is an example of a natural swamp system that has received wastewater for a long period (Chapter 2) but whose complex hydrology is not well known. The main objectives of this part of the study are to identify the key hydraulic and hydrological flow characteristics and to determine the overall water balance of the Nakivubo swamp. This will give the basic information for subsequent determination of the mass balances for nutrients and *Escherichia coli* in the system, and for the efficiency of the swamp as a waste water purification device.

Specifically, this study aimed to: (i) identify the overall basin geometry; (ii) determine the hydraulic profile of the system (flow paths of the waste water and average retention time); and (iii) quantify or estimate the water balance of the system through on-site data collection and application of a hydrodynamic simulation model.

3.2 Materials and methods

The water balance study was restricted to the lower Nakivubo swamp (area south of the railway line; Fig. 3.2) as this is biologically the most active part (Chapter 2).

3.2.1 Identification of the basin geometry

The basin geometry was determined from measurements of swamp vertical and horizontal profiles along the transects described in Chapter 2, and along the swamp edges. This was done to: (i) identify the nature of the swamp bottom, and the flow path, (ii) estimate the effective depth of water flow at each point (and therefore the cross-sectional area), to be used in the model; and (iii) correlate depth of flow with spatial variation in water quality and vegetation.

The study area, transects and measuring locations are shown in Fig. 3.2. The size of the individual portions of the study area were estimated using AUTOCAD (Chapter 2). Mat thicknesses, water depth, loose peat thickness and firm sediment level were measured at distances between measuring points of 25 to 50 m. Mat thickness and peat level were determined using a profiler (Fig. 3.3). Free water depth (b) and the mat thickness (a) were determined by carefully pushing the profiler through a hole dug through mat until it touched

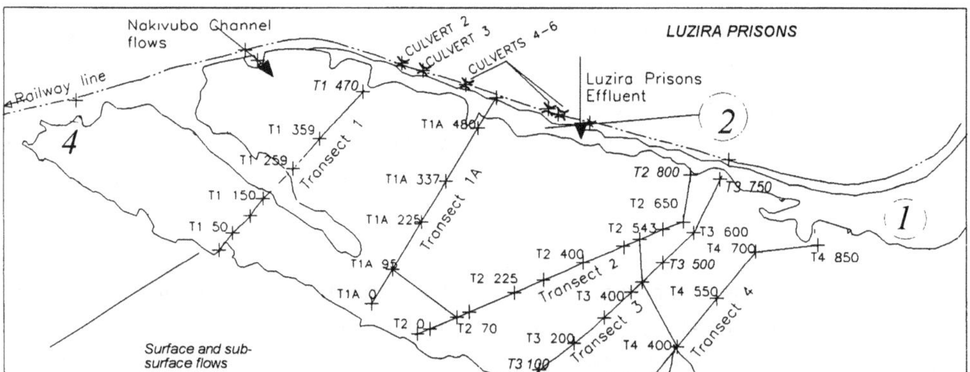

Fig. 3.2. Map of the lower Nakivubo swamp (study area) showing the sampling transects and locations. T1 to T4 refer to transects 1 to 4 as marked. The figure behind the transect code mentions the distance in metres from the western shore. Circled numbers 1 to 4 are the measuring locations for resistivity studies. The arrows show additional possible surface and sub-surface flows into the swamp.

the loose peat below and subtracting the length of the profiler portion above the mat from the total profiler length, L. The profiler was then carefully lifted upwards till its base plate was intercepted by the mat. Mat thickness was calculated as L minus above-mat length. The free water depth (b) was found by difference. Peat depth was estimated using a papyrus culm which was pushed into the sediment until a clear change in resistance between the soft peat and stiff sediment was met. The determined levels were later related to standardized sea level after a survey of the swamp was done.

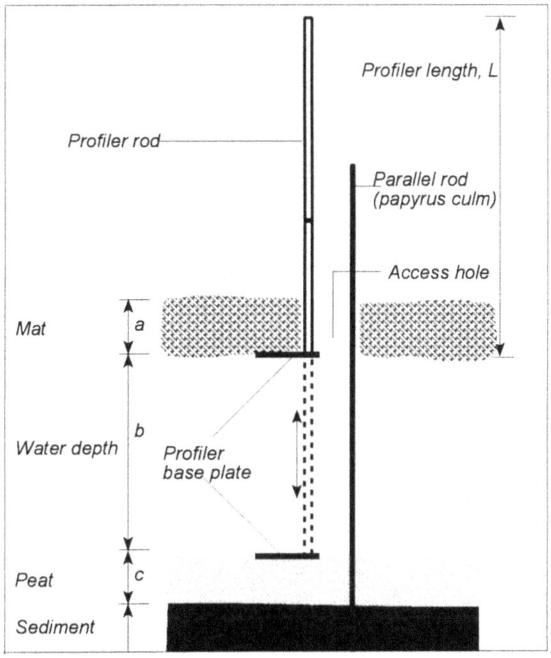

Fig. 3.3. Determination of swamp profiles using the profiler.

Because of variations in water levels due to seiches from the lake, surveying necessitated reading levels of fixed points. Graduated bamboo poles (water stages) were therefore driven into the firm sediments and their top levelled. The true water level at a particular time was then read as the level of the bamboo top minus the difference between this level and the water surface.

3.2.2 Water flow patterns in the swamp.

With a knowledge of the swamp bathimetry, electrical conductivity was used as a tracer for the determination of the major flow path of the wastewater in the swamp. EC was measured *in situ*, when samples were collected in sequence (from inlet to T4) or samples were kept under cool shades and measured within 7 hours of collection when simultaneous sampling was done. In the case of the latter, Makerere University students were stationed, one per sampling point to read water levels every 15 minutes and take water samples every 30 minutes or one hour. A portable conductivity meter (WTW) was used to read both EC and water temperature.

3.2.3 Water balance of the system

3.2.3.1 Description of water sources

i) Surface flows (Q_{into})

The lower Nakivubo swamp has a number of seasonal surface water/drainage streams flowing into it from the Luzira side (Fig.3.2). From Bukasa, no defined channels are present and the surface flow follows the natural topography. In addition, two major flow sources were described. These are the Nakivubo channel with wastewater from the city (entering the study area via the 9 m wide rectangular concrete culvert beneath the railway line) and the Luzira Prisons raw wastewater (which enters the swamp at culvert 6). A description of these wastewater channels can be found in Chapter 2.

Surface flows were measured using: (i) the area-velocity method (using an Ott-current meter) for culvert 1 of the major channel, (ii) the salt dilution method for the other minor culverts (2-5), and (iii) local water consumption and waste production estimates for the Luzira flows. In addition, a stage was put at the Nakivubo channel inlet and water levels recorded twice per day (7.00 a.m. and 7.00 p.m.). With a set of water levels and discharges measured during both dry weather flow periods and the rainy season, a discharge-rating curve was made. Water levels measured during the 17 months period over which the water balance is constructed were converted to equivalent inflows.

ii) Sub-surface flows (Q_s)

The seasonal sub-surface flow from Luzira reaches the swamp via culverts 2-6. Water from the Bukasa side reaches the swamp from a number of springs (spread along the edge) that are seasonally used as water abstraction points by the local population. The rest of the water enters the swamp as run-off.

For an estimation of this water contribution from the surrounding highlands into the valley, the watershed for the Bukasa side of the swamp was derived from a topographic map. Direct runoff and infiltration factors were estimated from curve number (CN) and infiltration charts according to Chow *et al* (1988) and sub-surface flows into the area as a function of precipitation. Luzira flows were in addition related to discharges in culverts 2 to 6.

A number of exploratory shallow bore-holes were augered (2 - 5 metres deep, using a 15 cm diameter hand auger) into the land adjoining either side of the swamp. This was done to determine the nature of the sub-surface soil strata around the swamp's edges and to provide an insight into sub-surface water flow characteristics from the swamp surroundings.

iii) Ground water flows (Q_g)

Given the complexity of the hydrological analysis of the system, only basic variables that support the overall study were studied. To determine whether the swamp water body is in significant communication with ground water the hydro-geology of the swamp was described and the infiltration capacity of the wetland soils determined. However, the installation of observation wells was beyond the capacity of this study. Hydro-geological records of the area closest to the swamp have been compiled earlier (Winters, 1922 a and b; Groves, 1929; Wayland, 1932; Harris, 1945) as they explored the possibility of ground water supply for Kampala.

To further confirm these earlier descriptions of the hydro-geology of the area, the resistivity of swamp soils was measured by the Vertical Electrical Sounding (VES) method according to Schlumberger's array (Dobrin, 1976). Sounding was done at four locations (see points 1 to 4 on Fig. 3.2). According to the array, the apparent resistivity of a soil formation, k [Ω.m], is given by:

$$k = G \frac{v}{I} , \tag{3.2}$$

where G = geometric factor for Schlumberger's array which depends on electrode separation, v = potential difference between two points [V], and I = current flowing through a layer [A]. The field results were interpreted by hydrogeologists at the Directorate of Water Development (DWD) Luzira.

Finally, the permeability of swamp soils was estimated using a varying head permeability test (Smith and Wheatcraft, 1993) on intact and undisturbed soil cores taken from the swamp at T1 150, T2 100, T2 225, T2 400 and T2 500 (Fig.3.2).

iv) Velocity measurements

Tracer tests using NaCl were made to determine the velocity of water flow in the swamp between the inlet and T1 during the rainy period (22nd and 23rd Nov. 1994). High flows (due to the rain) and the location of the test points ensured minimal effects by seiches. 10 kg of salt was mixed in 20 litres of the waste water and injected in a single pulse. At the same time, conductivity measurements at T1 259 were made every 5 minutes initially, increasing the frequency to every minute once EC started to rise. The average travel time for the 230 m distance was estimated from $\bar{t} = \frac{\Sigma\, c.t}{\Sigma\, c}$, (3.2a)

where c is the concentration and t is the time.

3.2.3.2 Hydro-meteorological data

i) Rainfall (R)

Monthly rainfall data for two stations within 8 km of the swamp (Makerere Weather Station and Kibira Road Station) and covering 30 years were collected from the Meteorological Department in Kampala. The assumption that this rainfall was evenly distributed over the study area and its catchment may create large errors in the short term, although analysis of the variance of means for annual totals showed no significant difference ($p=0.05$). The 30 year average data was used to describe the meteorology of the area (Chapter 2) and in the simulation model as an input.

For the water balance however, daily rainfall data during April 1995 to August 1996 was used. Hyetographs were plotted for these 17 months. The rainfall distribution into the system was later estimated for each section of the swamp (see section 3.4) by considering its catchment area.

ii) Evapotranspiration (ET)

Daily temperature (minimum and maximum), wind speed, relative humidity and sunshine hours for the 17 months (April 1995 -August 1996) were obtained from the Meteorological Department in Kampala, and applied in the determination of evapotranspiration.

ET determination is commonly achieved either by indirect measurement or through estimating using empirical methods. Direct measurement methods include: (i) tanks and lysimeters, (ii) field plots, and (iii) inflow-outflow measurements. Although there are controversies regarding the use of lysimeters (with respect to natural conditions of soil profile and moisture regimes, water application, plant root characteristics and net energy exchange), they give reliable ET measurements over a short time if installation satisfies certain minimum requirements (Singh, 1989). The third method only gives gross estimates of ET for large intervals like seasons or years and would not be applicable to the Nakivubo swamp whose flows are complicated by seiches and waves from Lake Victoria. However, in the constructed wetlands in Jinja (IHE-Delft/RIZA project), direct inflow-outflow-storage measurements were used to estimate ET in papyrus stands.

The various ET estimation methods can be classified into: (i) temperature methods, (ii) humidity methods, (iii) regression methods, (iv) radiation methods, (v)energy balance methods, (vi)mass transfer methods and (vii) combination methods (Saxton, 1982; Singh, 1989; Burman and Allen, 1990; Burman and Pochop, 1994). In this study the energy balance -combination method was used to estimate ET over the swamp.

To take into account the effect of aerodynamic resistance due to the wind and the canopy resistance, the Penman/Monteith combination method was used. The method employs the terms defined in Fig. 3.4. The major driving force for evapotranspiration is sunshine (Jensen *et al*, 1990; Deuver, 1990; Kadlec and Knight, 1996). The equation of the heat balance for the net incoming radiation R_n (MJ/m^2/d) (all terms per unit surface area) is:

$$R_n = \rho\lambda_m.ET + (U_o - U_i) + G + H_a + \Delta S, \qquad (3.3)$$

in which ρ = density of water (kg/m^3), λ_m is the latent heat of vaporization of water (MJ/kg) (2.453 MJ/kg at 20°C), U_i and U_o = heat content of the incoming and outgoing water, respectively (MJ/m^2/d), G = conductive heat transfer (MJ/m^2/d), H_a = convective heat transfer and ΔS = change in heat storage of a wetland (both in MJ/m^2/d). ET is in (m/d).

Since R_n is usually 10 - 20 MJ/m^2/d, compared to $(U_o - U_i)$ which is about 0.5 MJ/m^2/d (for a temperature difference between the influent and effluent of 5°C) and the peak storage changes are in the order of 0.5 MJ/m^2/d as well (Kadlec and Knight, 1996), these terms can be

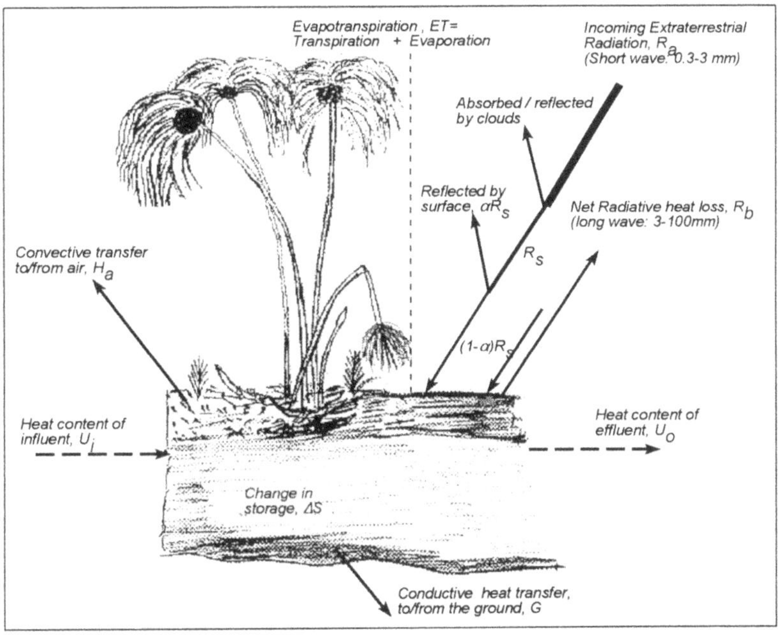

Fig. 3.4. Heat balance terms for a wetland.

neglected. The net incoming radiation R_n is thus simplified as:

$$R_n = \rho\lambda_m . ET + G + H_a \tag{3.4}$$

or

$$\lambda . ET = (R_n - G) - H_a \tag{3.5}$$

Where $\lambda = \rho\lambda_m$, or the volumetric latent heat of vaporization of water (2453 MJ/m^3 at 20°C). Equations 3.4 and 3.5 state that all (net) incoming radiation leads to evaporation except for the losses by conductive and convective heat transfer.

Applying the Penman-Monteith combination method (Shuttleworth, 1993):

$$\lambda . ET = \frac{\Delta}{\Delta + \gamma^*} (R_n - G) + \frac{\gamma}{\Delta + \gamma^*} (1710 - 6.85\ T) \frac{1}{r_a} [P_w^{sat}(T) - P_w^{sat}(T_{dp})], \tag{3.6}$$

for temperature T (°C). T_{dp}=dew point temperature (°C); Δ = slope of the vapour pressure curve (KPa/°C); $P_w^{sat}(T)$ =saturation water pressures at T (KPa). The psychrometric constant, γ is = $c_p P/0.622\lambda$ (kPa/°C); c_p is the specific heat of moist air at constant pressure (1.013 kJ/kg°C); $\gamma^* = \gamma(1 + r_c/r_a)$; r_a (s/m) = aerodynamic resistance given by (Jensen *et al* 1990):

$$r_a = \frac{\ln[(z_w - d)/z_{om}] \ \ln[(z_p - d)/z_{ov}]}{(0.41)^2 \ u_z} , \tag{3.7}$$

for wind velocity u_z (m/s). z_w = height of wind speed (u_z) measurement (m); z_{om} = roughness length for momentum transfer (=0.123h_c), h_c = height of crop (m); z_p is height of humidity and temperature measurements (m); z_{ov} = roughness length for vapour transfer (= 0.1 z_{om}) and d = displacement height for a crop (=2/3h_c; m). The crop surface resistance to vapour transfer, r_c, depends on soil saturation and plant type and attains a minimum value of 30 s/m for arable crops to 150 for forests. Grass has an equivalent value of 70 s/m. Equation 3.8 for non-clipped grass and alfalfa harvested only periodically, was used to estimate r_c (Shuttleworth, 1993):

$$r_c = 100/[0.75 \ln (100 h_c) - (0.7)] \cdot \tag{3.8}$$

Δ, is given (Kadlec and Knight, 1996) as:

$$\Delta = \frac{dP_w^{sat}}{dT} \approx \frac{P_w^{sat}(T) - P_w(T)}{T_w - T_{dp}} , \tag{3.9}$$

and T_{dp} was determined according to Kadlec and Knight (1996) from:

$$T_{dp} = \frac{5349.93}{19.0971 - \ln P_w(T)} - 273.16 . \tag{3.10}$$

The actual water pressure is $P_w(T) = P_w^{sat}(T) \ x \ RH$ in which RH is relative humidity. The average saturation water pressure, $P_w^{sat}(T)$ is the average of the saturation water pressures at maximum and minimum air temperatures $P_w^{sat}(T_{min})$ and $P_w^{sat}(T_{max})$ in equation 3.10 with T = ambient temperature (°C).

The net incoming radiation R_n in Equation 3.3 (also refer to Fig. 3.4) was calculated as:

$$R_n = (1 - \alpha)R_s - R_b, \tag{3.11}$$
$$= 0.77R_s - R_b$$

where α is the fraction of the solar radiation, R_S, reflected by the wetland (wetland albedo). For green crops, α = 0.20 - 0.25, say 0.23 (Burman and Pochop, 1994; Kadlec and Knight, 1996); in a wetland, there is no significant difference between vegetation types. R_b is the net outgoing long wave radiation (MJ/m^2/d).

R_S is given by Equation 3.12 (Shuttleworth, 1993) where the extraterrestrial radiation (MJ/m^2.d), R_a, depends on the latitude of an area and season of the year and S is the fraction of bright sunshine hours to the total hours in a day,

$$R_s = (0.25 + 0.5\frac{S}{100})R_a \quad (3.12)$$

For the study area, values for R_a were deduced from Jensen *et al.* (1990) (Table 3.1). R_b was computed as a function of atmospheric characteristics of cloud cover, absolute temperature and moisture content according to Equation 3.13 (Shuttleworth, 1993).

$$R_b = [0.33 + 0.67\frac{S}{100}][0.34 - 0.139\sqrt{P_w^{sat}(T_{dp})}]\ \sigma\ (T + 273.16)^4 , \quad (3.13)$$

where T is the ambient air temperature (°C), S is percent daily sunshine (-) and σ is Boltzman's constant = 4.903 $*10^{-9}$ MJ/m^2.d.

With the meteorological data of minimum and maximum temperature, relative humidity, sunshine hours and wind velocity, the daily ET in Equation 3.13 was calculated in a spreadsheet for April 1995 to August 1996. A sample output is shown in Appendix 3.1. Since G is usually negligible compared with R_n (Jensen *et al*, 1990), it was ignored in this calculation.

Table 3.1 Extraterrestrial radiation (R_a) at the equator 0° latitude (the Nakivubo swamp is located at 00°17' N - 00°19' N)

Month	Jan	Feb	Mar	Apr	May	Jun	Jul	Aug	Sept	Oct	Nov	Dec
R_a (MJ/m^2/d)	36.2	37.2	37.5	36.8	34.8	33.5	33.8	35.6	37.2	37.5	36.5	35.7

3.2.3.3 Seiches from the lake

Seiches are oscillations of a water body at its natural period (Pugh, 1987). The period depends on the physical dimensions of the basin, and on the periodicity and energy of the seiche producing force in the case of forced seiches (Chow, 1972). Seiches are caused either by a sudden gust of wind, tsunamis (submarine earth quakes) or by barometric pressure (storm surges). Once natural seiches are established, they can persist for several hours to weeks (Pugh, 1987), indicating least frictional dampening.

The natural period, T, of seiches with n multiple harmonics in a basin is given by the Merian formula (Eq. 3.14, Pugh, 1987) where L is the length of the basin [m], D is the depth [m] and g is the acceleration due to gravity [m/s^2].

$$T_n = \frac{2L}{n\sqrt{(gD)}} \quad (3.14)$$

For bays and harbours (including the Murchison Bay), the pivot node for oscillations is at the open entrance of the bay and the anti-node at the shore (Pugh, 1987). The period is then given by:

$$T_n = \frac{4L}{n\sqrt{(gD)}} \tag{3.15}$$

The theoretical period of seiches in Murchison Bay and in the whole Lake Victoria was calculated and compared with the observed oscillations at the swamp - lake interface. The results were used to formulate a series function which was used as an input into the model (DUtchFLOW or DUFLOW) to simulate longer term lake effects on water quality modifications by seiches.

Seiches were measured by recording water level variations. Level gauges (made of bamboo poles calibrated to the nearest cm, and driven into the firm clay/sediment) were put in the swamp at locations T1 50, T1 150, T1 259, T2 0, T2 225, T2 400, T2 600, T2 750, T3 200, T3 400, T4 0, T4 200 and T4 400 (Fig. 3.2). Measurements in the lake were hampered by the perpetual presence of the water hyacinth. By stationing one field assistant at each measuring point, it was possible to measure water levels and take water samples (for conductivity and nutrients measurements) simultaneously for all points. Readings were made every 15 minutes while samples were taken every 30 minutes for EC and every 1 hour for nutrient measurements. Measurements were made between 0900 hrs and 1600 hrs. At T4 400, an automatic logger fitted with a level sensor (accuracy $\pm$ 0.1 cm) , EC and temperature metres were also stationed and the difference in water level and conductivity over time recorded. Samples were kept under cool vegetation shades prior to measurement within 7 hours. EC was read using a WTW portable conductivity meter as described in Chapter 4.

3.2.4 Data presentation

All numerical data is presented as Mean $\pm$ SEM (Standard Error of Mean). Comparison of the means was done using ANOVA (Minitab, release 10).

3.3 Results and discussion

3.3.1 Basin geometry

The swamp mat-surface slope is very moderate although the basin profiles are very variable with respect to mat thickness, water depth and sediment thickness. Water depth (Fig. 3.5) for example changes from near zero at the edges of the swamp to about 3 m in the centre of the system and near the lake.

The width of the swamp varies from about 500 m to 850 m. Typical cross sections along the transects are shown in Fig. 3.6. The profile for T1 (closer to the inlet) shows a shallow water depth overlying a thick peat layer across the section. (In addition, T1 has an island at about

20 m width which acts as a barrier to free flow of water in the lateral direction. During rainy periods, a stream crosses this barrier eastwards in agreement with the swamp topography). As one moves towards the lake, the free water depth increases, the peat layer becomes thinner and the mat (especially in the *Miscanthidium* vegetated regions) increases. These profiles have an impact on the velocity of water flowing through the swamp, since they affect the cross-sectional area of and the resistance to flow.

The mat thickness is related to the vegetation type distribution as discussed in Chapter 2. Papyrus zones have thinner and more variable mat thicknesses compared to the thicker and more uniform mats of *Miscanthidium*. Thin mats in the papyrus zones are a result of a more loose mat-rhizome-root structure (Chapter 5) which facilitates the shedding of organic debris, thus leading to a high solids content in the water column and therefore a higher resistance to flow. On the other hand, the structure of *Miscanthidium* mats keeps organic debris from falling through thus becoming thicker and maintaining a much clearer water column. Thick mats also tend to reduce the free cross sectional area of flow (free water column, Fig. 3.6), and therefore increase flow velocities. Thinner free water layers and higher debris/particulate matter content in the water column near the swamp edges suggest less flow through these areas.

3.3.2 Water flow patterns

Electrical conductivity was used as a tracer to follow the flow path of wastewater through the swamp since the wastewater entering the swamp had a higher conductivity than the water from

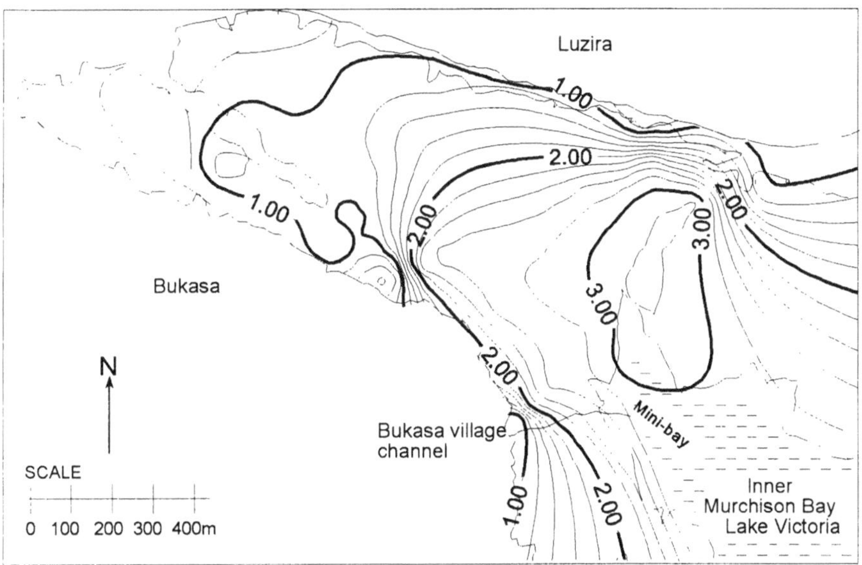

Fig. 3.5. Isolines of water depth distribution in the Nakivubo (m) swamp.

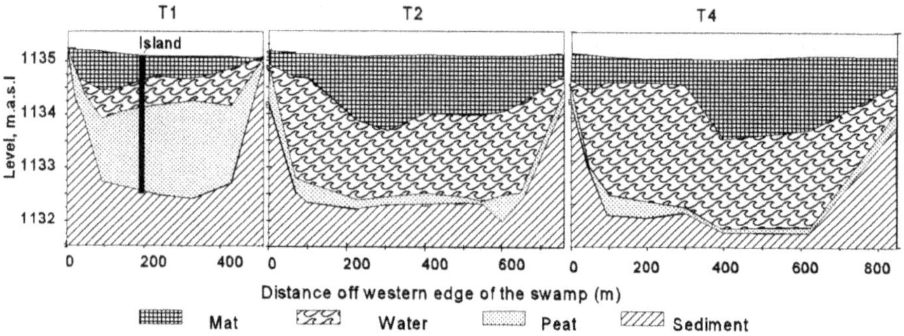

Fig. 3.6. Cross sections of swamp depths along transects 1, 2 and 4 (T1, T2 and T4).

the springs around the swamp (Nakivubo channel inlet = 438 ± 4 μS/cm; spring near Luzira range is 20 - 45 μS/cm).

Average EC profiles for T1 to T4 measured over two months in 1994 are shown in Fig. 3.7. Conductivity along T1 and T2 was lower for distances within 200 m from the western edge of the swamp (Bukasa). This corresponds to areas of lower mat thickness and water depth and

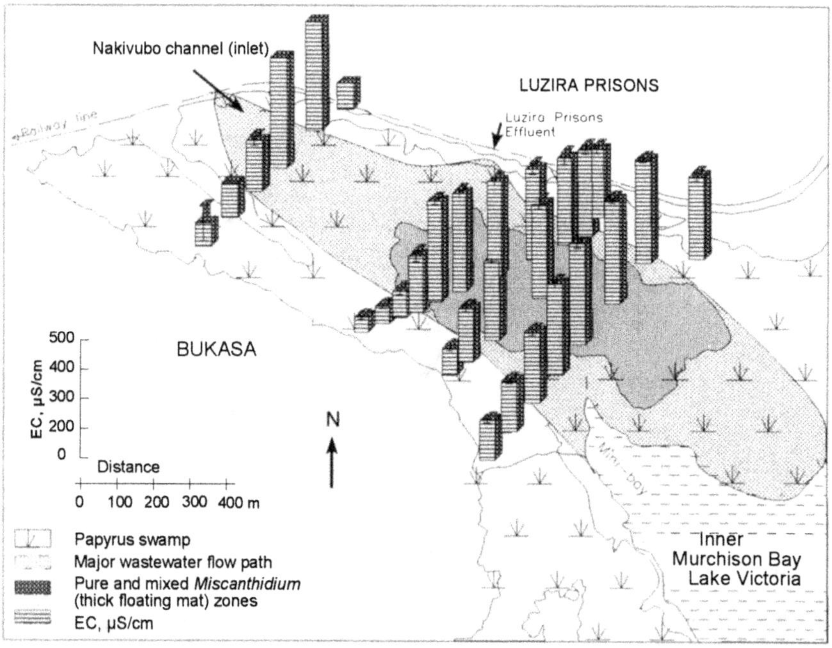

Fig. 3.7. Variation of EC along T1, T2, T3 and T4; n=7. March and August 1994. The major wastewater flow path is also shown.

indeed to the papyrus zones. Higher EC values were found in the centre of the swamp where the free water depth is greater (also corresponding to the mixed vegetation zone), and nearer to the eastern edge of the swamp (possibly due to the Luzira prisons' effluent and the general inclination of the swamp topography towards the east). Water samples from areas with high EC also had the smell and characteristic blackish colour of the influent when lake seiches did not occur. This was further indication that the bulk of the wastewater passed through this major flow path.

Lower EC values at the edges of the system could be explained either by assuming that wastewater does not reach the edges and hence they were not participating in wastewater purification or that sub-surface flow from the swamp edges was diluting the edge water and lowering EC. Since sub-surface flows were the result of rainfall, edge EC values during a typically dry season (January, 1994) are compared with those during a wet season (March, 1994; Fig. 3.8). The second half of December 1993 and January 1994 had an average rainfall of 0.7 mm/d compared to the average amount for March of 5.3 mm/d.

Figure 3.8 shows that during both dry and rainy periods, edge EC values are significantly lower than those in the main wastewater flow path (beyond 200 m from Bukasa in the figure). Edge EC values for T2 show very little variation during a particular period, although they are higher for the rainy period than for the dry period. The higher rainy-weather EC values are

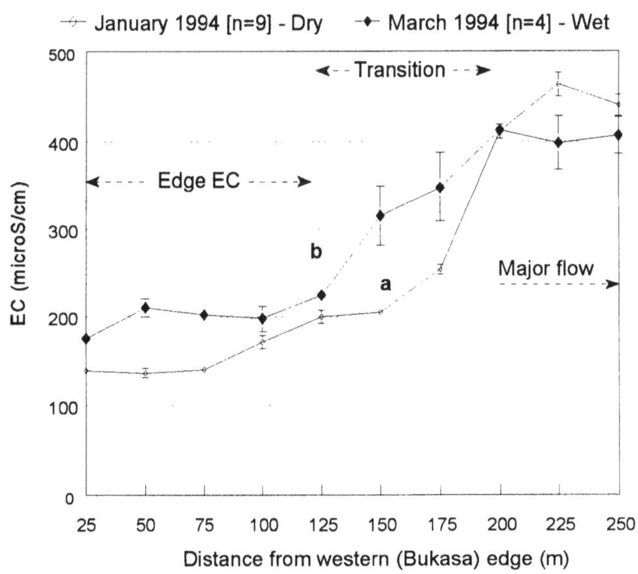

Fig. 3.8. EC on the western side of the swamp: T2 during the wet and dry seasons. EC during the wet season **b** is significantly higher than that of the dry season **a** for distances before 175 m ($p<0.001$, DF=82, F=21.8).

most likely a result of leaching and flushing of nutrients from plant material (Chapter 5) as rain water floods the mat. This effect appears to outweigh the dilution caused by rainwater which would reduce EC. The reduction of EC during the dry season could be due to plant uptake (Chapter 5), sediment sorption (Chapter 8) and/or dilution by sub-surface flows (which must be minimum in the dry period based on observations of dry pools and wells on the Bukasa edge of the swamp). Towards the main waste water flow path (beyond the distance of 200 m off the western edge of the swamp), the effect of rainfall on the dilution of Nakivubo Channel flows is depicted in Fig. 3.8. Conductivity for the rainy period were lower than those for the dry period.

Electrical conductivity values for the swamp edges (within 200 m from Bukasa and close to the railway line on the Luzira side) remained out of range of the influent (66-311 μS/cm). This suggested that the water had a preferential flow path in the centre of the system. The results of this study were later supported by nutrient and faecal coliform profiles (Chapters 4 and 6).

3.3.3 Velocity measurements

Using equation 3.2a , results from the tracer tests gave a travel time of wastewater from the inlet to T1 259 of 66 and 71 minutes (or average velocity = 0.056 m/s) on 2 separate occasions as shown in Fig. 3.9. Within the swamp (T2 and T3) however, the test failed because of dilution by the lake seiches, which change the whole flow pattern. Seiches are hardly noticed at T1 possibly due to the high flows at this point close to the inlet of the channel into the swamp and the dampening effects of the plant system on the seiche energy.

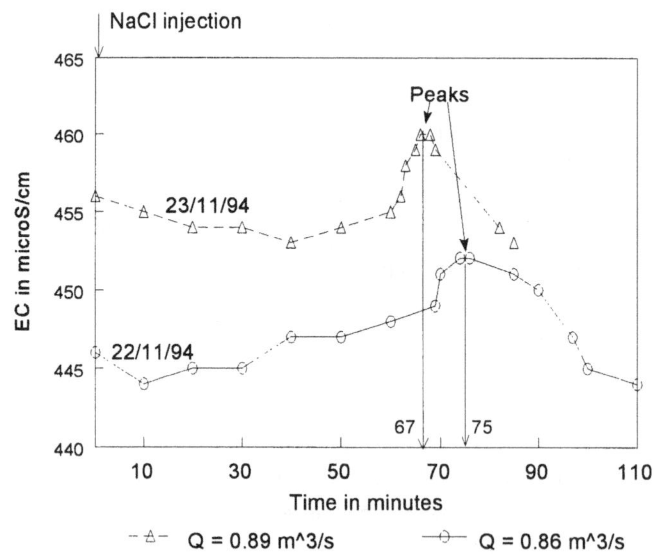

Fig. 3.9. Velocity determination in the swamp using NaCl. Salt injected at the inlet and EC measured at T1 259.

The flow velocity and the surveyed bottom slope (9.9 cm in 230 m) were used to estimate Manning's coefficient, *n*, according to: $n = \frac{1}{v} R^{2/3} S^{1/2}$, (3.16)

R = hydraulic radius $\approx H/2$ (H = flow depth, m; for large shallow basins, $R \approx H$; for floating swamps, the wetted perimeter is $2(H+W)$; W = flow width, m). The hydraulic radius, R = 0.30 m (during the measurements); the hydraulic gradient, S [-]= 0.099/230; and the velocity v= 0.056 m/s. Hence, $n \approx 0.17$. This range was used as an initial input for the model (Section 3.4). It is noted that this velocity corresponds to the upper range since the water was visibly channelized even in this comparatively short (230 m) stretch.

An estimate for the effective area of flow for the two discharges and velocities from A=Q/v gives 15.5 m^2 and 16.8 m^2 for 0.89 m^3/s and 0.86 m^3/s respectively. For H = 0.6 m , the effective flow width W is about 30 m (or only 10% of the total cross section of T1 beyond the island).This is the portion which will be in continuous contact with the bulk of the waste water, an indication of the channelization of flows in the swamp. Assuming a constant flow width for all discharges, then the range of flow velocities for this part of the swamp (for the influent discharges of 0.52 - 4.5 m^3/s - Section 3.3.4.1) would be 0.029 m/s - 0.07 m/s.

3.3.4 Water balance terms

i) Discharge in Nakivubo channel, rainfall and ET

Figure 3.10 shows the rating curve for the Nakivubo channel. Channel discharge is very highly correlated with depth of flow (r = 0.992, F = 2302, p<0.001, n = 37) and follows the regression equation: $Q[m^3/s] = 9.52H[m] - 2.95$. This equation was used to transform the recorded daily water levels of the channel into discharge values. The daily discharge was

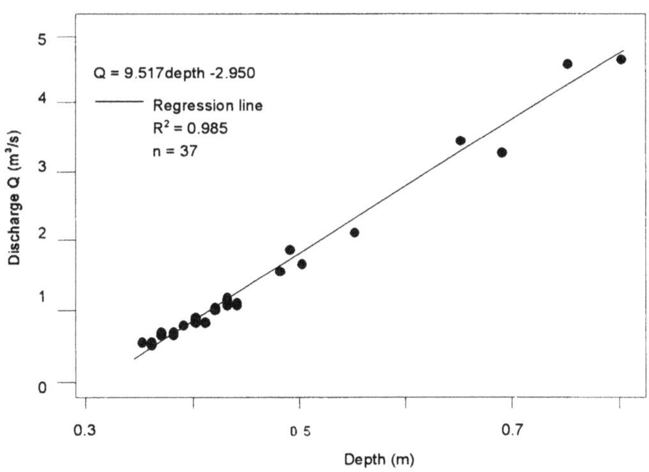

Fig. 3.10. Rating curve for Nakivubo channel.

taken as the average of morning and evening readings. Figure 3.11 shows the time series for this discharge, (averaged monthly for April 1995 - August 1996) ET and rainfall over the same period. Average daily ET experienced very little variation over time compared with rainfall and channel discharges (Fig. 3.11). The correlation between rainfall and channel discharges is depicted. Evapotranspiration is less variable over the year.

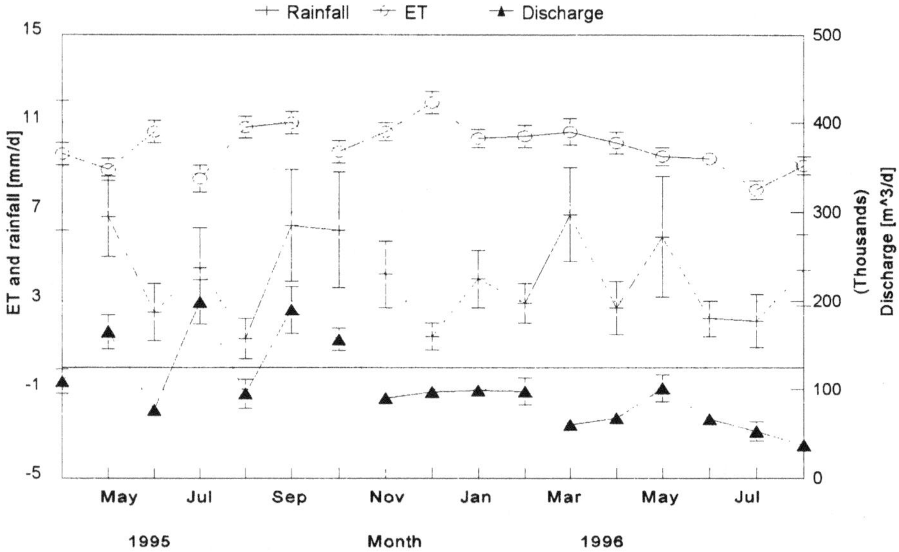

Fig. 3.11. Water balance terms for Nakivubo swamp: rainfall, Nakivubo Channel discharge and crop evapotranspiration [ET]. Vertical bars are $\pm$1SE.

Since the Nakivubo channel is also the drainage channel for central Kampala City, a cross correlation analysis for the whole data set showed a travel time of the water of less than 1 day; receding to dry weather flow levels possibly after about 4 days (Fig. 3.12). Finer resolution of the data suggests a travel time of about 0.4 days. On average therefore, rain water travels the approximately 8$\pm$4 km (surface runoff and channel flow) from the city centre including about 0.9 km of the upper swamp (Chapter 2) in about 10 hrs. High storm flows were maintained within the swamp and the turbid runoff water could be seen at the mini-bay (Fig. 3.2) within 18 to 30 hours after a storm.

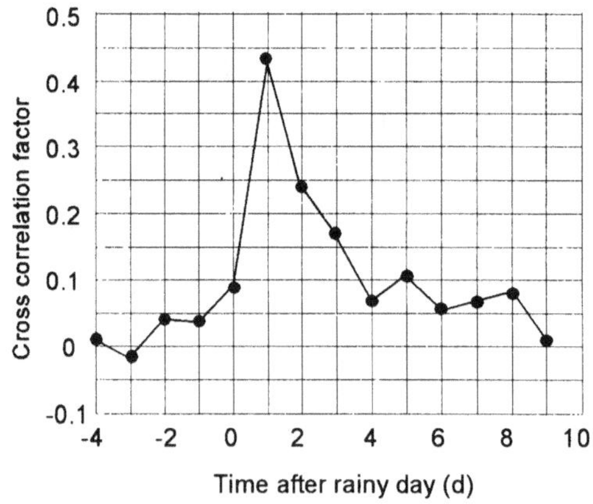

Fig. 3.12. Cross correlation of rain with the Nakivubo Channel discharge.

A summary of rainfall, channel discharge, evapotranspiration and open water evaporation data is given in Table 3.2. Total rainfall and open water evaporation (per unit area) are of the same order of magnitude. Evapotranspiration from the swamp is about 1.2 times (range 80% - 167%) higher than the open water evaporation/rainfall. Since the catchment area for rainfall is much larger than the swamp area (ratio ≈ 45:1), overall rainfall collection into the swamp is greater than evapotranspiration (rainfall in the catchment area of the Nakivubo channel is however already taken into account by the channel discharge measurements). The Nakivubo channel inflow is about 20 times as much as ET or direct rainfall, showing the significance of the channel flows to the water balance.

Table 3.2 Summary of rainfall, channel discharge potential evaporation and evapotranspiration data for Nakivubo swamp, 17 months averages

	Nakivubo channel Discharge, m^3/d	Nakivubo channel Discharge, mm/d	Rainfall, mm	Crop ET, mm/d	Potential open water Evaporation, mm/d
Mean (daily)	103575	90	4.3	5.0	4.1
Standard Error	11524	10	1.6	0.1	0.1
Minimum (daily)	9711	8	0.0	2.0	2.0
Maximum (daily)	522357	454	78.1	11.3	5.8
Sum (17 months)	51762826	45011	2218	3056	2103

ii) Runoff and subsurface flow from Luzira and Bukasa

Figure 3.13 shows the topography and the catchment area for Luzira and Bukasa hills. Using aerial photographs (1993; the Dept. of Mapping and Surveys - Entebbe) and estimations from physical surveys in the field, land use factors (Table 3.3) were derived to get a weighted runoff curve number (CN) according to Chow *et al* (1988).

The weighted curve number is therefore 8149/100 ≈ 81.5%. The same factors were used for both sides of the hill, taking into account the different watershed areas. Therefore the potential maximum retention (Chow *et al*, 1988), $S = 25.4/CN - 0.254 = 0.058$ m. The direct runoff

Table 3.3 Determination of Curve Number (CN) for swamp catchment area

Land use	Estimated % coverage	CN	% coverage x CN
Residential (25% impervious)	25	80	2000
Roads (dirt)	3	87	261
Cultivated land	40	88	3520
Open land (good cover)	32	74	2368
TOTAL	100		8149

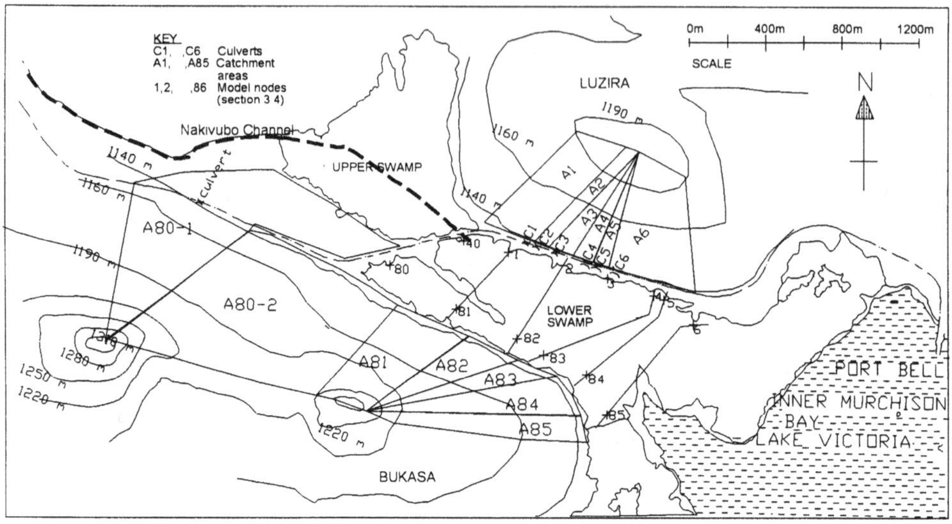

Fig. 3.13. Contributory areas for surface run-off (and sub-surface flow) for the lower Nakivubo swap. Flows into the upper swamp are mostly drained via the Nakivubo Channel.

P_e, is given by: $P_e = \dfrac{(3.937P - 0.2S)^2}{3.937P + 0.8S} * 0.0254$, where P is precipitation in mm. From these relationships, the total runoff from each area was calculated for individual rainfall events. The estimated catchment areas (Fig. 3.13; areas read using AUTOCAD, version 12) for Luzira and Bukasa catchment areas are 63.7 and 84.2 ha respectively. Daily runoff values as a function of rainfall were generated and used in the mass balance.

Table 3.4 gives a summary of the estimated runoff from Luzira and Bukasa from January 1995 to August 1996. Culvert flows from Luzira (C2-C6, Fig.3.13) are barely discernible during the dry season, except for culvert 3 which is a spring and culvert 6 which is an open sewer. During the wet season however, the culverts drain surface runoff and later (for a limited period) carry the base flow. Velocity measurements using the salt dilution method were rather erroneous and the average values could hardly be correlated with the rainfall. For example, during a 10-day rainy period in November 1994, the total measured culvert flow from Luzira was 97207 m^3/d while the total estimated rainfall on the Luzira side was 64718 m^3/d. Since

Table 3.4 Water flow characteristics for Luzira side culvert flows during a rainy period, n= 10. C2 to C6 are culverts 2 to 6 (Figure 3.13). All values are in m^3/d

Culvert	C2	C3	C4	C5	C6	sum:C2toC6	Rain
Mean±SE	1163±246	2673±210	1205±114	742±93	3936±380	9720±406	6471±2341
Range	365 - 2891	1700 - 3544	664 - 1847	405 - 1424	2575 - 6939	7904 - 12264	63 - 21913

the culvert flows are a result of the rainfall, this showed an error in velocity measurements of at least 33%. Therefore, the estimated runoff values were used in the water balance. From the water consumption values the dry weather flow for the prisons is estimated at 700 m^3/d.

3.3.5 Groundwater flows

i) Hydro-geology

The area of L. Victoria from the Luzira hills at Port Bell through Bukasa, and from Gaba to Entebbe (about 40 km to the west) is fringed by gneiss (Winters, 1922a and b; Wayland, 1932). In the valleys, including Nakivubo, the deduced (from literature and our findings) lithology of the geological formation is as shown in Fig. 3.14. The clays, probably derived from the decomposition of felspathic rocks and formed *in situ,* are impervious (Winters, 1922).

Early explorations revealed that the area consists mostly of crystalline rocks of impermeable nature with no sedimentary rocks overlying the formation and that the primary rocks are compact and not embedded by pervious strata. The rock formation is of such a nature as to preclude the possibility of a sub-artesian supply and therefore yields almost no water. This

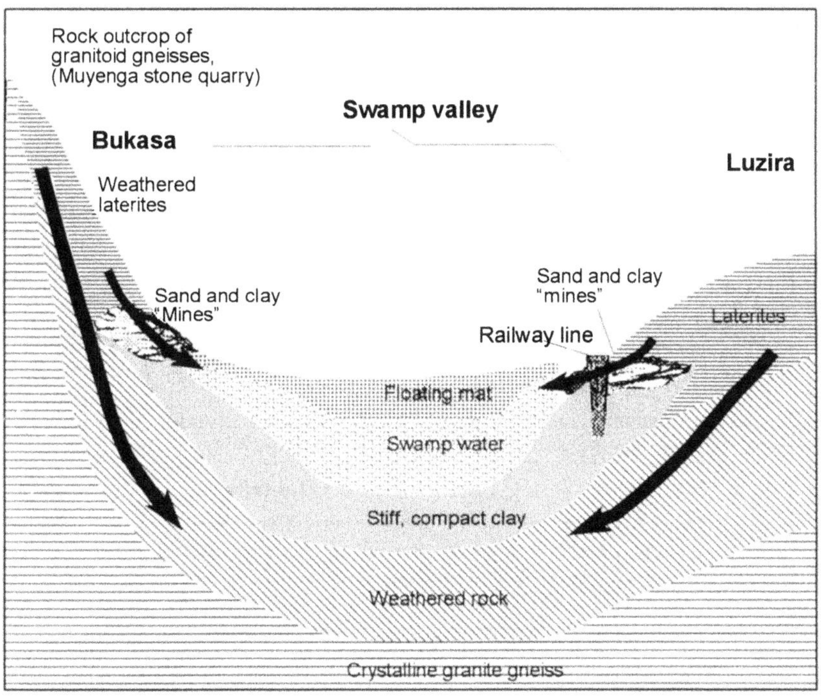

Fig. 3.14. Schematic presentation of the lithology of Nakivubo swamp. Black arrows indicate probable water flow paths. (Not to scale).

implies that the contribution of ground water to the swamp system is likely to be negligible. For the purposes of the water balance, seepage through the laterite soils into the weathered rocks is not relevant due to the thick overlying clay layer (9 -36 m; Table 3.6).

ii) Hydraulic conductivity measurements

Characteristics of cores from the swamp are given in Table 3.5. For all cores, clay (stiff and sometimes "dry") was encountered. The hydraulic conductivities for these shallow cores indicate the very low permeability typical of clay (Shaw, 1988). This further indicated negligible exchanges between the swamp water body and the shallow ground water in the laterite and weathered rock layers.

Table 3.5 Infiltration measurements

Location	Total sediment depth l (m)	Clay depth (m)	Q after 48 hrs (ml)	Final length H2 (m)	T2 -T1 (s)	$k = al(\ln H1 - \ln H2)/[(T2-T1)A]$	
						m/s	cm/hr
T2 100	0.35	0.25	15.3	0.988	172800	2.5E-08	8.9E-03
T2 225	0.34	0.24	61.2	0.951	172800	9.8E-08	3.5 E-02
T2 300	0.30	0.17	12.3	0.990	172800	1.7E-08	6.1E-03
T2 400	0.34	0.17	18.5	0.985	172800	2.9E-08	1.1E-02
T2 500	0.32	0.12	23.9	0.981	172800	3.6E-08	1.3E-02

Corer internal diameter = 0.04 m, area =1.257E-3 m^2, initial length, H1 = 1.0 m, a/A (smaller area/larger area was 1.0)

iii) Resistivity measurements

Table 3.6 shows that at all the points surveyed, the thin layer of top laterite soils (1 to 2 m) are underlain by thick clay aquiclude ranging from 9 to 35 m thick. This was also supported by observation at boreholes made along the edges of the swamp that had a mixture of clay and sand in the top 1 to 2 m. When the sand content allowed for deeper boring, thick compact clays were encountered within 2 to 3 metres depth.

Table 3.6 Layer resistivities and depths for the Nakivubo swamp area (Fig. 3.2). D (m) shows the depth, in metres, from the surface between which a particular material is found

Location (sites 1,2,3 and 4, Fig. 3.2)	Laterite/sandy soils		Clays		Weathered rock		Hard rock	
	k (Ωm)	D (m)	k (Ωm)	D (m)	k (Ωm)	D (m)	k (Ωm)	D (m)
1.Luzira 1	200	1	20	1-14	100	14-39	10E4	39 plus
2.Luzira 2	150	1	50	3-38	80	38 plus		
3.Bukasa 1	50	1	20	1-10	150	10-39	10E3	39 plus
4.Bukasa 2	150	2	15	2-27	250	27 plus		

3.3.6. Seiches from the lake

Seiche periods in the bay were estimated from water level variations like that given in Fig. 3.15a. for example. A compilation of logged water level readings at different times is shown in Table 3.7. Average periods per session were estimated from the number of cycles (peaks) divided by time. The period for Fig. 3.15a., for example, was 133 minutes (108 peaks in 14,400 minutes). The seiche period was 133.5 ± 1.0 minutes (range 130-137 minutes, n = 6).

Table 3.7 Seiche period determination for Murchison Bay and Lake Victoria

	Aug. '95	Nov.'95	Jan.-Feb.'96	Mar. '96	Apr.'96	May '96
Time (min)	3000	14400	14880	4080	6840	4320
No. of cycles	23	108	111	31	50	32
Period, T (min)	130	133	134	132	137	135
Max. amplitude (cm)	4.0	11.3	9.7	10.6	10.3	-

Gauff and Parkman (1988), on the other hand, measured the regularly changing velocities of water currents at 3 m depth, close to the water treatment plants at Gaba (Chapter 2) for 5 hours in December 1987. The following peak velocities were measured at following times: 0.05 m/s at 09:15 hrs; -0.067 m/s at 10:30 hrs; 0.033 m/s at 11:30 hrs; and -0.033 m/s at 12:45 hrs. From these results, the cycle period (from peak negative to peak negative or from peak positive to peak positive) is for both cases 2h 15min or 135 minutes, a value very close to our deductions from level measurements.

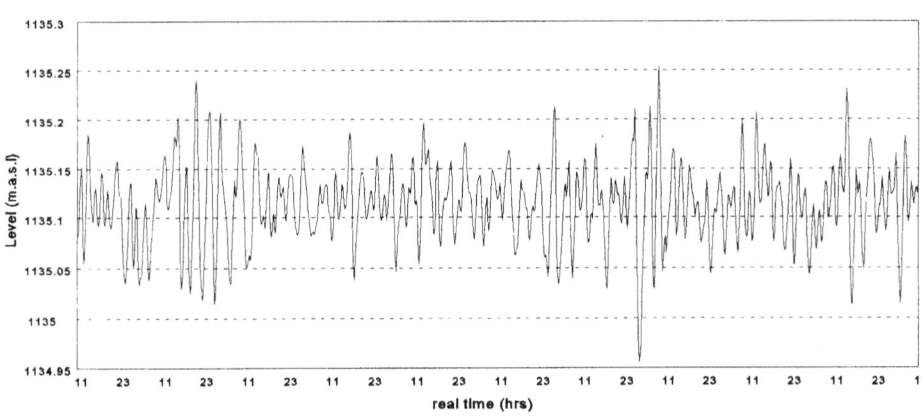

Fig. 3.15a. Water level time series at T4 400, 7 to 17 November 1995.

One can discern the dominant period generated by Lake Victoria, which is the strongest with the lowest frequency. It is visible on 8 Nov and 13 Nov. It has a periodicity of 225 min. With a free width of the lake between the island of about 200 km, and a periodicity of 225, eq.3.14 yields an average depth of 49m, which is close to the average depth in that part of the lake. The higher periods result from the waves that can develop in the part of the lake that is enclosed between the shore and the Sese islands (which also obeys eq. 3.14) and in the semi-enclosed Inner Murchison Bay (which obeys eq. 3.15). These waves can develop as a result of refraction of the main seiche from Lake Victoria, but can also be generated by local winds. The average period of 135 min is the result of the combined occurrence of these three waves. The periods of these waves can be accurately determined through spectral analysis.

Sample spectra of water level measurements are shown in Fig. 3.15b. The spectra of band width equal to 21 lags were chosen after studying the plots of several different maximum lags. The spectra show concentrations of variance in the harmonics for the following 'periods': 2.4, 3, 4, 6, 9, 15 and 50 for all seasons; 2.3, 2.8, 4.3, 6, 9 and 16 for the wet season and 2.4, 3, 4.5, 6, 9, 15 and 33 for the dry season. The major periods for all three plots occur at periods of 90, 135 and 225 minutes. Since hourly wind patterns are not available, it is not easy to relate these values to physical situations. These values however emphasize the likelihood

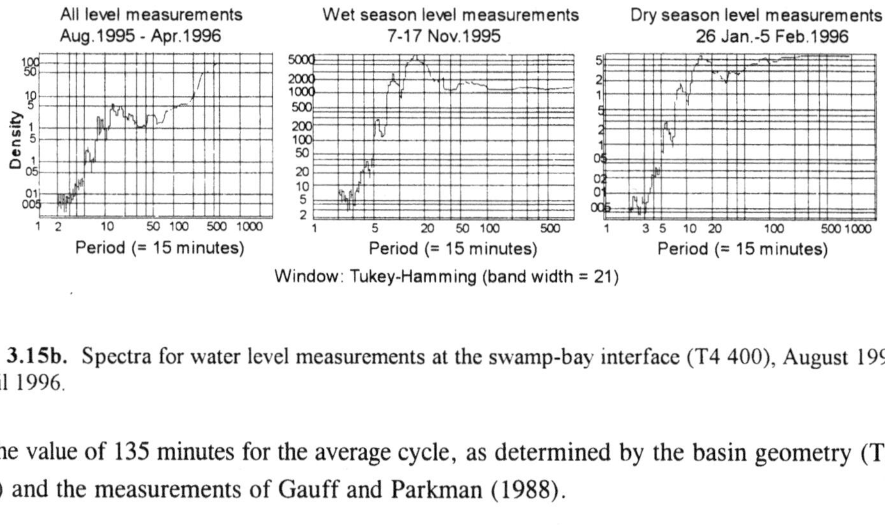

Fig. 3.15b. Spectra for water level measurements at the swamp-bay interface (T4 400), August 1995 to April 1996.

of the value of 135 minutes for the average cycle, as determined by the basin geometry (Table 3.7) and the measurements of Gauff and Parkman (1988).

Seiches in Nakivubo swamp tend to affect the purification capacity of the swamp by modifying the retention time of the water through build-up and flushing effects as well as resuspension and transport of sediment and mat materials. Considering a cycle period of 135 minutes and an average water level change of 0.1 m per cycle, the daily swamp - lake water exchange for the effective swamp area (785,000 $\pm$ 10,000 m^2; Chapter 2) with an average water depth of

2.5 m would be {(24*60)/135 [daily frequency]}*{0.1*785,000} = 837,300 m^3 . This amount is 8 times larger than the average daily wastewater flow of 103,575 m^3. For the total water volume in the swamp of approximately {1,150,000 m^2 *2.5 m} = 2,875,000 m^3, the turnover rate (period in which all the water in the swamp would have been replaced by a combination of lake water) would be about (2,875,000)/(103,575+837,300) ≈ 3.4 days. The swamp would therefore be completely flushed to lower EC levels of about 130 μS/cm within 3.4 days. This is however not the case. Although a very large volume of water is involved in the exchange process, the swamp is kept from complete washout possibly by considering that water is not replaced over the whole depth, but rather, the same water is frequently recycled in a "plug" or "piston" flow format. The exchange process on the lake side is also not complete, much of the same water re-enters the swamp.

For all measurements, a significant relationship between the water level and conductivity existed. Figures 3.16 and 3.17 show this relationship for November 1995 and January 1996. A dilution experiment for wastewater from the swamp with lake water was linear ($r = -1$, $n=20$, $p<0.001$), showing that over a short time interval, mixing swamp water with lake water results in dilution alone. In the swamp, deviation from a linear fit can be explained by temporal changes in conductivity of the wastewater in the Nakivubo channel (Chapter 4) and

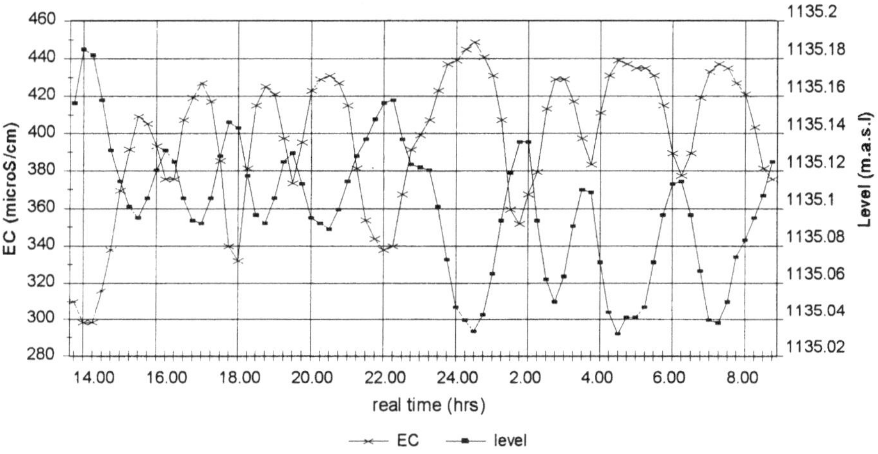

Fig. 3.16. Time series of water level and EC variation at T4 400, 7-8 th November 1995; $r^2 = 0.90$; EC = 1,026,006 - 903 Level ±17; $p< 0.001$, $n = 77$. Lake water EC ≈ 90 μS/cm.

system processes such as leaching, plant uptake and sorption onto sediment particles.
When the rise in water level was recorded simultaneously for several locations within the swamp, it was observed that currents reached locations along T1A. For short intervals and with a much dampened amplitude, currents even reached the eastern side of T1 (around T1 359). The western edges like T4 0 and T1 50 were least affected by the seiches. For example during one recording day, a water level change of 8.2 cm at T4 400 corresponded to a change of only 0.7 cm at both T1 50 and T4 0. This showed that the change in water level due to the

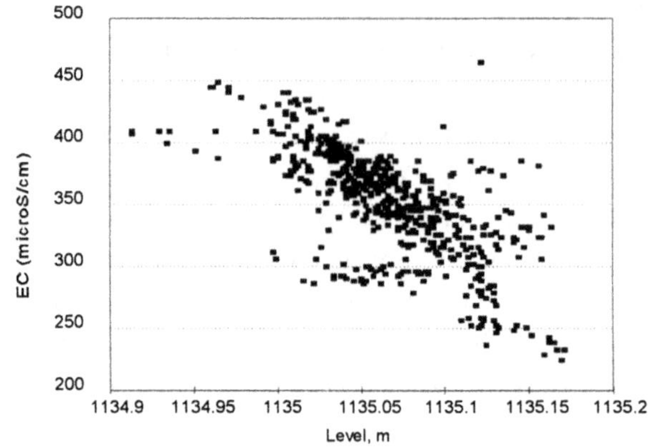

Fig. 3.17. Scatter plot of EC versus water level at T4 400, January 1996; r = -0.694, n = 581, p< 0.001; *EC = 798,402-703 level ± 31*

seiches occurs mainly in the middle portions of the swamp, and beyond T1, lakewards.

Figures 3.18 (A) and (B) show the temporal variation of water levels within the swamp along

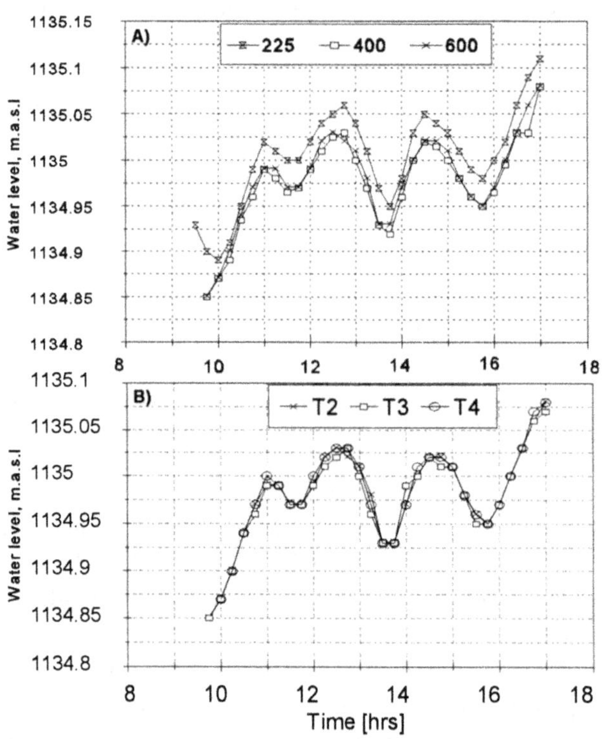

Fig. 3.18. Water level variations in the swamp due to seiches. (A) Simultaneous water level changes along T2. (B) Similar changes 600m from the western edge of the swamp at T2 600, T3 600 and T4 600. The water level rose by 25 cm in 7 hours.

T2 and at the 600 m points along T2, T3 and T4, respectively. These points, which are within the major flow path (Fig. 3.7), showed a similar response at all times. This shows that the effect of the seiches reaches deep into the swamp (within the central flow path) almost undampened. Dampening takes place near the edges of the swamp due to thicker peat layers, higher suspended solids loads in the water layer (near slurry) and shallower water depths at these locations (Fig.3.6).

Figures 3.18 (A) and (B) shows a water level change of over 20 cm or a volume increase of 157,000 m^3 in 7 hours for the major flow path alone. This is more than five times what can be accounted for by wastewater influent over the same period. Conductivity measurements in the swamp during seiche periods revealed that EC changes over the whole depth of a section when seiches enter the swamp. The likely transport mechanism of these seiches in the swamp is plug flow with axial dispersion. Along the longitudinal profile, there is limited mixing with the lake water closer to the railway, increasing towards the bay.

3.3.7 Discussion

Results show that wastewater flow in the swamp is not well distributed over the whole expanse/width of the swamp. There are zones of the system which seem to have no interaction with wastewater at all. This means that a lesser area of the swamp is involved in the treatment process. An extrapolation of these areas over the swamp shows that the effective swamp area is only about 785,000 m^2 out of a total study swamp area of 1,150,000 m^2. The cause for preferential flow in the system is mainly differential resistance due to deeper zones, spatial differences in vegetation density and types, and flow depth.

The lithology of the area indicates presumably negligible water exchanges with groundwater. Subsurface flows from the swamp edges are significant during rainy periods, although they primarily serve to increase EC concentrations near the swamp edges due to leaching from plant organs. During the dry season, there is reduced subsurface flow. Conductivity values are reduced too, possibly as a result of plant uptake and sediment sorption.

Figure 3.19 gives a summary of the water balance for the Nakivubo swamp over a 17 months period (April 1995 to August 1996). Open water evaporation for the swamp area is almost equal to direct rainfall. Shahin (1985) reported that evaporation from Lake Victoria is of the same order of magnitude as rainfall. However, evapotranspiration from the swamp is higher than direct rainfall over the swamp (ratio of ET to open water evaporation is 1.2 (range 0.8 - 1.6). Such values for littoral vegetation, whose above-ground biomass is well-watered and where meteorological conditions are favourable, are not uncommon. Ondok *et al* (1990) reported the work of different authors who found equally high ratios for littoral vegetation in Europe. According to these authors, Rodescu-Rosewald (1974) for example, found ratios for *Phragmites* in the Danube delta (using lysimeters according to Popov) to be between 1.0 and

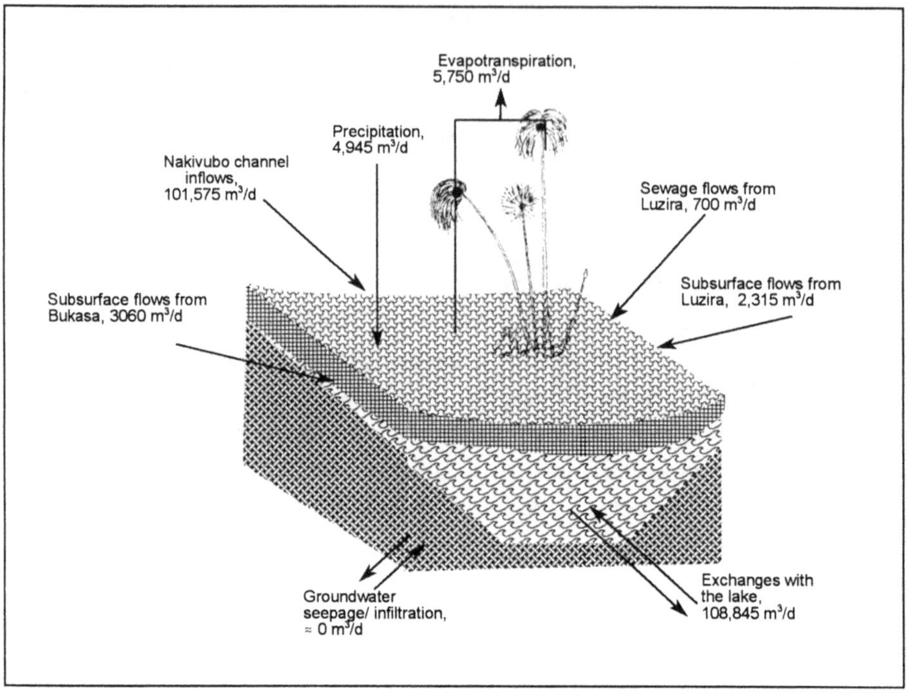

Fig. 3.19. Summary of the water balance for the Nakivubo swamp for the period April 1995 to August 1996.

7.0, depending on the season. Ratios for *Typha latifolia* determined by various authors ranged between 1.2 and 3.1. Okurut (pers. comm.), in a study on constructed wetlands in Jinja (Uganda), found evapotranspiration rates from papyrus to be 10 mm/d. Despite the relatively high evapotranspiration rates, Nakivubo swamp is kept from drying by both the large catchment area (50 km^2) of its main influent channel which drains up to 1200 mm/yr of runoff during the wet season waste water production from Kampala and the exchanges with the lake.

Seiches and currents from the lake have a large effect on the overall water balance and retention time of wastewater in the swamp. The oldest record on the presence of oscillations of Lake Victoria waters is by Stanley (1879) in Denny (1996). Stanley noted what he called positive and negative tides on the Inner Murchison Bay of Lake Victoria that had a period of 4 hours. This phenomena was also reported to occur in all inlets on the Ugandan coast. Our results, over a hundred years later (and with better time keeping facilities), are of the same order of magnitude (135 minutes). Figure 3.20 shows the effect of seiches on the water movement, over one cycle.

Another indicator of the little impact of seiches on the edges (especially the Bukasa side) and the role of rainfall on the water balance for these portions of the swamp is the drying out of these portions and the reattachment of the vegetation mats onto the substratum. This is sometimes followed by yellowing of plants if the dry period is prolonged.

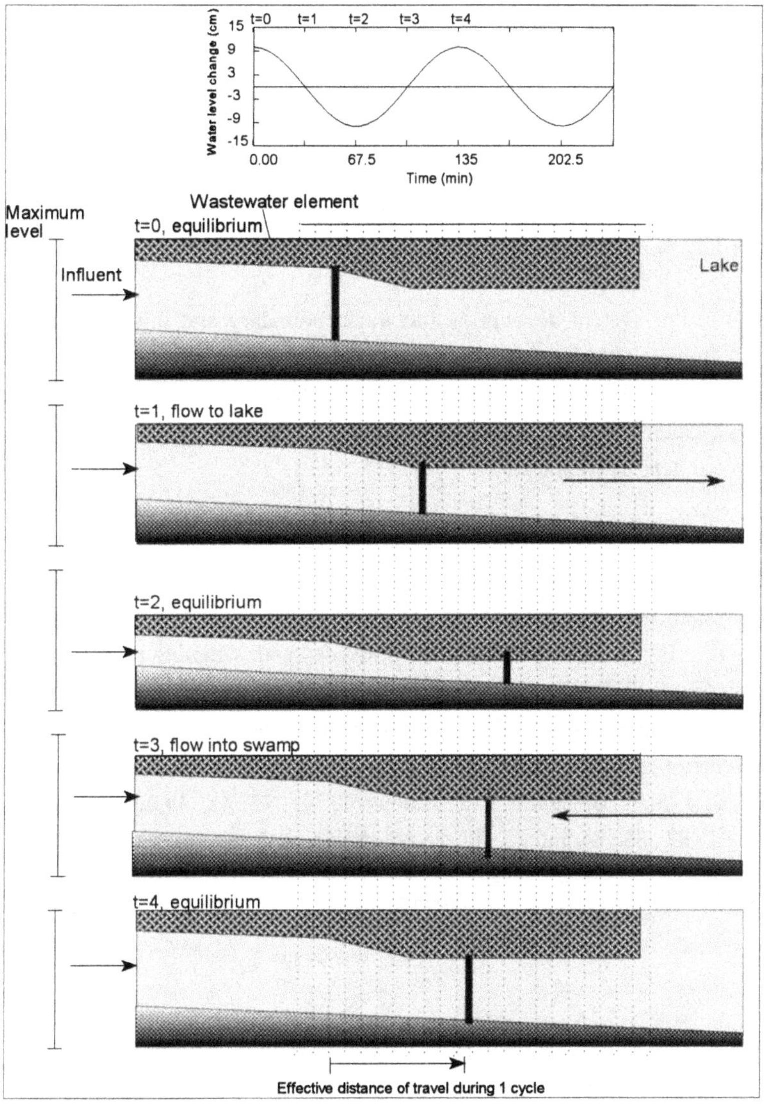

Fig. 3.20. View of effect of seiche movements into the Nakivubo swamp on the water level. The figure shows the average travel distance over one cycle (about 135 minutes) for the net inflow of Fig. 3.19 and flow velocities of 0.05 m/s for both wastewater and lake seiches.

Even over long periods, sub-surface flows do not contribute significantly to the water balance, due to the impervious nature of the swamp edges and bottom. The hydrogeological set-up, with thick underlying clay layers (9 -36 m) also limits exchanges with deep groundwater.

Section 3.4 discusses the hydrodynamic water model used and the overall water balance.

3.4. Modeling of water flows in the Nakivubo swamp

The use of models to predict and simulate water flows and quality was discussed under 3.1. In this study, a quasi-one dimensional flow/ quality simulation model DUFLOW (DUtch FLOW) was used to describe the characteristics of water flow in the swamp and their impact on waste water quality as it flows through the swamp. A brief description of the model structure is given in Appendix 3.5.

In DUFLOW, after system description (network, boundary and initial conditions, friction factors, etc.) flow calculations are done automatically. The quality part enables the user to define process descriptions for an overall tailor-made model to suit a specific problem of interest. This chapter discusses flow/quantity aspects. The quality aspects of DUFLOW will be handled separately in Chapter 9.

3.4.1 Input data

3.4.1.1 System schematization

The swamp system was schematized into different sections to correspond with transects and major flow zones (Fig. 3.21). Sectional nodes coincided with sampling points used in the study (compare with Fig. 3.2). Node 40 is the main swamp inlet, 3 is Luzira prisons effluent, 45 is the swamp-lake interface (T4 400), while 100, 101 and 102 are in the lake. Based on conductivity measurements, three major flow routes were identified namely Luzira with sections 1, 2, 3, 4, 5 and 6; Central with sections 40, 41, 42, 43, 44 and 45; and Bukasa with sections 80, 81, 82, 83, 84 and 85. These are shown in the network layout (Fig. 3.21). The

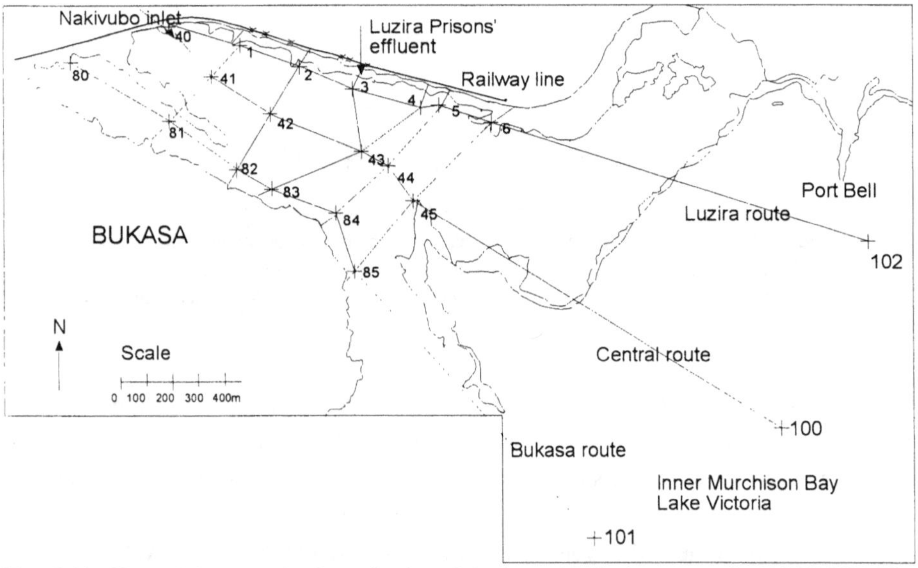

Fig. 3.21. Network layout and schematization of the lower Nakivubo swamp for DUFLOW. The numbers correspond to nodes.

network definition is presented in Table 3.5.1 (Appendix 3.5) while a description of the sections and their corresponding resistances to flow as used in the model are shown in Table 3.5.2 (Appendix 3.5).

3.4.1.2 Cross -sections

Survey data of section 3.2 was used to describe swamp bottom levels and cross sections. The surface level of the peat layer was used to represent the bottom level of any section. Differences between the channel and flooded plains including the mat, near the beginning of the swamp and the edges were taken into account by defining different flow and storage widths (flow width to storage width ratio). Table 3.5.3 in Appendix 3.5 gives this data.

3.4.1.3 Parameters and variables

The discharge in the system follows typical open channel flow formulae, namely either the Manning's equation or the Chezy formula. The former, represented by:

$Q = \frac{1}{n} R^{2/3} S^{1/2} bh$ was used.

In a wide and shallow floating swamp, the hydraulic radius ***R*** $\approx$ ***h/2***, where h is the depth of flow. ***S*** is the hydraulic slope and ***b*** is the width of flow of the water. Hence

$$h \propto \frac{Q^{3/5} n^{3/5}}{S^{3/10} b^{3/5}}$$

Hence the flow of water and the water level in the system will be a function of the resistance factor, the hydraulic gradient and the flow/storage width. The variables b, S and the resistance factor (1/n) as used in the model are shown in Tables 3.5.2 and 3.5.3 (Appendix 3.5).

3.4.1.4 Boundary conditions

Water level boundary conditions in the lake (nodes 100, 101 and 102) were selected. These actually corresponded to read values at node 445, extrapolated into the lake. At node 40, a water discharge boundary was used. A value of Q = 0.8 m^3/s was used since it was close to the average dry weather flow. Discharges from Bukasa and Luzira were considered as additional nodal inputs.

3.4.1.5 Initial conditions

Initial levels of 135.0 m were supplied at each node. Except for the swamp inlet (node 40, also a boundary condition) whose discharge was set at 0.8 m^3/s, all flows were initially set at 0 m^3/s.

3.4.2 Sensitivity analysis and model calibration

Sensitivity analysis was done to identify the parameters/variables for which the model is very sensitive. These were then used to calibrate the model in order to simulate with observed conditions. For the flow part of the model, Manning's *n*, flow width, storage width and the lake water level were confirmed to affect the flow characteristics.

Model n values were varied around the value of 0.17 (section 3.3.3; or a resistance factor of about 5 since in the model, it is 1/n that is used) depending on the morphology of a section. Using roughness factor values of 1,3,3,5,5,10 for sections 40 to 45 respectively, it was shown that as expected from Manning's equation of flow, changes to the flow and storage widths had a large impact on the flow especially in section 40.

It was assumed that there was more flow channelization in the first sections of the central flow path (40 and 41) compared to the others (43 and beyond). The selection of the flow width and storage width was therefore most uncertain in these regions. To determine the effect of these variables on the flow in the system, simulations were made for different ranges. It was seen that a very small flow width in section 40 led to high water levels at node 40 as expected from the Manning's equation. Figures 3.22a. and 3.22b. show the impact of flow width on the water levels in the central flow path. For Fig. 3.22a, the flow width at the end of section 40 was 150m compared to 20m for Fig. 3.22b. The rest of the variables were as presented in Appendix 3.5.4.

The impact of varying the storage width on the water levels in the central flow path is shown in Figures 3.23a. and 3.23b. For Fig. 3.23a, the storage width at the end of section 40 was 230m compared to 400m for Fig.3.23b. The rest of the variables were as presented in Appendix 3.5.4. The effect of a large storage basin is to dampen the peaks in water levels and velocities.

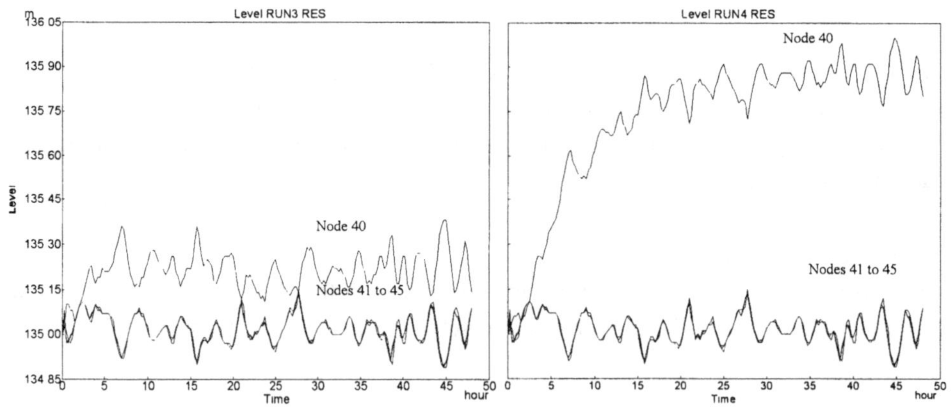

Fig. 3.22a. Fig. 3.22b.

Fig. 3.22. Effect of flow width on water levels along the major flow path. Fig. 3.22a., the flow width at the end of section 40 is 150m while the corresponding width in Fig. 3.22b. is 20m. This change did not affect downstream sections (nodes 41 to 45) for given friction factors. Velocity behaved similarly. affected.

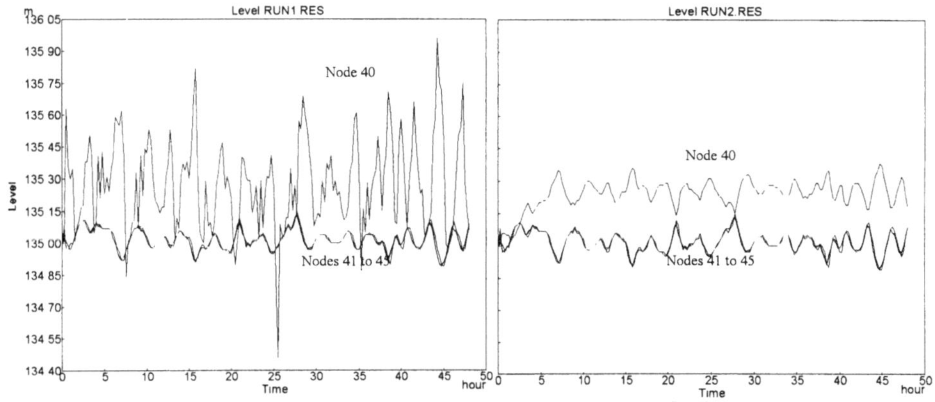

Fig. 3.23a. **Fig. 3.23b.**

Fig. 3.23. Effect of storage width on water levels along the major flow path. Fig. 3.23a., the storage width at the end of section 40 is 230 m while the corresponding width in Fig. 3.23b. is 400m. This change did not affect downstream sections (nodes 41 to 45) for given friction factors. The velocity was similarly affected.

The effect of a very high roughness factor in section 40 was shown, e.g. in Fig. 3.24. The roughness factor and the observed remarks are as shown in Table 3.8.

Model calibration was then done by adjusting the roughness factors for all sections until the observed levels at different sections were close to what was expected. Further work on calibration actually revealed that the water levels in the model were close to the expected levels for *1/n* values between 2 and 5. These values are well outside the conventional manning's n values which range from about 0.010 for very smooth channel to about 0.20 for very rough channels (Chow, 1972).This discrepany could be because of the nature of the swamp system with very large resistances from the bottom, the mat and at times from the roots of the

Table 3.8 Effect of roughness factor on water levels in the central flow path. A very low value indicates a very rough section.

Series	Roughness factor *(1/n)* in section 40	41	42	43	44	45	Remarks
A	0.01	1.0	3.0	5	5	10	Rising water level at 40, instability and low amplitude at 40.
B	5	0.1	0.1	0.1	0.1	0.1	Amplitude at 40 only about 3cm, amplitude gradient from 42 to 45. Level at 40 ≈level 41
C	1	0.5	0.5	1	1	1	Raise level at 41 (135.23 cm). Level difference between 40 and 45 about 20 cm. Amplitude at 40≈ 10 cm.
D	2	0.5	10	10	20	50	Higher amplitudes at 40 and 41, follows boundary 101, with lag at 41 due to rough section.
E	5	0.2	0.5	0.5	1	5	Fair variation at 40, and other nodes. Difference in levels between 40 and 45 about 5 cm

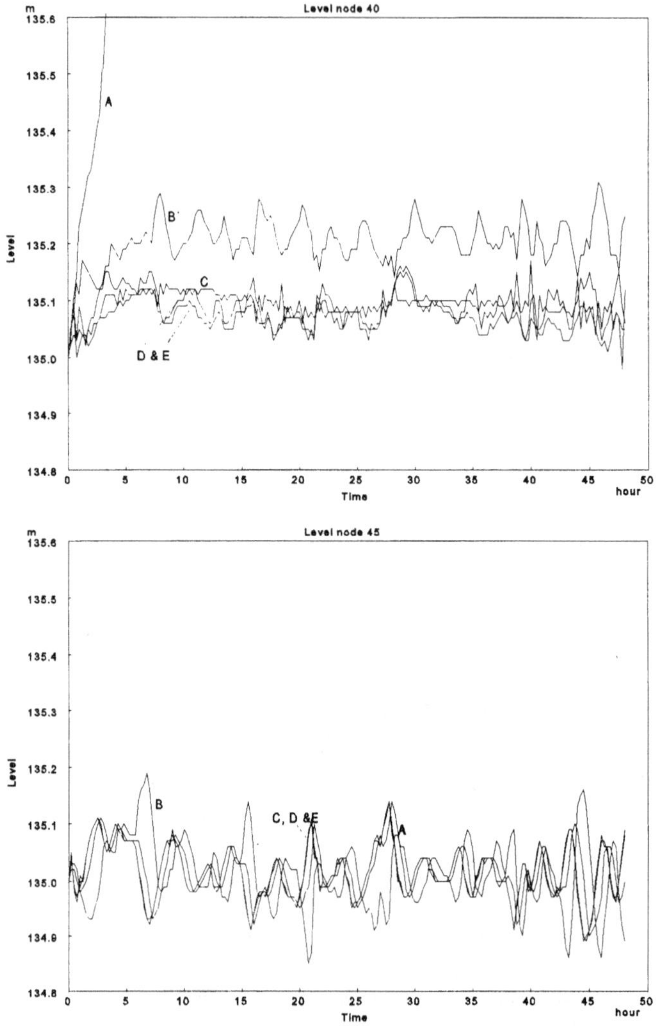

Fig. 3.24. Effect of the variation of the roughness factor on water levels within the central flow path. Less effect closer to the lake due to larger flow widths (Manning's equation).

vegetation. Hence the values used in the model were to fine tune it and may not have any physical significance. Values less than 1 in section 40 gave continuous rise in water levels at node 40, (a situation that would be likely to flood the city) irrespective of the downstream situation (Figures 3.24). It is also stressed that due to the non-free water surface in the swamp, higher resistances are to be expected than those of open channels.

The model was calibrated using the roughness factor by selecting different n values for the different sections and simulating the flows to ensure that:

- the water level at the inlet remains more or less at 135.1 m. During the dry season, there is very little variation in water levels at the inlet (usually ±3-5 cm) on any day;
- there are also less variations within section 41, unlike for sections further towards the lake (43-45); and
- the variations at node 45 are felt to almost the same degree and at nearly the same time as for nodes from 43 towards the lake.

There was a need to raise the roughness in section 41 to hinder backflow of lake water all the way to node 40. In the field, there appeared to be no direct effects on the water level at node 40 of water level changes in the lake (nodes 100, 101 and 102). After a number of runs, the situation shown in Fig. 3.25 (for the central flow path) was selected.

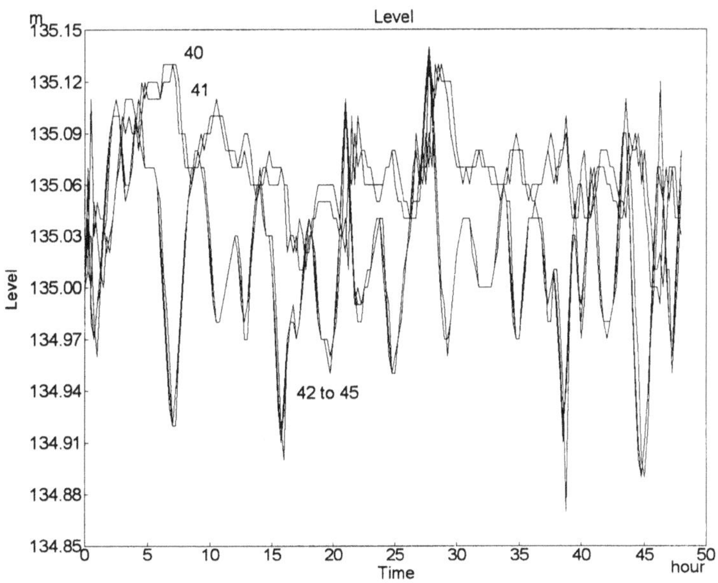

Fig. 3.25. Simulation of water levels in the central flow path within the Nakivubo swamp. There is little difference beyond node 42.

Figure 3.25 shows small differeneces in water levels and lower amplitudes at nodes 40 to 41. Nodes 43 through to 45 have larger amplitudes which are quite alike from node to node. This is very similar to what was seen from manual measurements in section 3.3.6 (Fig. 3.18), where water levels in the swamp measured concurrently every 15 minutes had nearly no variation in amplitude and frequency from node to node from the lake edge through to points along Transect 2.

3.4.3 Model validation

Model validation was done by a comparison of measured data at different nodes in the system with the simulated data. Data logger water level values at node 45 were again used for boundary conditions at nodes 100, 101 and 102. In addition, the simulated water velocities in section 40 were compared with those measured using the salt dilution method of section 3.3.2.

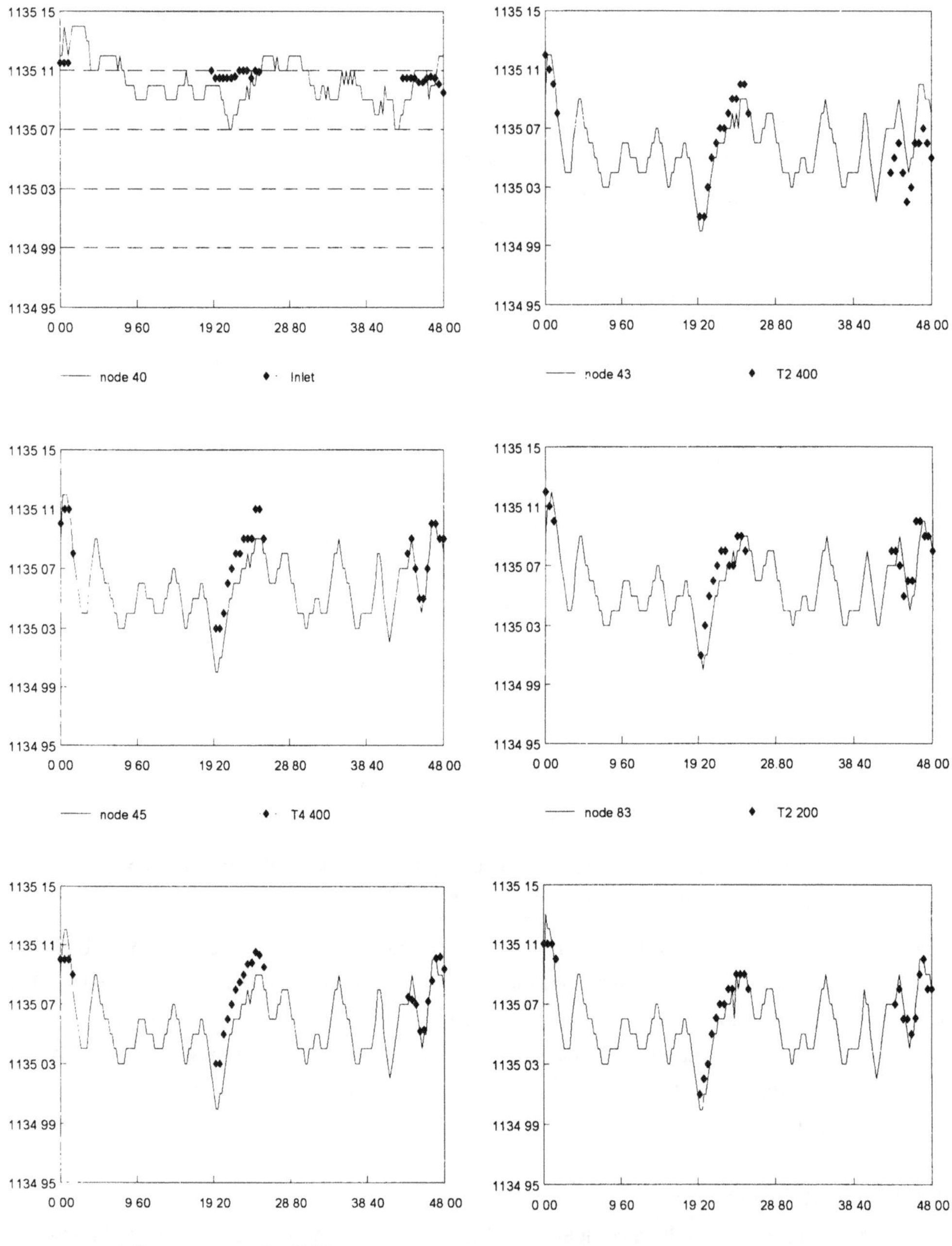

Fig. 3.26. Measured versus simulated values for water levels in the Nakivubo swamp during August 1995. The simulated values are continuous lines while the measured data are the triangular points.

Figure 3.26 shows good "correlations" between the measured and simulated water levels in the swamp. The fit for the inlet was however not so good and a check on the velocity in section 40 was therefore done. The simulated velocity gave an average value over section 40 of 0.017±0.001 m/s (Table 3.5.5, Appendix 3.5.4). The inlet discharge (0.8 m^3/s) used in the model, and the water flow section area of about 69.4 m^2 ([{9 x 0.42} +{150 x 0.9}]/2) gives a velocity of 0.012 m/s or about 70% and within the range of velocities in this section.

During the salt velocity measurements, the travel time of the plume in a channelized section (therefore a higher than average value) was 71 minutes. For the 230 m distance, the equivalent velocity was 0.05 m/s, (on the high side, but still within the range of simulated velocities where the highest value was 0.07 m/s).

Based on Fig. 3.26 and the above velocity check, it can be concluded that the model is a reasonable representation of the water flow characteristics in the Nakivubo swamp.

3.4.4 Characterization of water flows in the swamp

The simulations support other findings from water quality measurements that there are preferential water flow paths in the swamp and that seiches are propagated deep into the swamp. For average flow conditions, Table 3.5.5 in the Appendix shows that the water has a retention time of about 3.2 days in the central flow path, 4.9 days in the Luzira route and an even longer time of 17.5 days in the Bukasa route.

During the rainy period, water flows of 5.0 m^3/s as occasionally measured at the inlet led to simulated hydraulic retention times of about 1.0 days, 2.0 days and 0.7 days for the Central, Bukasa and Luzira routes respectively. Dry weather flow conditions with low discharges of 0.5 m^3/s (see also Fig. 3.9) at the inlet led to higher average retention times of about 5 days, 25 days and nearly 8 days for the Central, Bukasa and Luzira routes respectively.

The average discharges in the three routes namely Central,Bukasa and Luzira during average flow conditions are respectively, 0.55 ± 0.34 m^3/s, 0.17 ±0.13 m^3/s, 0.13 ± 0.08 m^3/s. This means that the central path carries over 65% of the total discharge. Considering the retention time of the water and the pollutant concentrations in the Luzira side of the swamp, then there seems to be little variation between the flow conditions in the Luzira and Central route, for purposes of performance evaluation. Summing up the Luzira and Central route discharges yields about 80% of the flow (also see Chapter 9).

3.4.5 Conclusion

It has been shown that the flows in the Nakivubo swamp can be reasonably simulated using DUFLOW. The flow in the system is clearly channelised with preferential flow sections of different velocities. The central flow path, which carries the bulk of the water in the system had the lowest retention times for both dry and average weather conditions. During the rains however, all three routes have very low retention times (as low as less than 1 day and Bukasa with about 2 days). This has negative impacts on the wastewater purification capacities of the swamp as elucidated in Chapters 4 to 9. The propagation of seiches deep into the swamp is likely to affect the dispersion of pollutants in the swamp and hence could affect its water treatment potential.

3.5 References

Boyt, F.L., Bayley, S.E. & Zoltek J.Jr. (1977). Removal of nutrients from treated municipal waste water by wetland vegetation. *Journal WPCF.* 5:789-799.

Brown, R.G. & Stark, J.R. (1989). Hydrologic and water quality characteristics of a wetland receiving waste water effluent in St.Joseph, Minnesota. *Wetlands*, Vol.9, 2:191-206.

Burman, P., and Pochop, L.O. (1994). *Evaporation, Evapotranspiration and Climatic Data.* Developments in Atmospheric Science, 22. Elsevier.

Burman, P., and Allen, R.G. (1990). *Evaporation, Evapotranspiration and Climatic Data. Developments in Atmospheric Science, 22.* Elsevier.

Chow Ven Te (1972). *Handbook of Applied Hydrology*. McGraw Hill.

Chow Ven Te, Maidment R.D. and Mays W.L. (1988). *Applied Hydrology*. McGraw Hill.

Denny, P. (1996). Gaia's Kidneys: Wetlands are our lifeblood. Inaugural address delivered on the occasion of the public acceptance of the chair of wetland and aquatic ecology at IHE-Delft, The Netherlands.

Deuver, J. M. (1990). Hydrology. In: *Wetlands and shallow continental water bodies, Vol 161-89.* Patten, C.B, Jorgenisen, E.S, and Dumont, H., (Eds.). SPB Academic Publishing.

Dobrin, MB. (1976). *Introduction to geophysical prospecting*. McGraw-Hill, New York.

Gaudet J. J. (1979). Seasonal changes in nutrients in a tropical swamp: North swamp, Lake Naivasha, Kenya. *Journal of Ecology.* Vol. 67, pp 953-981.

Gaudet, J.J., (1980). Nutrient dynamics of papyrus swamps. In *Proc. of first Int. Wetlands Confr.*, Bay Gopal (ed.). National Institute of Ecology and International Scientific publications, New Delhi-India. p.305-319.

Groves, A.W. (1929). Report on a visit to inspect excavations in progress at Luzira: Surficial geology. UGSM, unpublished report no. AWG/1. UGS,Entebbe.

Gauff/Parkman (1988). Kampala Water supply expansion Project. Final design report. I. water Supply Studies. Report prepared for the Ministry of Water and Mineral Development and National Water and Sewerage Corporation; Kampala - Uganda.

Jensen, M.E, Burman R.D. and Allen R.G. (Eds.) (1990). Evapotranspiration and irrigation water requirements. ASCE Manual. Report on Engineering Practice no.70. 332 p.

Hammer, D.E. & Kadlec R.H. (1986). A model for wetland surface water dynamics. *Wat. Res.*, Vol.22, 13:1951-1958.

Heut van H. (1991). Modeling water transport and phosphorus eutrophication in an interconnected lake system, A scenario study. Ph.D Thesis, Universiteit Twente.

Harris, N. (1945). Annual report for 1945: General geology. UGSM, unpublished report no. NH/34. UGSM, Entebbe.

Howell, P., Lock, M. and Cobb, S. (Eds.) (1988). The Jonglei Canal: impact and opportunity. Cambridge studies in applied ecology and resources management. Cambridge University Press.

Kadlec, R.H., and Knight, R.L. (1996). *Treatment Wetlands*. Lewis publishers.

Mitsch, W.J. & Gosselink, J.G. (1986). *Wetlands.* Van Nostrand Reinhold, New York.

Mitsch, W.J. & Reeder, C.W. (1991). Modeling nutrient retention of a freshwater coastal wetland: estimating the roles of primary productivity, sedimentation, resuspension and hydrology. *Ecological Modeling*, 54:151-187.

Ondok, J.P., Priban, K., and Kvet, J. (1990). Evapotranspiration in Littoral vegetation. In: *Guidelines of lake management,* UNEP and Intern. Lake Environ. Committee. Jorgensen, S.E. and Loffler, H. (Eds.); 5-11.

Pilgrim, D.H., Schulz, T.J. and Pilgrim, I.D. (1992). Tracer investigation of the flow patterns in two field scale constructed wetland units with subsurface flow. In *Wetland systems for water pollution control, Proceeding of the International Specialist Conference*. Sydney, Australia. 19.1-19.9.

Pugh, T.D. (1987). *Tides, seiches and mean sea-level.* A Handbook for Engineers and Scientists. John Wiley and sons. 442 pp.

Saxton, K.E. (1982). Evapotranspiration. In: *Hydrologic modeling of small watersheds.* Haan, C.T., Johnson H.P., and Bakensiek D.L., (Eds.). Amer.Soc. of Agr. Engs.SCE,

Shahin, M. (1985). *Hydrology of the Nile* Basin. Elsevier.

Shaw, E.M. (1991). *Hydrology in Practice.* Chapman and Hall.

Shuttleworth, W.J. (1993).Evaporation. In: *Handbook of Applied Hydrology*, Maidment R.D. (Ed.). McGraw Inc. Chapter 4.

Singh, P.V. (1989). *Hydrologic systems. Volume II. Watershed Modeling.* Prentice Hall. Inc.

Smith, L., and Wheatcraft, W.S. (1993). Groundwater flow. In: *Handbook of Applied Hydrology*, Maidment R.D. (Ed.). McGraw Inc. Chapter 6.

Walton, R., Davis, J.E., Martin, H. T., and Chapman, R.S. (1996). Hydrology of the Black Swamp Wetland on the Cache River, Arkansas. *Wetlands*,16 no 3: 279-287.

Wayland, E.J. (1932). Notes on Gaba Peninsula and Kiruba Island. UGSM, unpublished report no. EJW/8. UGS,Entebbe.

Winters, R.J. (1922a). Report on the possibilities of an underground water supply for Kampala. UGSM, unpublished report no. RJW/1. UGSM, Entebbe.

Winters, R.J. (1922b). Notes on the geology of Kampala with reference to the possibilities of an underground water supply. UGSM, unpublished report no. RJW/2. UGSM, Entebbe

Chapter 4

Water quality of the Nakivubo swamp

Kansiime, F. and Nalubega, M.

4.1 Introduction

Industrialization and population increase have resulted in more wastewater production which has in turn accelerated eutrophication of surface waters and spreading of water borne diseases. The diseases are responsible for millions of deaths each year in developing countries (UNEP, 1995). Conventional wastewater treatment is practised in both developed and developing countries. However, only a few of the conventional wastewater treatment plants are well equipped to remove the nutrients (Brix, 1987) and none has been reported in developing countries.

High costs of conventional wastewater treatment and stringent rules on effluent quality standards for major plant growth nutrients (N&P), have revived interest in use of wetlands for wastewater treatment especially in regions where the cost of land does not preclude land extensive treatment systems like lagoons. As a result, in the last decade most attention has been directed towards the use of wetlands in controlling water pollution through treatment of municipal, agricultural and industrial wastewater as revealed by a number of publications and proceedings (Reddy and Smith, 1987; Hammer, 1989; Cooper and Findlater, 1990; Bavor and Mitchell, 1994; Kadlec and Brix, 1995; Haberl *et al.*, 1997). In these conferences, wetlands have been recognised as potentially cost effective systems for purifying wastewater because of the following advantages compared to conventional treatment systems: low operational and maintenance costs, low energy requirements and no need for skilled labour. Furthermore, these systems can be established where wastewater is produced, can treat very small flows previously untreated and can be developed locally for a wide range of wastewater volumes to be handled. However, wastewater treatment using wetlands requires more land and may be expensive where wetlands are not already existing, offers less control of treatment systems processes, can harbour pests and can act as breeding grounds of mosquitoes.

Although wetlands have been used as convenient wastewater discharge points as early as when sewage was collected and conveyed into 'nature' (Richardson and Davis, 1987; Kadlec and Knight, 1996), the use of wetlands for nutrient removal remains controversial, despite the large international research effort on the topic. Much research has been devoted to understanding physical, biological, and chemical processes in constructed wetlands (Hammer, 1989; Cooper and Findlater, 1990; Bavor and Mitchell, 1994). One of the likely drawbacks for most of these studies carried out on linearly configured constructed wetlands is that extrapolation of results from such studies into a more complex environment may end up in poor prediction of nutrient removal (Howard-Williams, 1985; Cooke, 1994). Natural wetlands are characterized by extensive variability in functional components which make it virtually impossible to predict their response to wastewater application and translate the results from one geographic region to another (Hammer and Knight, 1994; Kadlec and Knight, 1996).

Howard-Williams (1985), reviewed the role of nutrient storage compartments in mediating the fluxes of nutrients through wetlands. These compartments have a limited storage capacity and if this is exceeded, it leads to breakdown of nutrient retention. Thus, wetlands may have a limited capacity in improving the quality of water passing through them. The limited capacity of such systems may not become apparent if studies to demonstrate the efficiency of such systems are of relatively short duration. Furthermore, performance of natural wetlands has been reported to depend on applied loading rates and natural conditions (Brodick *et al.*, 1988; Hammer and Knight, 1994). This may be attributed to gradual changes in plant species composition and accumulation of pollutants which in turn depend on wetland soil type, hydrology of the area and flow through of the effluent. Lijklema (1990) reported that continuous addition of fresh organic matter to a wetland may decrease the absorption capacity of the sediment for phosphate. This may subsequently lead to release and export of ortho-phosphate from the sediment of natural wetlands as observed by Gehrels and Mulamoottil (1989). Because of the occasional export of accumulated organic sludge and embedded nutrients into adjacent open waters, wetlands have been viewed as large holding tanks or septic tanks (Gaudet, 1976).

Nakivubo swamp has been receiving waste water for over 30 years now. To assess the overall removal of pollutants in the swamp and an indication of the processes therein, conductivity (EC) was used as a tracer. Other conversions and transformations that could lead to increase or decrease in EC (e.g., precipitation, pH changes, redox reactions, and degradation of organic matter) were assumed to be small compared to the value of the wastewater. The lake water has a background EC of about 90 μS/cm whereas the wastewater in the swamp exhibits an average conductivity of 437 μS/cm. This situation made it possible for us to a certain extent trace the pathways of the waste water. This was further supported by faecal coliforms numbers and the physical appearance of vegetation (lush or stunted growth) in the swamp especially papyrus.

Whereas EC may act as an alternative to chloride, sodium or any other conservative material, nutrients and *E. coli,* the variables of interest in this study, certainly will undergo changes in their concentration other than transport and dilution. In addition nutrients may not behave conservatively within the wetland. Nutrients may be gained or lost by chemical reactions, by biological uptake or metabolic activity and adsorption to desorption from suspended matter.

Although individual processes, such as nutrient uptake by plants, phosphate uptake in the sediment, cycling and export of nutrients, nitrification, denitrification, and pathogens die-off will be described in more detail in later chapters, in this chapter the overall removal rates are introduced. The approach used was the assessment of the decrease of P:EC and N:EC ratios along the swamp. Since velocities are not uniform over the width of the swamp, reaction times are expected to vary accordingly. However also, reaction (removal) rates are not necessarily the same everywhere in the swamp and may be due to a function of vegetation, the depth of water under the floating mat, the thickness of the mat and local water velocity. Henceforth processes in the swamp were assessed in view of the major dominant vegetation types in the swamp.

Previous studies in the Nakivubo Swamp- Murchison Bay area concentrated on the wastewater purification capability of the swamp (Kizito, 1986; Taylor, 1991) and the possible impact of the discharge of pollutants on the operations of the Gaba water works (WHO, 1970; Gauff, 1988). However, most of these studies were based on data of inflow and outflow of the swamp ('black box approach') and some of the conclusions from these studies were that Nakivubo swamp purifies wastewater flowing through it and that the raw water quality in the Murchison Bay is acceptable for abstraction at Gaba water works. There is lack of information on the quality of water in the swamp and the transport and processes taking place therein. The fact that there is a possibility of pollutant overloading of the swamp and eventual contamination and local eutrophication of the Inner Murchison Bay, where the city's water supply is abstracted, necessitated this study. The primary objective was to assess the variation of pH, temperature, dissolved oxygen (DO), nitrogen and phosphorus (temporal and spatial) as wastewater carried by Nakivubo channel flows through Nakivubo wetland before entering Lake Victoria at Inner Murchison Bay.

4.2 Materials and methods

4.2.1 Study area

The study was carried out in the lower Nakivubo swamp and Inner Murchison Bay (Fig. 4.1). Nakivubo channel connects the upper and lower swamp at the railway embankment. This is the point that was designated as the Nakivubo channel inflows. Besides the wastewater carried by Nakivubo channel, Luzira prison discharges raw sewage of variable strength and flow into the swamp. The major part of the study concentrated on the lower swamp, which has a well-defined inflow.

Samples for water quality assessment were taken from the Nakivubo channel flows, in the swamp along and across transects and in the Inner Murchison Bay. Physical and chemical variables, namely electrical conductivity (EC), pH, dissolved oxygen (DO) and temperature were monitored in addition to nitrogen (total Kjeldahl nitrogen, ammonium nitrogen, nitrate nitrogen and nitrite nitrogen) and phosphorus (total phosphorus, total reactive phosphorus and total dissolved phosphorus).

4.2.2 General sample collection, handling and preservation

Water samples for laboratory analysis, were collected in 500 ml plastic containers previously washed with dilute HCl and rinsed with distilled water. Before sample collection, bottles were

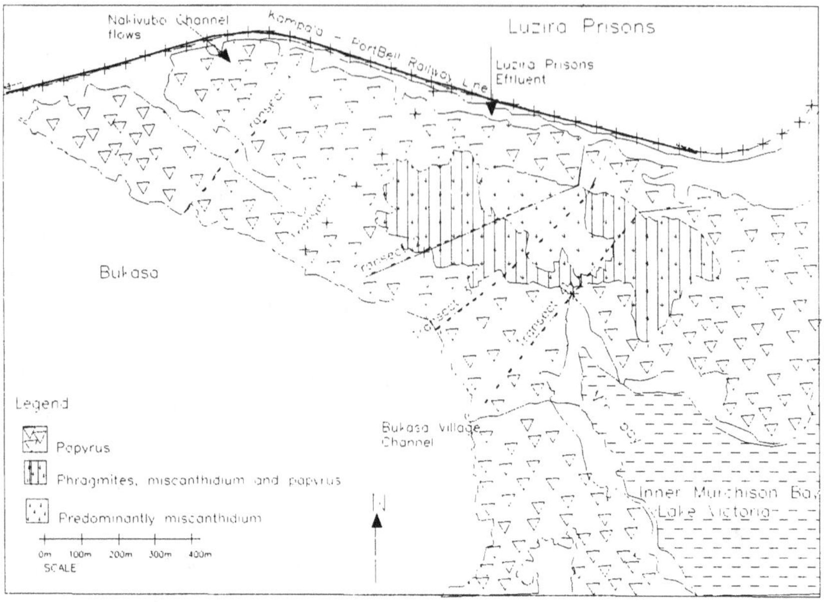

Fig. 4.1. Map of the lower Nakivubo swamp showing transects and dominant vegetation.

rinsed several times with the water sample to be collected. After collection, samples were stored in vegetation shades in the swamp and thereafter transported within six hours to the laboratory for analysis. Samples not analysed immediately were preserved and stored in a deep freezer at - 20 °C. Samples were in most cases collected between 8.30 a.m. and 1.00 p.m.

4.2.3 Field measurements

Inflows into the Nakivubo swamp

Samples of the influent into the swamp were taken at the railway embankment (Nakivubo channel flows) and at Luzira drain (Fig. 4.1). Water samples were collected just below the surface (10 - 20 cm). In order to establish diurnal variation in the quality of wastewater at the railway embankment, measurements were made *in situ* (or in the laboratory) from selected sites every 20 to 30 minutes for 8 hours a day and for at least a month. A data logger was also placed at transect 1 to monitor variation of conductivity and temperature for several days.

Sampling in the swamp

The quality of water in the swamp was measured in transverse and longitudinal directions using access transects 1 (T1) to 4 (T4) cut across the swamp (Fig. 4.1). During the preparation of the transects papyrus culm bundles were laid on the path to ease movement along the transect. However, for transect 1, the last half of the transect had open pools of deep water making it difficult to lay papyrus bundles to make a walkable path. Because of this, the channel was dug and a canoe was used to move from one location to another. The transects' direction, measured in metres, is marked from Bukasa towards the Luzira prison's side of the swamp. T1-259 denotes 259 m along transect 1.

Depending on the site and aim of monitoring, different sampling techniques were used to collect water samples. In zones dominated by papyrus an opening for sample collection was made by driving a graduated pole in the papyrus mat to the desired depth. Thereafter, the pole was gently withdrawn and a papyrus culm to which a plastic sampling tube (with its lower end wrapped in a papyrus umbel) was bound, was placed in the hole. The umbel served to filter out mud and other debris from the water sample. The upper end of the plastic tubing was connected to a hand pump which pumped out water from a given depth into a wide mouthed plastic bottle. The bottle was filled and allowed to overflow until conductivity, pH, and temperature readings had stabilized. After the readings a sample for laboratory analysis was also collected.

The *Miscanthidium* mat was thick and solidly packed. This made the drawing of water from the mat difficult. Furthermore, there was water inflow from lateral and overlying areas of the open hole. In such situations, to avoid mixing of waters, separate openings were made for each sampling depth. A free water column existed below the mat and water could be pumped out easily. Subsequent samples in the water column were taken from the same hole.

Wastewater flow patterns

Conductivity and faecal coliforms were used as natural tracers to locate the major flow paths of wastewater. This was possible because the conductivity of wastewater entering the Nakivubo swamp via the Nakivubo channel and Luzira drain was found to be much higher than the water at the edges of the swamp (not under the influence of the wastewater), or that of adjacent Lake Victoria. Similarly, faecal coliform numbers at the edge of the swamp were very low. So, high conductivity measurements and high numbers of faecal coliforms in the Nakivubo swamp, were associated with point source input of ions and pollutants and in this particular case, the Nakivubo channel and Luzira drain.

Some sites were more intensively studied than others. The criteria for their selection were first: location within the preferred flow path of wastewater and secondly dominance of the site by papyrus or *Miscanthidium* vegetation. Lastly undisturbed areas (no trampling or clearing of vegetation) where plants were healthy-looking were preferred.

Vertical profiles of water quality in the swamp compartments.

To assess the variation of water quality in different swamp compartments two sites (under the influence of wastewater), were selected: one in a *Miscanthidium* dominated area and another in a papyrus area. The *Miscanthidium* and papyrus sites were at 400 m along transects 2 and 4 respectively. The sites and sampling points were selected randomly. Water was taken from

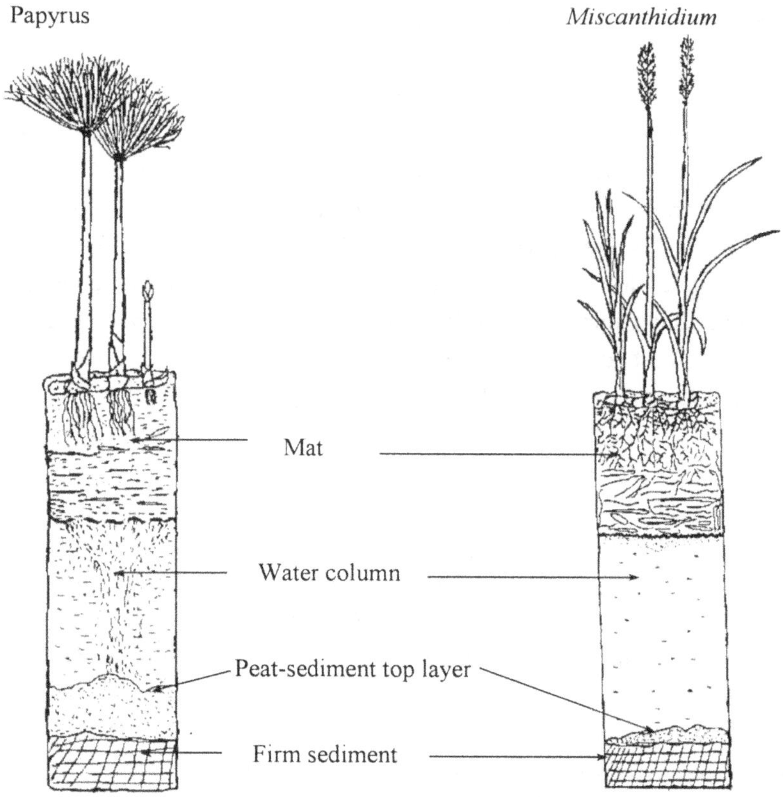

Fig. 4.2. Schematic representation of vertical compartments of sample collection in the dominant vegetation types.

the surface, mat, free water and peat/sediment top layer (Fig. 4.2).

Variation of water quality in Inner Murchison Bay

Samples in Inner Murchison Bay were taken at approximately 500 m, 1000 m, and 2500 m in the open bay perpendicular to the swamp-lake interface at T4-400. In the early sampling periods, these sites could be approached from the open bay at Gaba. However, at a later stage the bay was blocked by the water hyacinth and navigation from Gaba became impossible. Therefore a canoe was used along a channel dug from the swamp edge (T4-400) to the open water. The canoe had to be forced through the water hyacinth to reach all the sampling sites. In some cases when it was difficult to penetrate to the 2500 m location, samples were collected only up to 1000 m. Samples were taken from the subsurface between 20 and 30 cm.

Variation of water quality along the longitudinal axis in the swamp.

To assess the longitudinal variation of water quality in the swamp samples were taken from the railway embankment, at different locations in the swamp at T2 (T2-100 and T2-400) and T4 (T4-200 and T4-400). Samples were taken between 9.00 a.m and 1.00 p.m for 4 days during a dry season. For each sampling date, samples were taken at the same time. To cater for dilution (due to lake seiches), values of conductivity and nutrients were collected for baseline lake values. The ratio of nutrients to conductivity (both corrected for lake value), were used to assess the removal of nutrients over the longitudinal section of the swamp.

Variation of water quality and water levels in the swamp.
For monitoring water level variations in the swamp resulting from seiches in the lake, graduated bamboo poles were installed at different locations in the swamp (Chapter 3). To relate the effect of water levels on water quality, changes in water levels were recorded and samples taken at the same time by different people stationed at the following locations: Inlet, T2-200, T2-400, T4-200 and T4-400. Samples were only taken during the day and for at least six - eight hours on a given day.

4.2.4 Measurement of physical and chemical variables

Physical and chemical variables were measured with Wissenschaftlich-Technische Werkstätten (WTW) microprocessor meters. Electrical conductivity and water temperature were measured with a WTW LF 96 conductivity meter, pH with a WTW pH meter (PH 196T) and dissolved oxygen with an oximeter (OXI 96). In most cases measurements were made *in situ*, between 8.30 a.m. and 1.00 p.m. Whenever it was not possible to take measurements in the field, samples were kept in vegetation shades and transported to the laboratory for measurement within six hours. The calibration of portable meters was according to the manufacturer's instructions.

4.2.5 Chemical analyses

Generally, methods used for chemical analyses of water samples were as described in Standard Methods (APHA, 1992). In the laboratory, turbid samples were filtered immediately through a Whatman general purpose filter paper (no 91,GP), to reduce turbidity. Samples not analysed immediately were preserved in a deep freezer at - 20 °C. Samples from the deep freezer were allowed to warm to room temperature before analysis.

Ammonium concentration was determined by the direct nesslerisation method. Nitrate and nitrite were determined by the cadmium reduction and diazotisation methods respectively. Kjeldahl nitrogen was determined by the macro-Kjeldahl method. Total reactive phosphorus and total phosphorus were determined by ascorbic acid and persulfate digestion methods respectively. All chemical analyses were carried out as described in Standard Methods (APHA, 1992). Measurement of concentrations of given variables was done spectrophotometrically. Spectrophotometry was with a Hach (DR/2000). Precision and accuracy of the spectrophotometer and the procedures for sample analysis were counterchecked by determining the concentration of known standards.

4.2.6 Statistical analyses

Statistical analysis was carried out using MINITAB software (MINITAB Release 10 for Windows). Box plots (Fig. 4.3) were used to depict trends in the data. The width of the box does not represent anything. The distance between the upper and lower quartile (the interquartile range) is the measure of the spread of the distribution (i.e shows how much the value varies). Whiskers are lines which extend from the top and bottom of the box to adjacent maximum and minimum values within 1.5 of the interquartile range. If r is the interquartile range, the maximum observation is the one that is less than or equal to the upper quartile plus 1.5r. The minimum observation is the one greater than or equal to the lower quartile minus 1.5r. Values which are higher or lower than 1.5r are interpreted as outliers and in the statistical box plot are represented by asterisks.

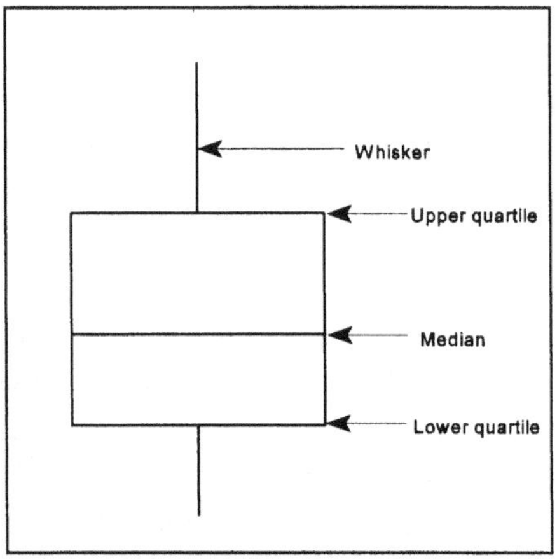

Fig. 4.3. Features of a box plot which show statistical distribution of a given variable.

Comparison of variables was performed using analysis of variance (ANOVA). Following ANOVA, differences among means were tested using Tukey's and Dunnett's multiple comparisons. The family error rate (experiment-wise error) was set at 5% and the individual error rate (compromise-wise error) was automatically adjusted by the statistical programme depending on the number of comparisons. All data were first tested for homogeneity of variances and normality using both Bartlett's and Levene's tests.

4.3 Results

The physical and chemical characteristics of wastewater carried by the Nakivubo channel and Luzira drain are presented in Table 4.1.

Nakivubo channel wastewater was in most cases hypoxic while conductivity was lower than that at Luzira drain. Concentrations of nutrients (nitrogen and phosphorus) were higher for Luzira drain (Fig. 4.4). However, considering the average flows (both dry and rainy weather) into the swamp, the nitrogen load discharged into the Nakivubo swamp was 1000 kg N/d (91%) whereas Luzira prisons contributed only 101.6 kg N/d (9%). The phosphorus loads were 77.54 kg P/d (80%) and 16.85 kg P/d (20%) respectively.

Table 4.1. Average values ± standard error of the mean and range of physical and chemical variables of major inflows into Nakivubo swamp. Numbers in parentheses represent sample size.

Location		EC (μS/cm)	DO (mg/l)	Temp (°C)	pH
Inlet	Mean	437 ± 24 (36)	0.2 ± 0.04 (15)	23.1 ± 0.1 (32)	6.8 ± 0.2 (20)
	Range	380 - 488	0 - 0.5	21.3 - 24.0	6.7 - 7.1
Luzira	Mean	1092 ± 81 (11)	3.4 ± 1.1 (3)	25.3 ± 0.4 (11)	7.2 ± 0.1 (8)
	Range	752 - 1680	1.3 - 4.9	23.6 - 28	6.6 - 7.6

EC = electrical conductivity; DO = dissolved oxygen; Temp = temperature.

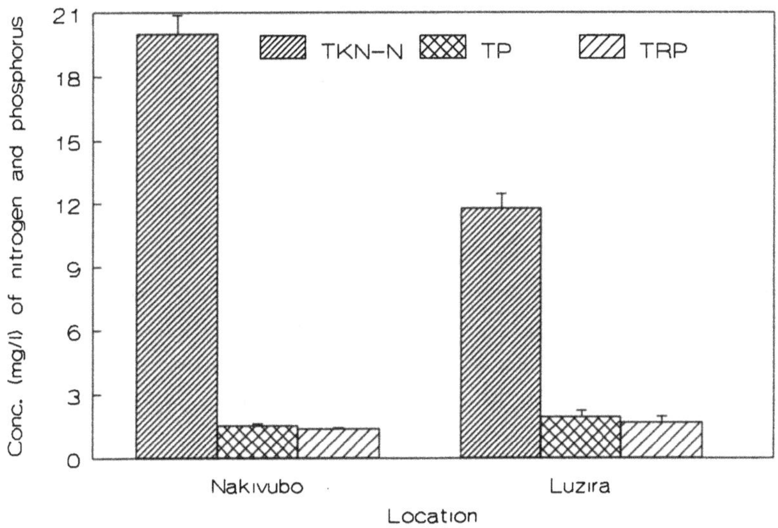

Fig. 4.4. Concentrations of nitrogen and phosphorus in the major wastewater inflows into the Nakivubo swamp. Note: The actual values of Luzira are ten times higher than indicated on the scale. Vertical bars indicate standard error of the mean.

Diurnal variation of conductivity and temperature for three continuous days at transect 1 is presented in Fig.4.5. The travel time between inlet and transect 1 was approximately one hour (as determined by the salt dilution technique, Chapter 3). Since the concentration of pollutants were more or less the same, readings at this transect reflected the diurnal pattern of the inflow at the railway embankment. Maximum values of conductivity and temperature were observed around 4 o'clock in the morning and minimum values at 16.00 h. The daily maxima and minima corresponded to the cycle of discharge of wastewater into the channel, taking into account the lag distance between the city and the swamp.

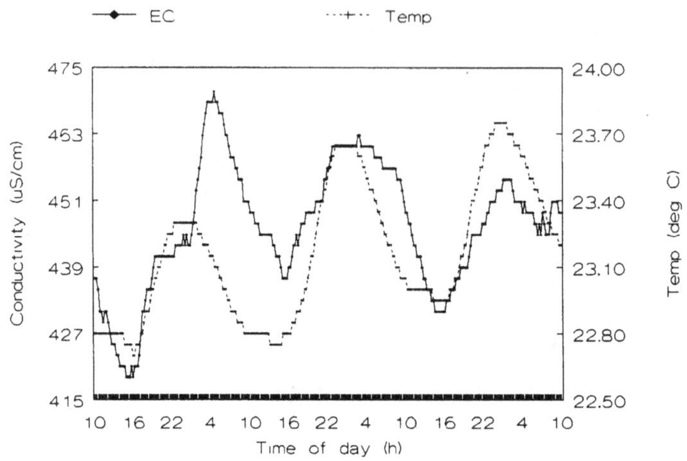

Fig. 4.5. Variation of conductivity (EC) and temperature (Temp) at transect one (T1-350 m).

Individual conductivity values measured intensively for 11 days at the swamp inlet (from 8.00 a.m to 5 p.m) were significantly different from each other ($p < 0.001$), implying temporal variation in quality from the discharge point(s). Conductivity varied between 412 - 449 μS/cm. The lowest conductivity (354 μS/cm) was recorded during a rain event. Temperature and pH did not vary much during the day and in some cases it remained constant for the eight hours of monitoring. Temperature varied between 22.5 - 23.9 °C and pH from 6.7 - 6.8. The diurnal variation of conductivity was in rhythm with that of nutrients (Fig. 4.6), however, monitoring of nutrients could be done only during day time.

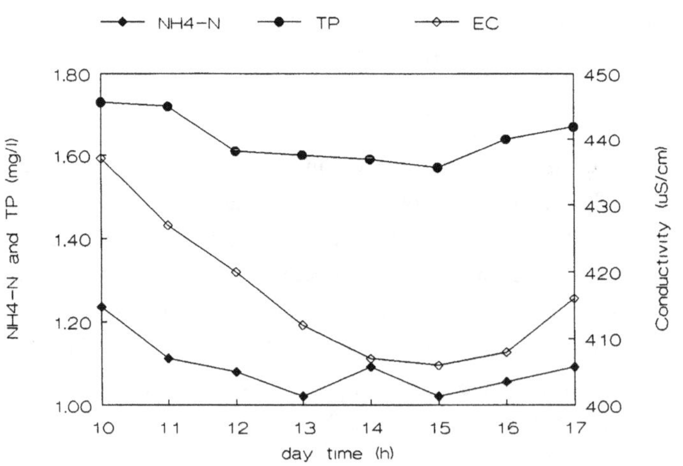

Fig. 4.6. Diurnal variation of ammonium-nitrogen (NH4-N), total phosphorus (TP) and conductivity during the day at the swamp inlet. Note: Ammonia concentration is ten times higher than that indicated on the scale.

4.3.1 Spatial and temporal variation of water quality in the swamp

Water quality in the swamp was variable both in time and space depending on the wastewater flow paths. Trends across the swamp were generally the same for all four transects and thus only the results of transect 2 are presented. Conductivity and faecal coliforms, used as natural tracers to locate the flow paths of wastewater, along transect 2 depicted a similar trend (Fig. 4.7) being maximum in the middle of the swamp and lowest at the edge. Analysis of variance for variables (conductivity, temperature, pH, ammonium nitrogen and total phosphorus) measured along the transect resulted in five distinct reaches: (i) 0 - 150 m, (ii) 175 - 200 m, (iii) 225 - 300 m, (iv) 400 - 625 m and (v) 675 - 800 m, the first (Bukasa side) and fifth (Luzira side) portions are at the edges of the swamp.

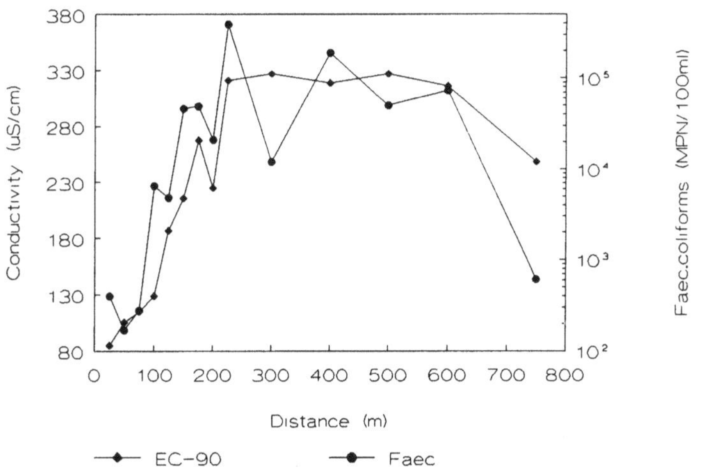

Fig. 4.7. Variation of conductivity, (EC-90) and faecal coliforms along transect 2. Note conductivity was corrected for the lake value (90 μS/cm).

Conductivity, temperature and pH were lower at the edges and higher in the middle of the swamp (Fig. 4.8). Similar trends were observed for ammonium- nitrogen and total phosphorus, the highest value in the middle of the swamp, dominated by *Miscanthidium*, (Fig. 4.9) Nitrate and nitrite nitrogen concentrations were in most cases less than 1% of the total dissolved inorganic nitrogen.

The combination of variation of faecal coliforms across transect 2 and of other variables within the 5 reaches of the transect, suggests that most of the wastewater flows through the central part of the swamp between 100 and 650 m along the transect. The above trend was also observed for other transects and levels of pollutants were lower at the edge of the swamp and highest in the middle. Dissolved oxygen was generally very low ($\leq$ 0.2) or absent and no trend was observed.

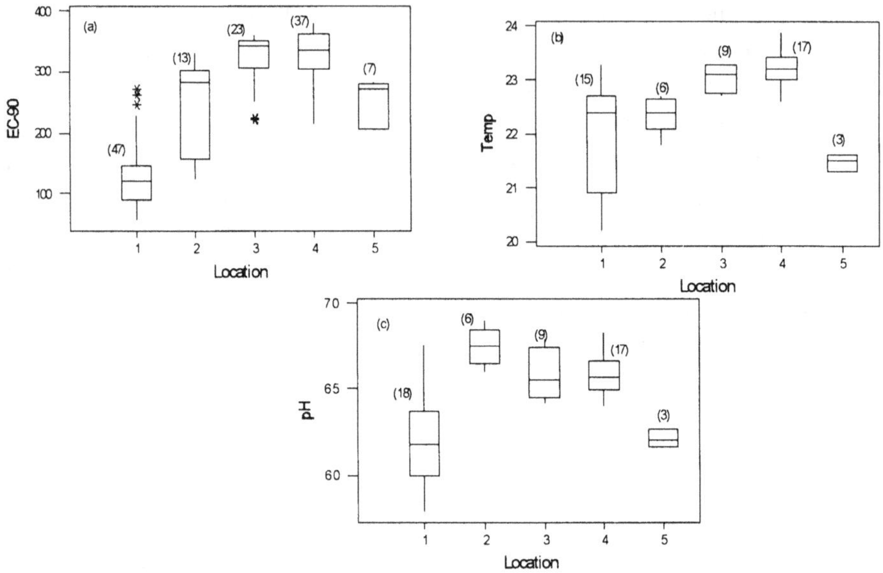

Fig. 4.8. Variation of conductivity (EC-90; μs/cm; graph a), temperature (graph b), pH (graph c), at different locations along transect 2. Location 1 = distance 0-50 m, 2 = 175 - 200 m, 3 = 225 - 300 m; 4 = 400 - 600 m. and 5 = 700 - 800 m from the western edge of the swamp. Numbers in parentheses represent sample sizes.

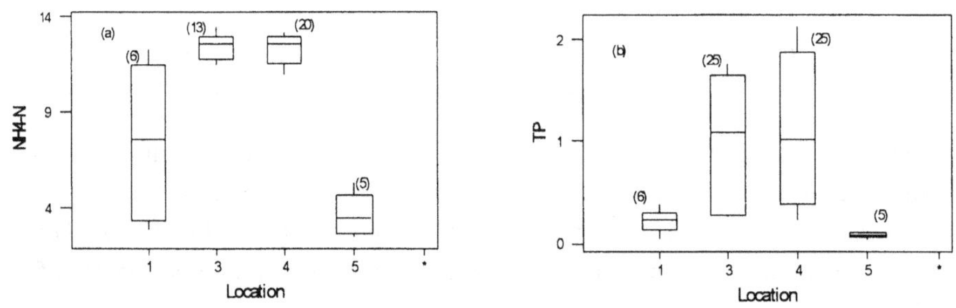

Fig. 4.9. Variation of ammonium-nitrogen (mg/l, graph a) and total phosphorus (mg/l, graph b) along Transect 2. Locations are as in Fig. 4.8.

4.3.2 Vertical profiles of water quality in different compartments in the swamp

To assess vertical profiles of water quality in the swamp, samples were taken at different depths from locations dominated by papyrus (T4-400) and by *Miscanthidium* (T2-400).

In the papyrus site, values for conductivity, temperature and pH were not significantly different over the vertical profile (p = 0.458, 0.183, 0.811, respectively), Fig. 4.10. Similarly concentrations of ammonium-nitrogen and total phosphorus were not significantly different (p = 0.248 and 0.175, respectively), Fig.4.11.

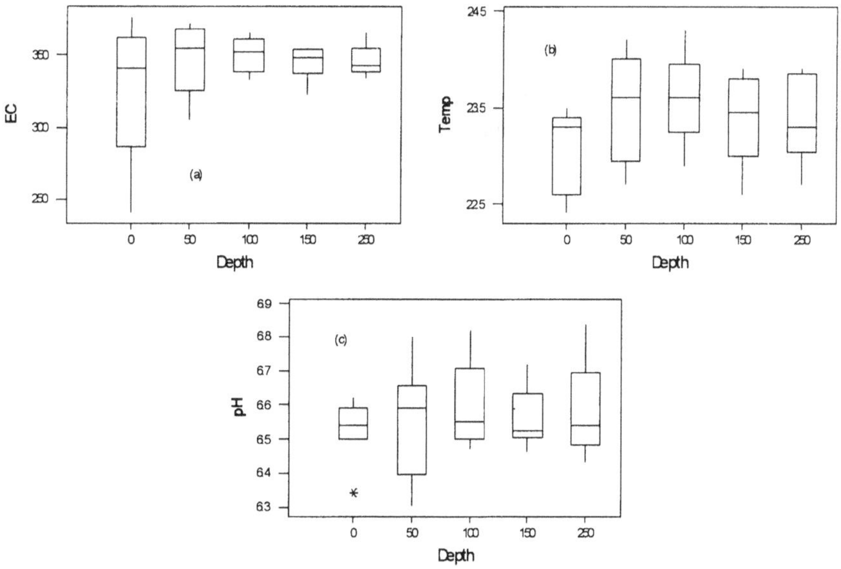

Fig. 4.10. Variation of conductivity (EC, μs/cm, graph a), temperature, graph b) and pH (graph, c) at different depths in papyrus vegetation at T4-400. Depth 0 = surface, 50 = mat, 100- 150 = water column and 250 = peat/sediment top layer. Sample size = 9.

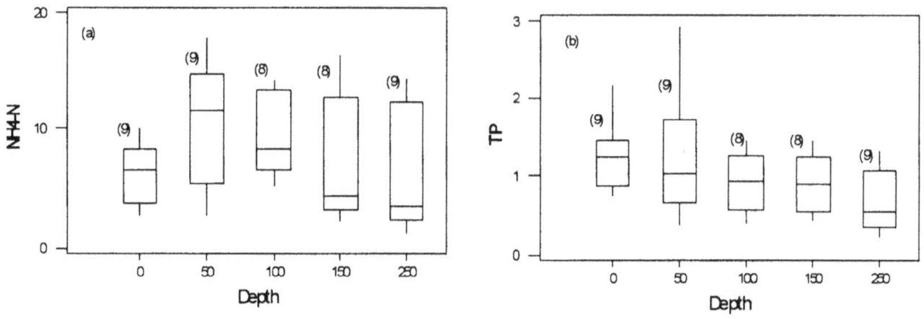

Fig. 4.11. Variation of ammonium nitrogen (NH_4-N; mg/l, graph a) and total phosphorus (TP; mg/l, graph b), at different depths (in centimeters from the surface) in papyrus vegetation at T4-400. Depth 0 = surface, 50 cm = mat; 100 - 150 = water column and 250 = Peat/sediment top layer.

In contrast, the water quality variables in the different compartments of *Miscanthidium* were different. Conductivity and pH values were significantly lower in the mat and highest in the water column and peat/sediment top layer, which were the same (Fig. 4.12).

Temperature was higher in the water column and sludge/peat layer, followed by the top mat and lowest in the mid-mat (Fig. 4.12, graph b). The high acidity (pH as low as 4.3) in the *Miscanthidium* mat was not recorded elsewhere in the Nakivubo swamp. Ammonium-nitrogen was lowest in the mat and more or less the same in the water column and sludge/peat layer (Fig. 4.13, graph a). However, total phosphorus depicted an opposite trend (Fig. 4.13, graph b). Total reactive phosphorus was lower in the top mat and the sludge/peat layer and highest in the water column.

Dissolved oxygen was generally low in all swamp compartments being zero in the water column and mid mat and 0.5 - 0.9 mg/l at the surface. No particular trends were observed.

Comparison of vertical profiles in the two vegetation types, indicated that nutrients' concentrations were higher in the mat of papyrus than of *Miscanthidium*. The water column of *Miscanthidium* had the highest concentrations.

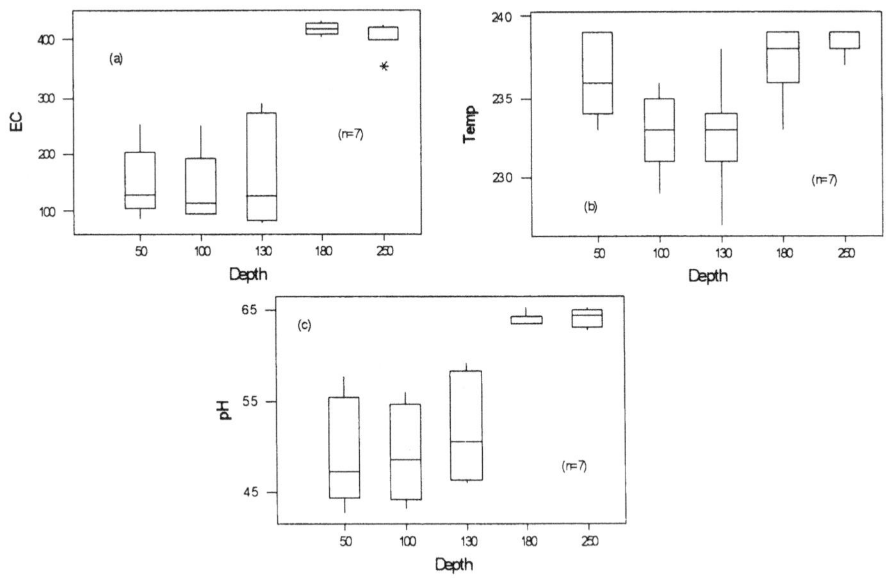

Fig. 4.12. Variation of conductivity (EC μs/cm, graph a), temperature (graph, b), and pH at different depths from the surface (in cm) in *Miscanthidium* vegetation at T2-400. Depth 50 - 130 = mat, 180 = water column and 250 = peat/sediment top layer.

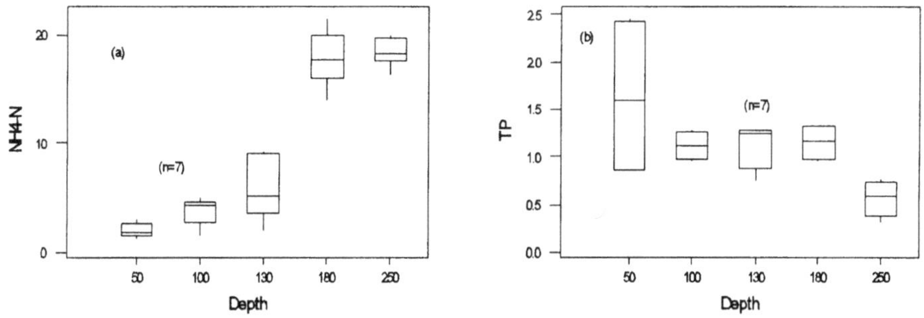

Fig. 4.13. Variation of ammonium-nitrogen (NH4-N; graph) and total phosphorus (TP; mg/l, graph b), at different depths (in centimeters from the surface) in *Miscanthidium* vegetation at T2-400. Depth 50 - 130 = mat; 180 = water column and 250 = peat/sediment top layer.

4.3.3 Variation of water quality in the Inner Murchison Bay

Temperature, pH, DO increased from the swamp interface to the open waters of the bay (Fig. 4.14). Water was still hypoxic at 750 m offshore from the swamp interface. Conductivity, ammonium-nitrogen and total phosphorus concentrations sharply decreased in the open waters of the bay, just 1 km offshore from T4-400 (Fig.4.15). However, when the water hyacinth was resident in the bay, the conductivity values and nutrient concentrations were high, whereas dissolved oxygen concentration and temperature values were low.

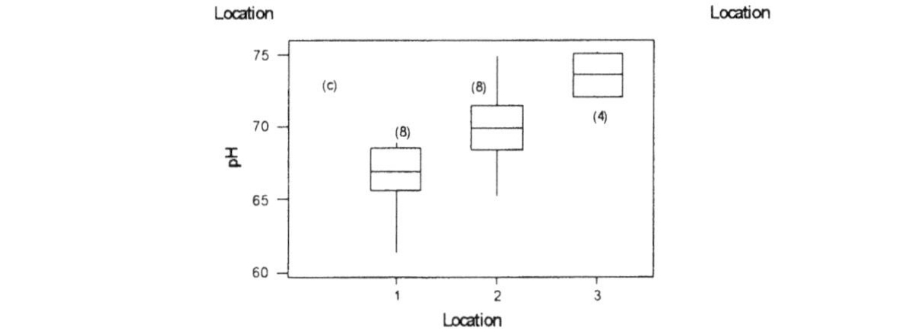

Fig. 4.14. Variation of dissolved oxygen (DO; mg/l , graph a); temperature (Temp, graph b) and pH (graph c) at given locations in Inner Murchison Bay. Locations 1, 2 and 3 are 750 m, 1500 m and 2500 m respectively from the lake swamp interface at T4-400.

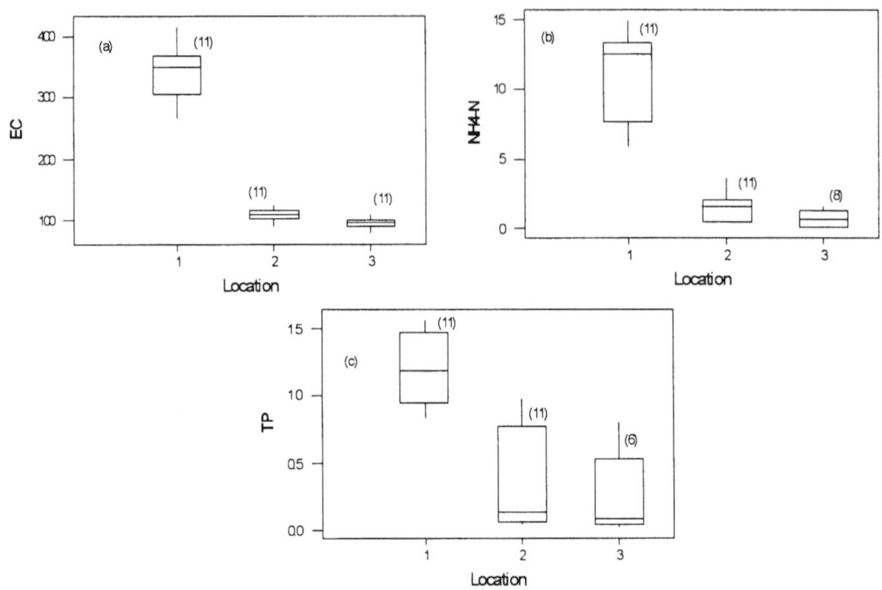

Fig. 4.15. Variation of conductivity (EC; μS/cm, graph a), ammonium-nitrogen (NH4-N; mg/l, graph b) and total phosphorus (TP; mg/l, graph c) at given locations in Inner Murchison Bay. Locations 1, 2 and 3 are 750 m, 1500 m and 2500 m respectively from the lake-swamp interface at T-400.

4.3.4 Variation of water quality along the longitudinal axis of the swamp

The variation of water quality along the horizontal axis of the swamp was monitored by simultaneously taking water samples at the swamp inlet and at different locations in the swamp. The results are presented as the ratio of nutrients to conductivity (both corrected for lake base values), to differentiate between retention or transformation of pollutants and dilution by lake water. Samples were taken from areas dominated by *Miscanthidium* and papyrus vegetation. The decrease in ratio of nutrients to EC was higher in papyrus than in *Miscanthidium* (Fig. 4.16).

The ammonia-nitrogen to conductivity ratio decreased to 75.5% and 53.3 % respectively from the swamp inlet towards respectively the *Miscanthidium* and papyrus zones along Transect 4. For total reactive phosphorus, the ratios decreased to 18.9% and 53.5% for the two respective zones.

Overall, lower concentration of nutrients, values of pH and temperature were found at the edges of the swamp, whilst high levels were found in areas dominated by *Miscanthidium* located in the main wastewater path. Conductivity and temperature values, recorded in the

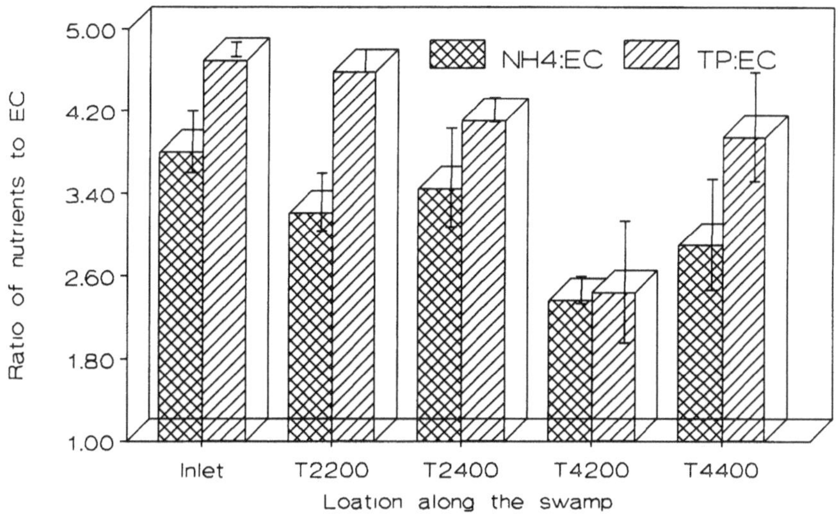

Fig. 4.16. Variation of the ratio of ammonium-nitrogen (NH4-N :EC; (*100), and total phosphorus (TP :EC; (*1000), along the swamp. Inlet = Nakivubo channel swamp inlet; T2200 = 200 m along transect 2. Distance 200 is located in papyrus whereas 400 is in *Miscanthidium* vegetation.

swamp water under *Miscanthidium*, were in most cases similar to that at the swamp inlet (Fig. 4.1). The smell of water at the locations was always similar. Furthermore, the colour of water during storm events was similar for the two locations.

4.3.5 Variation of water quality in the swamp and the effect of lake seiches

The pattern of diurnal variation of water quality at the swamp inlet was not reciprocated in the swamp pattern beyond transect 1. Instead, the variation in the swamp was more dependent on the movement of lake water in and out of the swamp. Rising water levels were accompanied by a rise in water temperature and a drop in electrical conductivity. Lake water moving into the swamp in some cases transported reasonably oxygenated water (up to 5.8 mg / l). However, in most of the times there was little inflow of DO rich water which maybe attributed to the duration of the seiches and presence of water hyacinth in the bay. The transported oxygen remained in the water column as no oxygen was detected in the swamp mat. This could be due to oxygen consumption by microorganisms during the mineralisation of organic matter in the mat. Movement of lake water into the swamp also resulted in dilution of nutrients in the water column.

4.4 Discussion

The high nutrient load into the Nakivubo swamp, through the Nakivubo channel, originates from the discharge of secondary treated effluent of Bugolobi sewage works and rainfall run-off from the city of Kampala (Taylor, 1991; Matagi, 1993; SOE, 1994). The variation in water quality in the inflow, which showed daily maxima and minima, reflects diurnal variation in the discharge of domestic and industrial wastewater into the channel. The wastewater seems to be discharged into the channel in the late evening and reaches the railway embankment at around 4 o'clock in the morning. Strong sewage conveyed by Luzira drain, is due to lack of pretreatment of the wastewater discharged into the swamp.

The variation in water quality in the swamp indicates that wastewater is not evenly transported to all parts of the swamp as it passes through. The bulk of the water funnels through the middle part, whereas the edges of the swamp receive low wastewater flow or none at all. The low concentration of nutrients and low values of physical variables at the edge of the swamp could be due to the natural topography making less wastewater flowing into this area. Nakivubo swamp vegetation is attached and emergent on the landward side and at the edges whereas it is deep and has a floating mat on the lake-ward side. The floating nature of the swamp influences the horizontal and vertical transport of pollutants in the swamp (Howard-Williams and Gaudet, 1985), in most cases resulting in canalisation (Gaudet, 1976; Howard-Williams, 1985).

Closer to the edges and where papyrus is rooted or overlying a thin layer of highly turbid water, the loose rhizomatous papyrus structure and debris continuously falling out of the mat onto the sediment via the water column (Gaudet, 1976; Denny, 1985), may increase resistance to flow and reduce the velocity of water, possibly increasing the retention time of water in these areas. The higher retention time and easy penetration of wastewater into the mat could have resulted in nutrient uptake by papyrus plants and hence explaining the low nutrient concentrations in the water column in these areas.

On the contrary, the compact and solid nature of the *Miscanthidium* mat with its low hydraulic conductivity may have impeded penetration of wastewater into the mat. The vertical movement of water into the mat appears to be only by capillary action and through evapotranspiration loss from the plants and mat surface, which keeps the mat soggy. The slow movement of water in the mat accompanied by ion uptake by *Miscanthidium* plants and dilution from rain could explain the low levels of conductivity and nutrients in the mat (Gaudet, 1976; 1979; Verhoeven, 1986).

The observed low levels of oxygen in the Nakivubo swamp are in the range reported for other papyrus swamps (Gaudet, 1976; Howard-Williams and Gaudet, 1985) and other types of natural wetlands receiving sewage effluent (Cooke, 1994; Osborne and Totome, 1994). This is attributed to the high oxygen demand exerted by decomposing organic matter in the swamp. This implies that, the little oxygen reported to be released by wetland plants (Brix, 1993), is consumed in the oxidation of decomposing organic matter.

The pH values of water in the Nakivubo swamp were lower compared to the inflows and in the Inner Murchison Bay - Lake Victoria, but within the range reported for other papyrus swamps in East Africa (Gaudet, 1976; 1979; Howard-Williams and Gaudet, 1985) and other natural wetlands / marshes (Howard Williams, 1979; Cooke, 1994; Kadlec and Knight, 1996; Thompson and Hamilton; 1983; Verhoeven, 1986). This is attributed to the formation of humic and fulvic acids during organic matter mineralisation.

Miscanthidium has been reported to tolerate low pH environments and to colonise areas that are too acidic for other macrophytes (Denny, 1993), to be drought intolerant (Thompson and Hamilton, 1983) and to precede papyrus in the hydrosere (Lind and Visser, 1962). The observations of this study, suggest that *Miscanthidium* mat induces low pH, resulting from the compactness of the mat, which in addition to impeding mixing of mat and swamp water, traps the produced humic and fulvic acids in the interlaced mat structure. These compounds formed from lignin and phenolic molecules, have strong ion exchange capabilities and can remain in the system for 2000 to 3000 years (Verhoeven, 1986). Furthermore, the acidity may suppress microbial activity, resulting in further accumulation of undecomposed dead plant material, explaining the thick mat observed in *Miscanthidium* zones. The morphology and ecological differences between papyrus and *Miscanthidium* will be expounded further in Chapter five.

Variation of water quality along the length of swamp revealed that there was a decrease in the concentration of nutrients with respect to the base values in the lake, implying the retention of nutrients in the swamp. Overall, the decrease was higher in the areas dominated by papyrus compared to those under *Miscanthidium*. Channelisation accompanied by limited interaction between wastewater and the mat in which plants are rooted, have been reported to reduce nutrient stripping by natural wetlands (Viner, 1970; Gaudet, 1978). Less interaction between the wastewater and the mat may explain the high concentrations of nutrients observed in the water column under the mat of *Miscanthidium*.

The sharp decline in the concentration of pollutants from the swamp interface to the open waters of the Inner Murchison Bay suggests that mixing and dilution in the lake is responsible for the low levels of chemical variables and faecal coliforms observed in the open waters of the bay. Values of physical variables and coliforms recorded far offshore site (2500 m from the swamp edge), were in the range earlier reported (WHO, 1970; Kizito, 1986; Gauff, 1988; Taylor, 1991). The ratio of N: P (> 12) as reported by Jorgensen *et al.*, (1988) for fresh water lakes, suggest that phosphorus may be limiting the productivity in the Inner Murchison Bay.

Diurnal variation of water quality in the swamp, indicated that the lake influences significantly the quantity and quality of water in the swamp. The seiche movements of water into the swamp, diluted the pollutants. In some cases, there was a strong correlation between lake water inflow into the swamp and the dilution of pollutants. For some other data sets, the correlation was weak, hence making it difficult to predict the actual dilution of pollutants by lake water. Depending on the season and duration of the seiches, variable amounts of oxygen are transported into the swamp. The maximum oxygen concentration recorded was 5.8 mg/l. The effect of this oxygen in the nitrification of ammonia will be discussed in Chapter 7. No oxygen was transported into the swamp (during seiches) when the water hyacinth was resident in the bay.

4.5 Conclusions

Wastewater carried by the Nakivubo channel and the Luzira drain is not evenly distributed in the swamp and the water flows in preferential paths. Water quality in the swamp varied depending on the flow paths of the wastewater and the vegetation type. Low concentrations of nutrients were recorded in the zones dominated by papyrus. Dilution and mixing in the Inner Murchison Bay-Lake Victoria generally may explain further observed reductions of nutrients and faecal coliforms at the swamp-lake interface. The lake has a significant impact on water quantity and quality in the swamp.

4.6 References

APHA, (1992). *Standard Methods for the Examination of Water and Wastewater*, Washington, DC.

Bavor, H. J. and Mitchell, D. S . (eds.) (1994). Wetland systems in water pollution control. *Wat. Sci. Tech,. 29*: (4)

Brix, H. (1987). Treatment of wastewater in the rizhosphere of wetland plants: the root - zone method. *Wat. Sci. Tech., 19:* 107-118.

Brix. H. (1993). Macrophyte-mediated oxygen transfer in wetlands: transport mechanisms and rates. *In:* Moshiri, G. A. (ed.). *Constructed wetlands for water quality improvement.* CRC Press, Inc. Boca Raton.

Brodrick, S.J.,Cullen, P. and Maher, W. (1988). Denitrification in a natural wetland receiving secondary treated effluent. *Wat. Sci. Res., 22:* 431- 439.

Cooke, J. G. (1994). Nutrient transformations in a natural wetland receiving sewage effluent and the implications for waste treatment. *Wat. Sci. Tech,.* 29: 209 - 217.

Cooper P. F. and Findlater, B. C. (1990). *Constructed wetlands for water pollution control.* Pergamon Press, Oxford. 605 pp.

Denny, P. (1985). Wetland vegetation and associated life forms. In: Denny, P. (ed.). *The ecology and management of African wetland vegetation.* Dr. W. Junk Publishers, Dordrecht. pp 1-18.

Denny, P. (1989). *Wetlands.* In: UNEP. Report IX; Strategic planning in Uganda.

Denny. P. (1993). Wetlands of Africa. *In*: Whigham, D. F; Dykyjová, D. F. And Hejný S. (eds.). *Wetlands of the world, I. inventory, ecology and management.* Kluwer

Academic Publishers, Dordrecht. pp 1- 46.

Gaudet, J. J. (1976). Nutrient relationships in detritus of a tropical swamp. *Arch. Hydrobiol., 78:* 213 - 239.

Gaudet, J. J. (1978). Effects of a tropical swamp on water quality. *Ver. int. Verein. Limnol., 20*: 2202-2206.

Gaudet, J. J. (1979). Seasonal changes in nutrients in a tropical swamp: North swamp, Lake Naivasha, Kenya. *J. Ecol., 67*: 953-981.

Gehrels, J. and Mulamoottil, G. (1989). The transformation and export of phosphorus from wetlands. *Hydrobiological processes, 3:* 365 - 370.

Gauff, H. P. (1988). *Kampala water supply expansion project. Water quality studies.* Report prepared for the Ministry of Water and Mineral Development and National Water and Sewerage Corporation; Kampala - Uganda.

Hammer, D. A. (1989). *Constructed wetlands for wastewater treatment.* Lewis Publishers, Inc. Michigan.

Hammer, D. A. and Knight, R. L. (1994). Designing constructed wetlands for nitrogen removal. *Wat. Sci. Tech., 29*: 15 - 27.

Haberl, R., Perfler, R., Laber, J. and Cooper, P. (eds.) (1997). Wetlands systems for water pollution control. *Wat. Sci. Tech. 35*: (5)

Howard-Williams, C. (1979). The aquatic environment: II. Chemical and physical characteristics of the lake Kilwa swamps. *In*: M. Kalk, A. J., McLachlan and Howard-Williams, C. (eds.). *Lake Chilwa studies, Monographiae biologicae.* Vol. 35, Dr W. Junk, The Hague. pp 231 - 245.

Howard-Williams, C. (1985). Cycling and retention of nitrogen and phosphorus in wetlands: a theoretical and applied perspective. *Freshwater Biology, 15*: 391 - 431.

Howard-Williams, C. and Gaudet, J. J. (1985). The structure and functioning of African swamps. *In:* Denny, P. (ed.). *Ecology and management of African wetland vegetation,* Dr. Junk Publishers, Dordrecht. pp 153 - 175.

Jorgensen, S. E. and Vollenweider, R. A. (1988). *Guidelines of Lake management: Vol 1. Principles of lake management.* International Lake Environment Committee and United Nations Environment Programme.

Kadlec, R. H. and Brix, H. (eds.) (1995). Wetland systems in water pollution control. *Wat. Sci. Tech., 32*: (3)

Kadlec, R. H. And Knight, R. L. (1996). *Treatment wetlands.* Lewis, Publishers, Boca,Raton.

Kizito, Y. S. (1986). *The evaluation of pollution levels in Nakivubo Channel, Kampala-Uganda.* M.Sc. Thesis, Makerere University, Uganda.

Lijklema, L. J. (1990). Mass balances and cycling of nitrogen and phosphorus in flooded soils. *In*: Oorschot, M. M. P. and Verhoeven, J. J. A. (eds.) *The Utrecht plant Ecology News report.* pp. 44 - 51.

Lind, E. M., and Visser, S. A. (1962). A study of a swamp at the north end of Lake Victoria. *J. Ecol., 50*: 599-613.

Matagi, S. V. (1993). *The effect of pollution on faunal distribution in Bwaise and Nakivubo channels, Kampala.* M.Sc. Thesis. Makerere University, Kampala, Uganda.

Osborne, P. L. and Totome, R .G. (1994). Long-term impacts of sewage disposal on a tropical wetland. *Wat. Sci. Tech., 29*: (4) 111 - 117.

Reddy, K. R. and Smith, W. H. (1987). *Aquatic plants for wastewater treatment and resource recovery* . Magnolia Publishing Inc. Orlando - Florida.

Richardson, C.J. and Davis, J.A. (1987). Natural and artificial wetland ecosystems: ecological opportunities and limitations. *In:* Reddy, K. R. and Smith W.H. (eds.) *Aquatic plants for wastewater treatment and resource recovery*. Magnolia Publishing Inc. Orlando - Florida pp 819 - 854.

Sasser, C. E., Gosselink, J. G., and Shaffer, G. P. (1991). Distribution of nitrogen and phosphorus in a Louisiana freshwater floating marsh, *Aquatic Botany*. *41*: 317-331.

SOE (1994). *State of Environment for Uganda*. Ministry of Natural Resources, Kampala - Uganda.

Taylor, A. R. D. (1991). Status of Nakivubo and Luzira swamps in Kampala. Report prepared for the Ministry of Environment Protection, Natural Wetlands conservation and management programme, Kampala- Uganda.

Thompson, K. and Hamilton, A. C. (1983). Peatlands and swamps of the African continent. In: Gore, A. J. P. (ed.). Mires: Swamp, Bog, Fen and Moor. Ecosystems of the world series, Vol 4 B (Regional studies). Elsevier, Amsterdam. pp 331 - 373.

UNEP, (1995). Water Quality of World River Basins. UNEP Environment Library No 14. Nairobi.

Verhoeven J. T. A. (1986). Nutrient dynamics in minerotrophic peat mires. *Aquat. Bot.*, 25: 117-137.

Viner, A. B. (1970). *Ecological chemistry of a tropical African Lake*. PhD Thesis, University of London.

Wetzel, R. G. (1993). Constructed wetlands: scientific foundations are critical. In: Moshiri, G. A. (ed.) *Constructed wetlands for water quality improvement*, Lewis Publishers, Boca Raton, pp. 3 - 7.

WHO/Resources Group London (1970). Master plans for Water Supply and Sewerage for greater Kampala and Jinja Areas (1970 - 2000). Special studies Reports ``C'' and ``D'' Lake investigations. Special Studies report ``B'' Water quality.

Chapter 5

Nutrient transformations by dominant vegetation types in the Nakivubo swamp.

Kansiime, F. and Nalubega M.

5.1 Introduction

Nakivubo swamp is dominated by the aquatic macrophytes *Cyperus papyrus* and *Miscanthidium violaceum* (Fig. 4.1, Chapter 4). The swamp has been receiving wastewater from the city of Kampala for more than 30 years now. Plant uptake is one of the potentially major processes that remove nutrients from wastewater discharged into the swamp.

Uptake by macrophytes (both in natural and artificial wetlands) has been recognised by several authors as an important process in removing nutrients from wastewater (Gaudet, 1977; Chale, 1987; Breen, 1990; Rogers *et al.*, 1991; Wetzel, 1993a). Considering mass balances, plant uptake, has been reported to vary between 16 - 90% and 21 - 75% of influent nitrogen and phosphorus loads respectively (Reddy and Debusk, 1987; Breen, 1990; Kadlec and Knight, 1996). Nutrient uptake is directly related to the growth rate, as the nutrients are transformed and used up in the production of new cells during plant growth. However, uptake varies from one plant species another depending on their physiological adaptations and phenology (Denny, 1985). Plants with high nutrient uptake and high standing crop are usually preferred for wastewater renovation (Chambers and McComb, 1994) as high amounts of nutrients can be removed through harvesting.

The high productivity of macrophytes accompanied by high nutrient uptake, make these plants act as stores of nutrients (Gaudet, 1977; Thompson *et al.*, 1979; Wetzel, 1993a), which are eventually recycled into the system through processes of leaching, decomposition and mineralisation (Howard-Williams, 1985; Verhoeven, 1986; Denny, 1987; Sasser *et al.*, 1991). When biomass production in wetlands exceeds the rate of decomposition, peat accumulation occurs (Howard-Williams, 1979). The nutrients trapped in swamp compartments may dissolve into the water or be deposited as peat or sediment and eventually be exported to the nearby water bodies (Howard-Williams, 1985; Richardson and Davis, 1987). Part of the recycled nutrients may again be taken up by growing plants (Sasser *et al.*, 1991).

To minimise internal nutrient cycling and eventual export of nutrients from the system, harvesting is one of the options to remove nutrients from wetlands (Muthuri *et al.*, 1989; Suzuki *et al.*, 1989; van Oorschot, 1990; Kadlec and Knight, 1996). Harvesting is especially important for phosphorus removal as there is no equivalent for denitrification to eliminate phosphorus by transforming it into volatile substances.

In addition to nutrient uptake by plants, removal mechanisms of nutrients from wastewater include physical and bio-chemical processes such as adsorption, and nitrification/ denitrification whereby wetland plants indirectly provide attachment sites for microorganisms involved in nutrient cycling and removal (Gersberg *et al.*, 1986; Brix and Scheirup, 1989; Kadlec and Knight, 1996).

The papyrus vegetation in the Nakivubo swamp varies in terms of lushness. Since healthy plants in the swamp are found in the main flow path of the wastewater, their lushness may be attributed to nutrient uptake from the water in which the plants are anchored. Hence, the nutrient content of plants growing in the main flow path of the wastewater is expected to be higher than those not affected. In chapter 4 it was observed that the water quality in the Nakivubo swamp was influenced by the vegetation type. The nutrient concentrations were higher in the water column under *Miscanthidium* than in that under papyrus where there was better contact between the plants and wastewater. In addition to other factors like mat thickness, nutrient uptake is suggested to be one of the causes for the observed differences in water quality. To get a better insight into the role played by the two vegetation types in nutrient uptake and processes taking place therein, this study was carried out to:
(i) determine the nutrient content of representative plant parts and biomass production in the Nakivubo swamp, and correlate these with the wastewater flow patterns;
(ii) determine nutrient uptake, storage and biomass production, of the two dominant vegetation types;
(iii) assess the development of the *Miscanthidium* mat and the acidity therein; and
(iv) assess the importance of the leaching process in nutrient cycling.

This was done to provide an insight in the nutrient uptake and storage capacity of the Nakivubo swamp thus assisting in planning and managing of the swamp and its vegetation for wastewater treatment purposes. Since the vegetation in the Nakivubo swamp (especially papyrus) has several anthropogenic uses like thatching of houses, basket making, act as a source of fuel, a habitat of wildlife, it combines both characteristics of conservation and resource recovery which makes it an ecosystem of particular value.

5.2 Materials and methods

5.2.1 Field studies

Study area and site selection

This study was carried out in the Nakivubo swamp at different locations along Transects 1, 2 and 4 (Fig. 4.1). In the text T1-150 refers to 150 m along Transect 1 from the western edge (Bukasa side) of the swamp. Study sites along a given Transect were selected in view of the wastewater flow patterns (Chapter 4) and represent *Miscanthidium* and *Cyperus papyrus* vegetation.

Wastewater flow patterns in the swamp were monitored as described in Chapters 3 and 4. To correlate wastewater discharges with nutrient content of the above-ground (aerial) plant parts and biomass production, study plots (1 m^2 each) were randomly selected at locations T1-50, T1-100, T1-259 and T1-359, T4-100, T4-300 and T4-400. At T4-400, two plots were selected, one in papyrus monoculture (400P1) and the other at the interface between papyrus and *Miscanthidium* (400P2). The aim of selecting these sites was to assess the effect of mat thickness on wastewater penetration into the mat and on biomass production. Furthermore, plant species growing with the dominant vegetation were observed for their diversity to get an idea on the interspecies competition. All shoots from each plot were harvested just above the rhizome, taken to the laboratory and oven dried (103 ± 2 °C) to constant weight for dry weight determination.

Samples for nutrient analysis (N and P) were sub-sampled from the dried plants of each plot. Total nitrogen and total phosphorus in dried plant parts were determined according to the method of Novozamsky *et al.* (1983).

Biomass productivity

Biomass productivity in the Nakivubo swamp was assessed using either the culm-girth and dry weight relationship (Denny, 1984; Muthuri *et al.*, 1989) or by the harvest method.

To assess the growth rate of papyrus, a relationship between the girth of a culm unit (culm + umbel) and the dry weight of the culm and umbel (i.e. shoot) was first established by measurement of at least five plant units at different locations along transects 1 to 4. Locations were randomly selected at 100 - 150 m intervals along each Transect in order to cover a large area of the swamp. At each location, the plant culm girth and umbel length were measured. The girth was measured just above the sheathing scales. Thereafter, plants were harvested and taken to the laboratory for oven drying and dry weight determination.

Plots (at T1-150, T1-259, T2-100, T4-100 and T4-400) were used to monitor regrowth of papyrus plants and biomass production over time. All plants in 2 m x 2 m plots were harvested and allowed to grow again. Regrowth was monitored every month for a period of 5 months. During the regrowth, the increase in biomass was estimated using the established culm-girth and culm-unit dry weight relationship.

Since it was not easy to accurately estimate the biomass production of *Miscanthidium* by establishing a relationship between plant part(s) and total above-ground plant dry weight, the harvest method was used. In this method all the above-ground biomass is harvested. With this method the productivity of papyrus and *Miscanthidium* could easily be compared. Two plots, one in papyrus (T2-270) and the other in *Miscanthidium* (T2-400), were monitored. Monitoring was carried out for a period of eight months. At each location, a plot of 10 x 10 m was demarcated, all the plants were harvested just above the rhizome level and allowed to regenerate. Open areas of two metres were left between the plots to prevent fire attack and trampling by people. At each harvesting time, two plots of 1 m^2 each, were randomly selected and all plants harvested. The plants were taken for oven drying to determine dry weight. Harvesting of plants was done every three to four weeks for a period of 8 months.

Nutrient uptake and storage in different plant organs

To assess the effect of age (juvenile and mature) on nutrient uptake and storage by papyrus and *Miscanthidium*, plants were collected from the field and separated into different organs (roots, rhizome, culm, stalk, umbel, leaves and inflorescence). Papyrus juvenile plants were those with umbels just out of the scale leaves, whereas mature ones were well grown and the umbel was about to flower. For *Miscanthidium*, juvenile plants were ones whose stalk was still embedded in the leaves while for mature ones the stalk was out of the leaves and well developed. Nitrogen and phosphorus content of different plant parts was determined according to Novozamsky *et al.* (1983).

5.2.2 Pilot potted experiments

Miscanthidium mat morphology and acidification

To understand the development of *Miscanthidium* mat and the low pH values observed under it, pH measurements were made in the field. Also a block of a *Miscanthidium* mat (1 m x 1 m x 1.5 m deep) was cut out and lifted from the swamp and its different structural layers examined. A pilot controlled experiment was set up to check field observations (pH variation and root/mat development) in an enclosed compound at Makerere University Institute of Environment and Natural Resources (MUIENR), Kampala - Uganda.

In this experiment, stalks of *Miscanthidium* clones from the swamp were planted into four plastic buckets (10 l) filled with garden soil. A control with only garden soil was also used. Each bucket containing plants (the inner/top bucket) had perforations at the bottom and was placed in another bucket (the outer/lower bucket) containing supporting stones which kept the inner bucket suspended in the outer one. The outer bucket maintained a water column thus making plants in the inner bucket simulate the floating mat of *Miscanthidium* in the Nakivubo swamp. Wastewater was discontinuously added to the inner bucket; it seeped through the soil and finally into the outer bucket up to the outlet level of the outer bucket which was slightly above the bottom water level of the inner bucket. The water level maintained in the outer bucket kept the soil and plants wet simulating the soggy mat in the swamp. Once every week, 5 litres of wastewater were added to the inner bucket, which in turn displaced the water in the outer bucket. The wastewater which was added to the plants was collected from the water column under the *Miscanthidium* mat in the Nakivubo swamp at T2-400. Weekly, before and after changing the water, pH was measured in the water (outer buckets) and interstitial water (inner buckets). Tap water was added daily to replace that lost through evapotranspiration. At the end of the experiment, the soil was removed and roots examined.

Nutrient uptake by papyrus and Miscanthidium violaceum

To assess the nutrient uptake of papyrus and *Miscanthidium* under similar conditions and to carry out a mass balance, a pilot experiment was set up in an enclosed compound at MUIENR. Triplicates of both papyrus and *Miscanthidium* clones were grown for one month in plastic buckets (10 l) fed weekly (seven day retention time) with fresh wastewater collected from the Nakivubo swamp influent (Fig. 4.1). A control with only the wastewater was also monitored. The plants were acclimatised for three weeks. During the experimental period, and each time prior to feeding the plants, conductivity, dissolved oxygen, pH, NO_2^-, NO_3^-, NH_4^+, TRP and TP of fresh wastewater were measured. At the end of each week, samples were again taken from the buckets for the same measurements and the water replaced by fresh wastewater. Plant characteristics (shoot density, recruitment and physical appearance) were also monitored. Water lost by evapotranspiration was replaced every day with tap water.

Five litres of wastewater were added weekly to the papyrus buckets and a control whereas *Miscanthidium* buckets contained only four litres. Less water was added to *Miscanthidium* plants because, during the acclimatisation period, the plants did not grow well (as noticed from the rate of recruitment of young shoots and recovery of planted clones) when the head of water was more than 5 cm above the rhizome. On the contrary, papyrus plants easily adapted in the water even when the head of water was 15 cm above the rhizomes.

At the end of the experiment, plants were harvested, weighed and divided into different parts and oven dried to determine dry weight. Sub-samples were dried again and used in the determination of N and P in the different plant parts. The increase in biomass was deduced from the initial and final weights of plants.

5.2.3 Laboratory studies

Leaching of nutrients from plants

Senescing whole plants were obtained from T2-100 (papyrus) and at T2-400 (*Miscanthidium*). In the laboratory, papyrus was separated into umbel, culm, scales, rhizome and roots. *Miscanthidium* was separated into shoots and below-ground material (roots, rhizome and peat). The different plant parts were air dried (23 - 25 °C) for 3 weeks.

The air-dried plant organs were cut into fragments of approximately 1 cm. 10 g of the fragments were sub-sampled and put in 1 litre bottles. To minimise air intrusion, a known volume of distilled water was added to leave just enough space for the glass stoppers. The samples were incubated in the dark, at room temperature (23 - 25 °C). Sub-samples were oven-dried at 103 $\pm$ 2 °C for 24 hours for dry weight determination.

Water samples from the bottles were taken after 1, 3, 5, 14, 28 and 40 days for measurement of conductivity, temperature, pH, and analysis for NH_4^+, NO_3^-, NO_2^- , DRP (Dissolved Reactive Phosphorus) and TP (Total Phosphorus). The sample bottles were topped to original levels with a measured volume of distilled water. Prior to each sampling, the bottles were manually shaken for 30 seconds and allowed to stand for 2 minutes. All analyses were carried out as described in Chapter 4.

5.3 Results

5.3.1 Field studies

Relationship between culm girth and culm-unit dry weight for papyrus

The relationship between the culm-girth and culm-unit dry weight was established as:

$$\log W = 0.033 + 2.197 \log G \tag{5.1}$$

where, W is dry weight (g) of a culm unit and G is culm girth (cm) of the culm. A strong correlation ($r^2 = 0.813$, $P < 0.001$) was found between log culm-girth and log dry weight (Fig. 5.1).

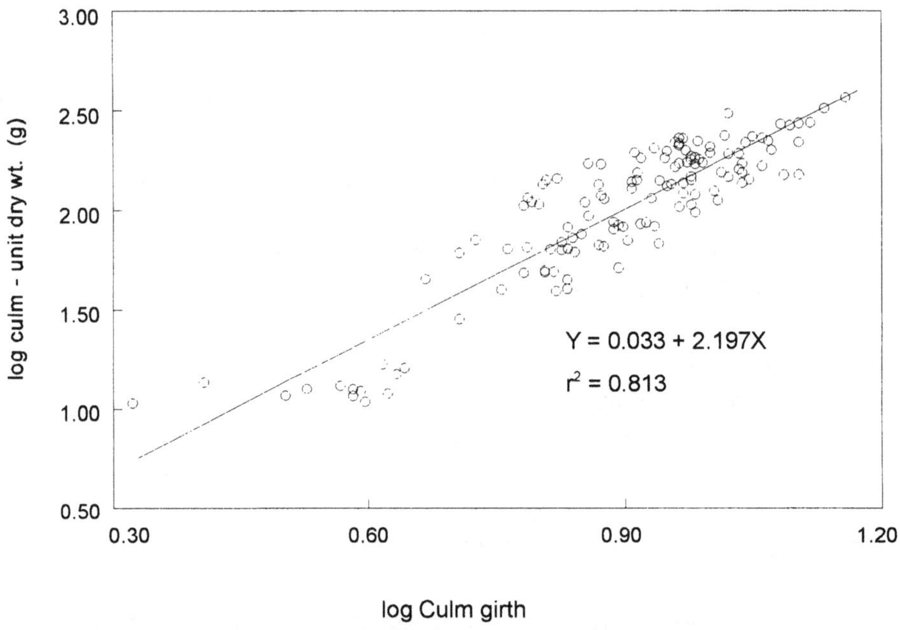

Fig. 5.1. Relationship between culm girth and culm-unit dry weight of papyrus in the Nakivubo swamp.

Aerial biomass production and nutrient content in the Nakivubo swamp

Data on culm girth, umbel length and observations on associated vegetation along Transect 1 are presented in Table 5.1. Along Transect 1, there was a general increase in culm girth and umbel length both of which became maximum in the region under the influence of wastewater. Species diversity, was higher in areas not influenced by wastewater.

Table 5.1. Culm girth, umbel length of papyrus and associated vegetation types along Transect 1. Values are for 3 replicates ± standard error.

Distance (m)	Culm girth (cm)		Umbel length (cm)		Associated vegetation types
	Mean ± SE	Range	Mean ± SE	Range	
50	7.43 ± 0.63	6.8 - 8.7	34.9 ±10.1	21.0- 54.5	*Vossia cuspidata* *Ipomoea rubens* *Leersia hexandra*
100	8.0 ±0.87	6.6 - 9.6	46.4 ± 11.3	25.3 - 64	*Ipomoea rubens* *Leersia hexandra*
259	10.8 ± 0.17	10.5 - 11.5	59.2± 8.2	51 - 75.5	*Enhydra fluctuans*
359	11.1 ± 1.0	9.6 - 13.1	52.5 ±6.0	44 - 46	*Enhydra fluctuans*

N and P contents (as percentage of dry weight) in above-ground papyrus plant parts, and biomass per unit surface along Transect 1 are shown in Fig.5.2. Nutrient concentrations in above-ground plant parts and in the swamp water as well as aerial biomass per unit area increased from the Bukasa to the Luzira side of the swamp. The biomass at T1-259 and T-359 (sites under pronounced influence of wastewater, see Chapter 4) was no significantly different

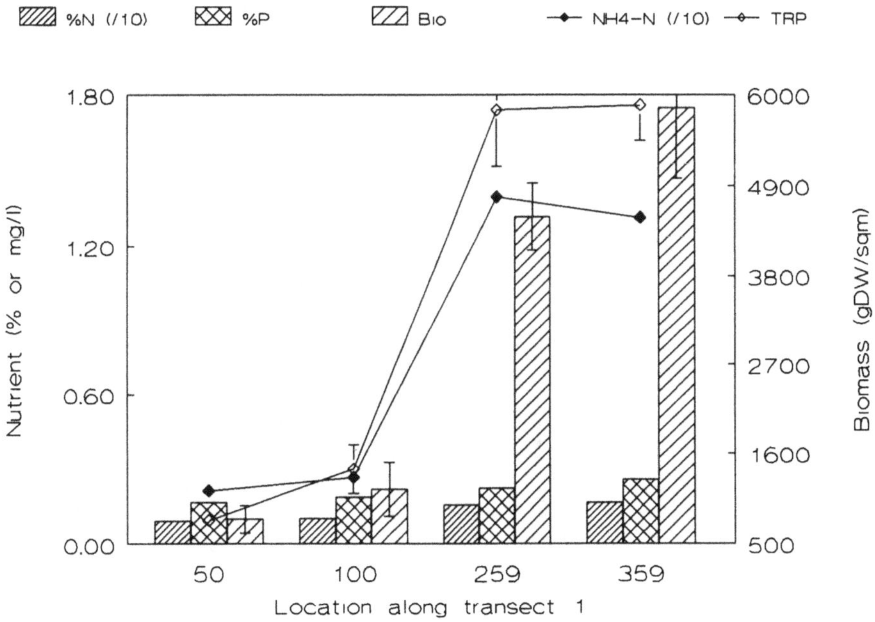

Fig. 5.2. NH_4-N (mg/l) and TRP (mg/l) in swamp water; TN (% DW) and TP (% DW) in above ground papyrus parts and aerial biomass, (g DW/ per square metre; sqm) at indicated distances (m) along Transect 1. Bars indicate standard error. n= 3 for aerial biomass and 5 for nutrients.

($P > 0.05$) but were approx. 5 times higher than that at 50 and 100 m ($P < 0.001$) which inturn were identical.

N and P content (% dry wt) in above-ground plant parts, and the aerial biomass per square metre along Transect 4, similarly increased from the Bukasa side of the swamp (Fig. 5.3). The biomass of papyrus found at the interface with *Miscanthidium* (Fig. 5.3, location 400P2) was lower than that observed at the same area where papyrus was growing as a monoculture (location 400P1). In the monoculture vegetation, papyrus was growing on a thin loose mat (50 cm on average) with some open water spots which could have allowed better contact between plants and the wastewater. At the former site (interface, location 400P2), the mat is thick (1 m on average) and papyrus growth was poor (stunted). The plant parts (especially the umbels) were growing poorly and had a yellowish colour. The creeper *Ipomoea rubens* was abundant and always intertwined with papyrus culms.

The aerial biomass of *Miscanthidium* at T-400 (400Mis) was three times lower than that of papyrus (Fig. 5.3), but higher than for papyrus observed in areas not under the influence of wastewater (T1-50). At location 400Mis (Fig.5.3), the mat in which the plants are anchored is thick (1.3 m on average) and emergent. Papyrus under pronounced influence of wastewater

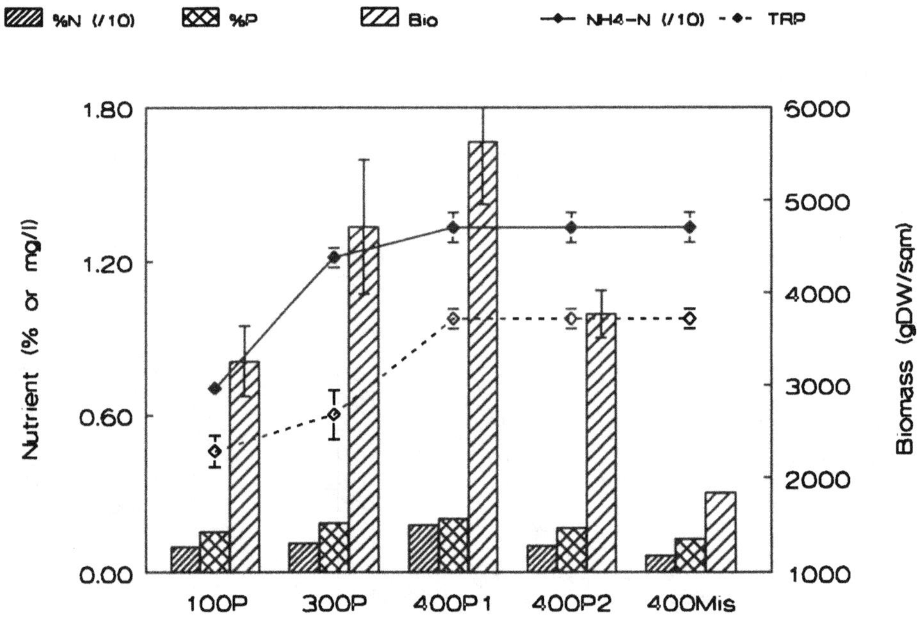

Fig. 5.3. NH_4-N (mg/l) and TRP (mg/l) in swamp water; TN (% DW) and TP (% DW) in above ground papyrus parts and aerial biomass (Bio) at indicated distances (m) along Transect 4. P after the distance denotes a papyrus dominated vegetation whereas Mis denotes *Miscanthidium* dominated vegetation. Pi is in a papyrus monoculture while P2 is at the interface between papyrus and *Miscanthidium*.

and that close to the swamp inlet, at T1 (T1-259 and T1-359) exhibited higher biomass, compared to that observed close to the lake, at T4 (T1-300 and T4-400). Similarly nutrient concentration depicted the same gradient decreasing lakeward.

Growth rates and aerial ceiling biomass

The growth of papyrus at harvested plots in the Nakivubo swamp, is depicted in Fig. 5.4. The asymptotic maximum biomass density is proportional to the extent of exposure to wastewater underneath the mat. High growth rates were recorded at sites located in the main flow path of the wastewater namely; T1-259 (36 g DW / m^2 /d), T2-100 (38 g DW / m^2 /d) and T4-400 (31 g Dw / m^2 /d). Similarly biomass density was higher at these locations (Fig 5.4). Lowest growth rates were recorded at sites (T1-150 and T4-100) not under the influence of the

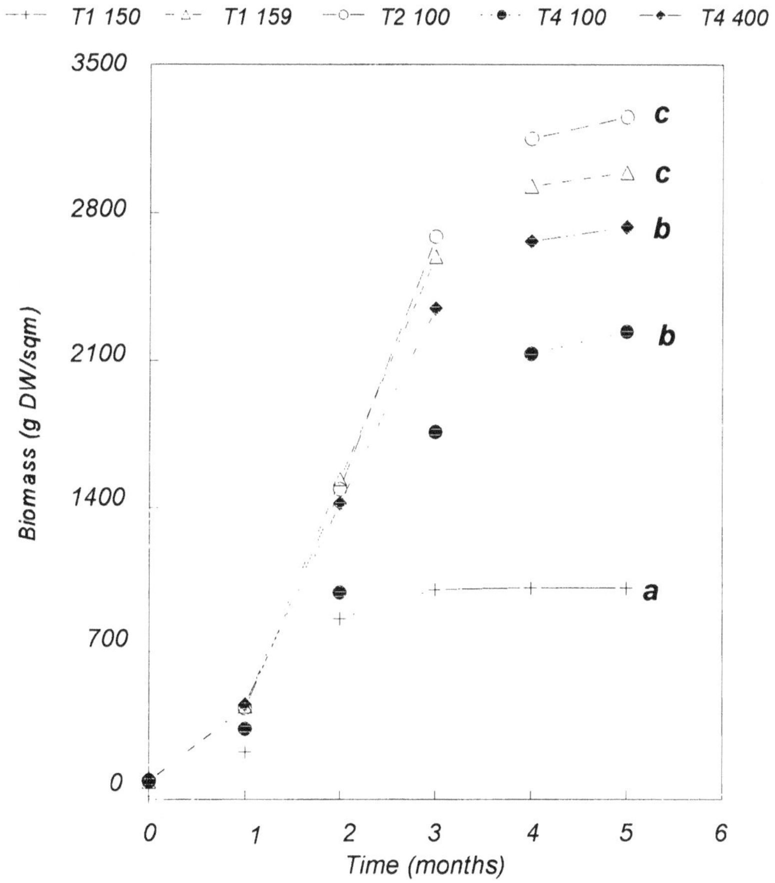

Fig. 5.4. Aerial biomass production and fitted logistic growth curves of papyrus at different locations in the Nakivubo swamp. Biomass was estimated using a culm girth and culm-unit dry weight relationship. Growth curves with the same letters are not significantly different from one another (p =0.05). T1-150 represent 150 m along Transect one.

DW / m^2 /d). Location T1-150 attained it's low asymptotic maximum density (1012 g DW / m^2) within 2 months whereas at other sites the maximum density was attained after 4 - 5 months.

Overall, papyrus vegetation under the influence of wastewater had higher biomass production. The predicted asymptotic maximum density was higher at locations under the influence of wastewater. The highest biomass density were recorded for papyrus sites close to the swamp inlet (T1-259; 2991 g DW / m^2 and T2-100; 3252 g DW / m^2).

The growth rate of *Miscanthidium* of 11.8 g/m^2 /d (Fig. 5.5) was lower than that recorded at all the papyrus sites including those not under the main flow paths of the wastewater.

The harvest method confirmed higher biomass density by papyrus compared to *Miscanthidium* (Fig. 5.5) and ditto for the growth rates. For the first two months after cutting the site, the aerial biomass of the two plant types were comparable and higher productivity of papyrus became apparent in the third month and remained significantly higher till the final harvest. *Miscanthidium* plants grew evenly and densely whereas papyrus, despite its high productivity, had some open patches which may be related to the growth and expansion of papyrus rhizomes. The maximum aerial biomass recorded during the experimental period was higher for papyrus (4490 g DW /m^2) compared to *Miscanthidium* (1109 g DW /m^2).
The Juvenile plants of papyrus and *Miscanthidium* had higher concentrations of P and N in

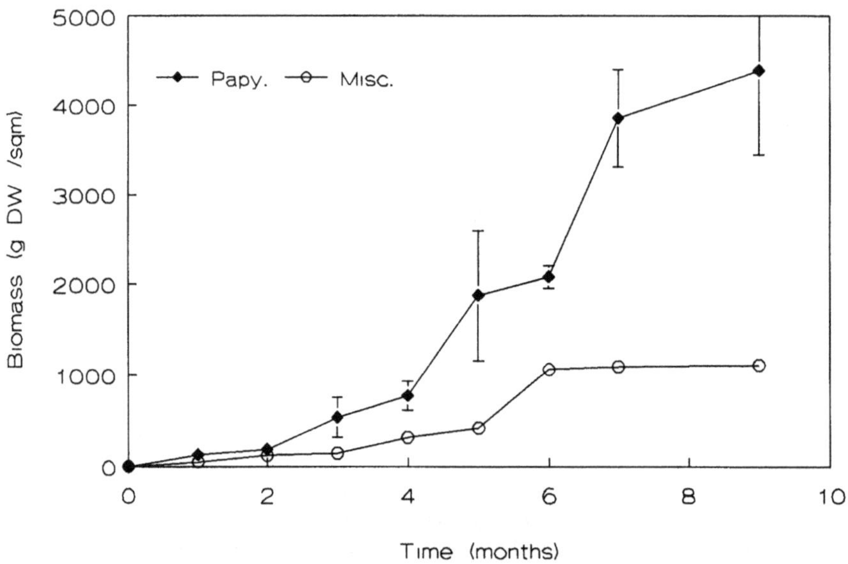

Fig. 5.5. Aerial biomass production of papyrus and *Miscanthidium* in the Nakivubo swamp as measured by the harvest method.

The Juvenile plants of papyrus and *Miscanthidium* had higher concentrations of P and N in their organs compared to the mature ones (Figs. 5.6 & 5.7). The roots and culms of mature plants of papyrus had more or less the same concentration of P but this was significantly lower than observed in the rhizome and umbel. For juvenile plants, P content was more or less the same in all organs except for the roots whose value was lowest. There was no significant difference between mature and juvenile plants for N, suggesting this nutrient is utilised by both young and matures plants to the same degree. The concentrations in the culms were almost half that recorded in the rhizomes and umbels.

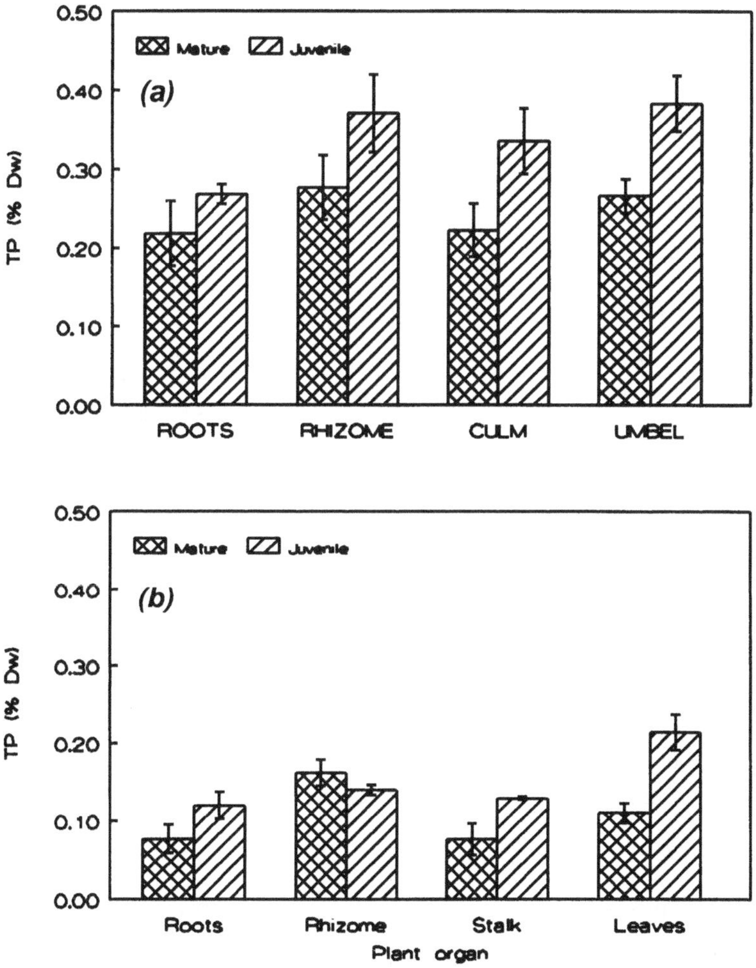

Fig. 5.6. Total phosphorus in mature and juvenile papyrus (a) and *Miscanthidium* (b) organs, (n=7).

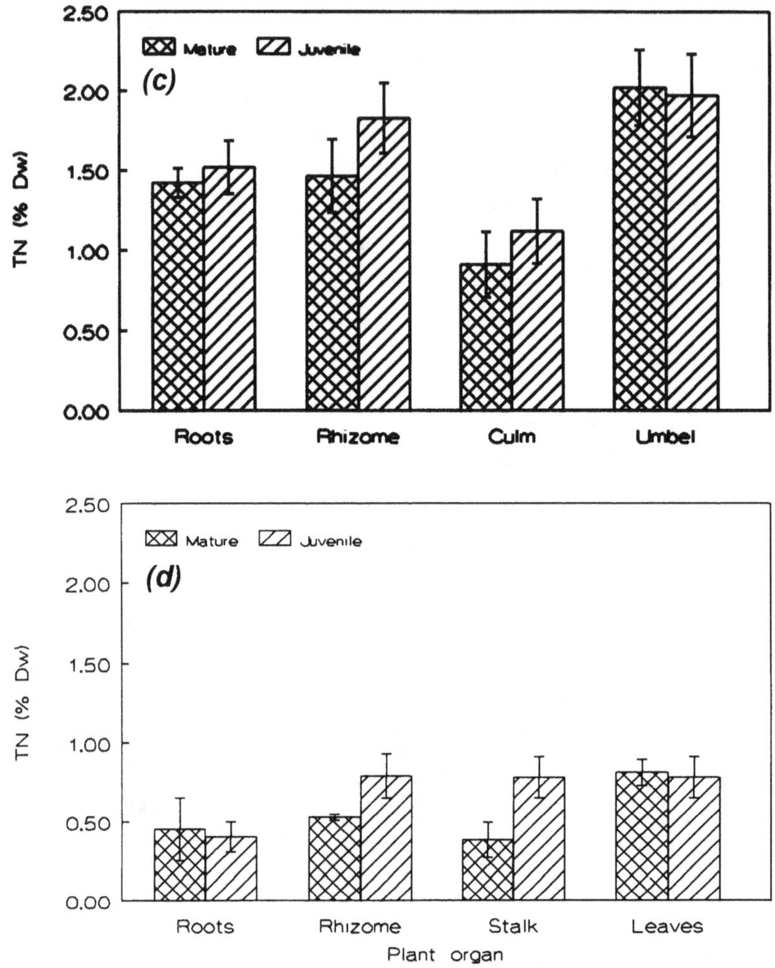

Fig. 5.7. Total nitrogen in mature and juvenile papyrus (c) and *Miscanthidium* (d) organs, (n=7).

For *Miscanthidium*, the leaves of juvenile plants had highest concentration of P, whereas the roots and stalks of mature plants had the lowest. Concentration of P was not significantly different in the roots and stalk. The concentration of N in *Miscanthidium* leaves was more or less the same for both age groups, but in other organs, it was generally lower for mature plants.

Generally, for both vegetation types, the roots had the lowest concentrations of N and P compared to the other organs whose concentrations were more or less the same, suggesting that nutrient translocation from the roots is limited by the nutrient uptake rate of roots to other plant parts. Furthermore, papyrus organs had higher concentrations of N &P compared to *Miscanthidium*.

Considering all measurements made in the Nakivubo swamp, during the study period, papyrus plants had a significantly higher phosphorus content ($p = 0.005$) than *Miscanthidium* (Fig. 5.8). Nitrogen content on average was two times higher in papyrus plants than in *Miscanthidium* ($p < 0.0001$).

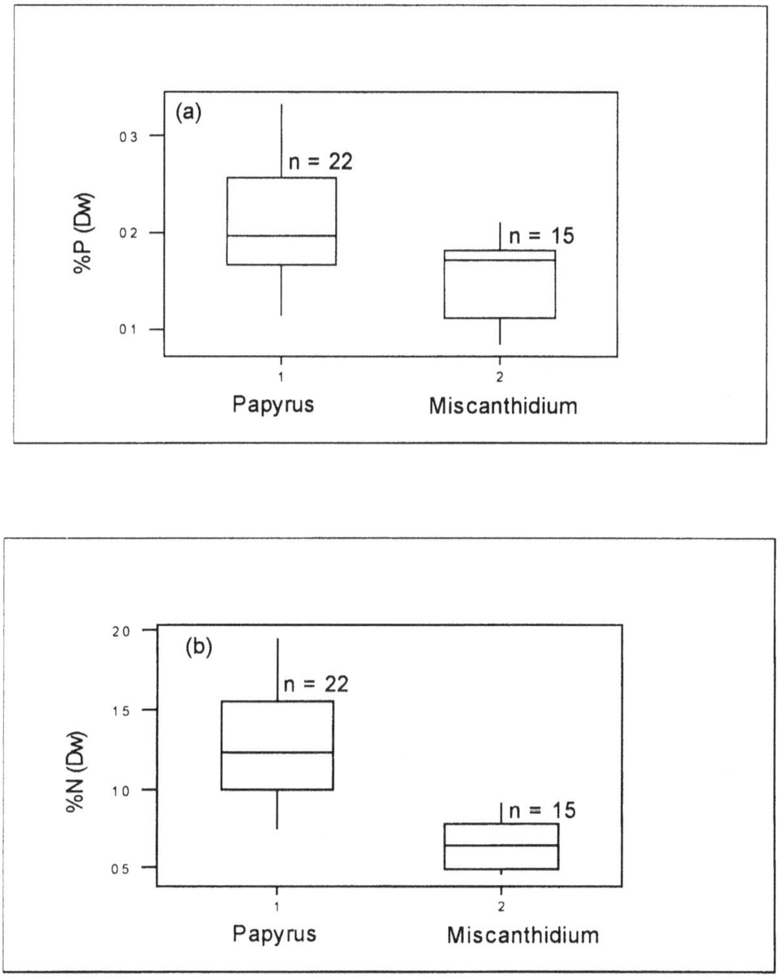

Fig. 5.8. Total phosphorus (a) and total nitrogen (b) content of papyrus and *Miscanthidium* in the Nakivubo swamp.

Development of Miscanthidium mat in the Nakivubo swamp.

Papyrus and *Miscanthidium* in the Nakivubo swamp have different mat morphology. The *Miscanthidium* mat was 1.5 m thick on average while that of papyrus was 0.5 m and always loose. The bulk density was higher in the *Miscanthidium* (69.7 ± 5.7 kg/m^3, n= 7) than in papyrus (52.1 ± 4.7 kg/m^3, n= 7).

In the field it was possible to cut out and lift a rectangular block (about 1 m x 1 m x 1.5 m deep) from an intact *Miscanthidium* mat (Fig 5.9). Close examination of the mat in the field revealed several layers. The top layer was made up of plant remains and coarse organic peat (greyish-black in colour), whereas the subsequent lower layers were brown in colour with decomposing and undecomposed plant and debris/peat trapped in between.

Fig. 5.9. Side view of a block of *Miscanthidium* mat lifted out of Nakivubo swamp. Total length = 1. 5 m. The white mark along the block is a measuring tape.

The papyrus mat had more loosely attached roots and rhizomes and the detritus was always falling to the swamp bottom. It was not possible to obtain an intact part of a papyrus mat as it would disintegrate when being lifted out. In *Miscanthidium*, even the plant debris at the interface between the mat and the water column was intact with negligible material falling out/sedimenting during removal. This explains why the peat/sediment accumulation on the bottom is lower in *Miscanthidium* dominated areas as compared to papyrus ones (Chapter 8).

5.3.2 Pilot studies

Acidification and development of the Miscanthidium mat

During growth of *Miscanthidium* in an enclosed compound at MUIENR, new shoots of plants sprouted fast from planted stalks. Some grew to their full length of about 2 metres within a period of three months. The plants were green and healthy and started flowering in the fourth month. At the end of the experiment, the roots had developed extremely well, the soil was embedded in the interlaced roots and some of the roots had passed through the perforations of the bucket in which they were growing, into the outer bucket. After removing the soil, the roots were found to be interlaced into each other (especially adventitious roots, Fig. 5.10).

Fig. 5.10. Appearance of Miscanthidium roots after four months of growth in soil as a substratum. Not the interlacing roots especially the adventitious roots.

During the growth period the pH of the experimental buckets in which the plants were growing fluctuated between 6.8 - 7.8 (Fig. 5.11), and was not significantly different from the control although a slight lowering of the pH value in the planted buckets was observed.

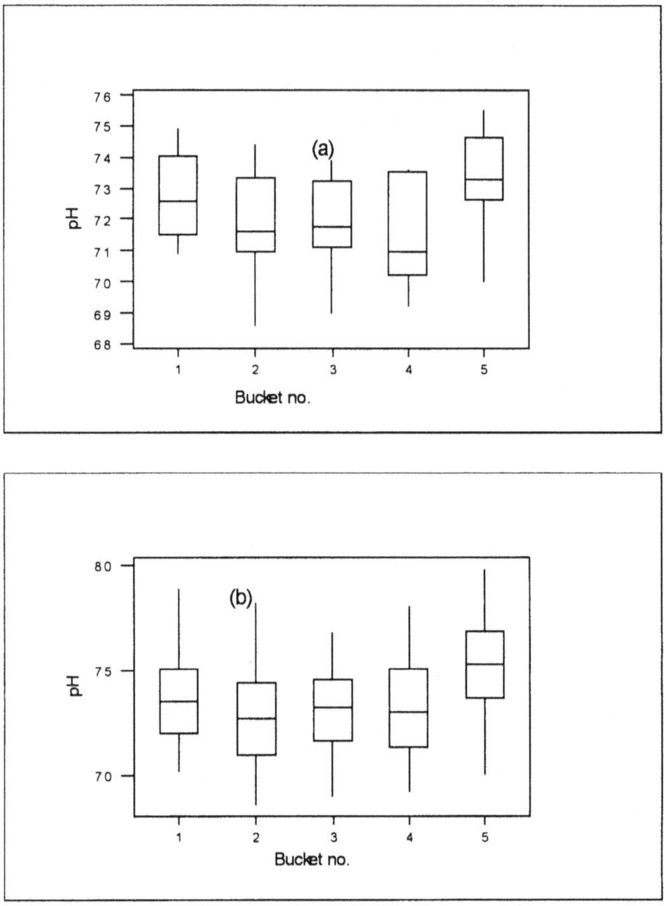

Fig. 5.11. Variation of pH in buckets planted with *Miscanthidium* over the experimental period. Graph (a) represents the inner buckets containing plants grown on soil whereas graph (b) is for the outer bucket, which maintained the water column below the inner bucket. Buckets 1 - 4 = replicates and 5 = control.

Nutrient uptake by papyrus and Miscanthidium plants in a pilot experiment.

During the initialisation period, it was observed that *Miscanthidium* with shoots trimmed off could not acclimatise well when grown in a water media, which led to most plants rotting away. Even when whole plant units (roots and shoots) were used, *Miscanthidium* did not grow well when the head of water was 5 cm above the base of the shoots. This implied that *Miscanthidium* cannot tolerate flooding conditions especially during the early stages of growth. However, cut stalks could grow well in the soil (section 5.3.2 above). Papyrus plant clones on the other hand acclimatised easily in water in the absence of soil, even when the head of water was 20 cm above the rhizomes level. Characteristics of plants used in the experiment and biomass changes are shown in Table 5.2. Recruitment was higher in *Miscanthidium* than with papyrus buckets.

Table 5.2 Bucket characteristics before and after the experiment.

Bucket	Initial no. of plants (clones)	Final no. of plants	% total shoot weight	Wet (dry) weight gain, (g)
Papyrus 1	8 (2)	8 (3 dead)	46	331 (54)
Papyrus 2	9 (2)	12 (1 dead)	45	556 (78)
Papyrus 3	8 (2)	14 (1 dead)	44	182 (29)
Miscanthidium 1	15 (3)	23	64	74 (17)
Miscanthidium 2	27 (4)	37	65	377 (85)
Miscanthidium 3	24 (4)	37	67	120 (26)
Control	-	-	-	-

The roots of papyrus only interlaced at a later stage, when there was no room for expansion in the buckets. The papyrus roots were growing straight at an early stage of development implying that the observed interlacing was because of the small bucket volume which could not allow all the new roots to expand freely downwards. Unrestricted root development in papyrus is shown in Fig. 5.12.Papyrus plants grew very well and by the third week of the experimental period the roots had covered the entire width of the bucket, whereas those of *Miscanthidium* covered about half the width but interlaced each other (Fig. 5.13).

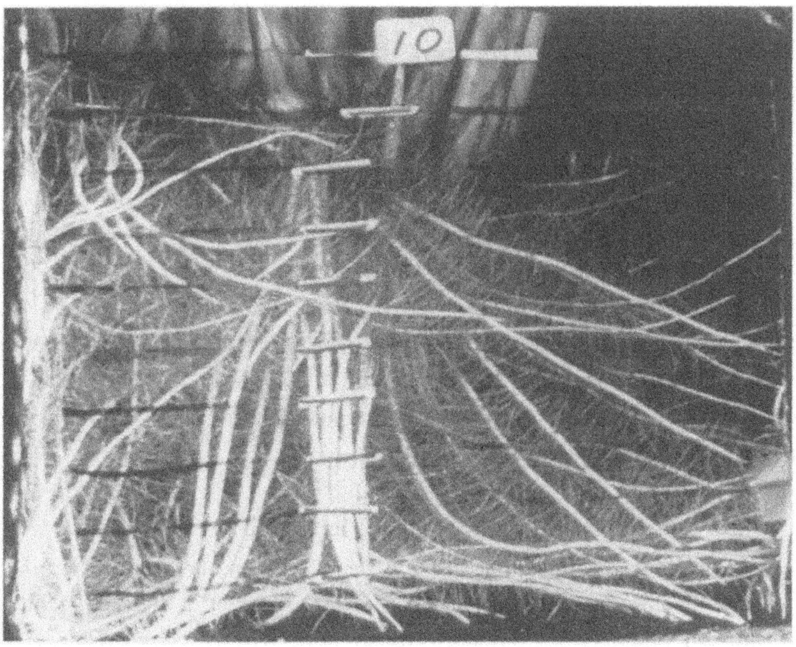

Fig. 5.12. Appearance of papyrus roots when expansion is unrestricted.
Note: roots start curving after touching the bottom or sides of the container.

Fig.5.13. Appearance of *Miscanthidium* roots after 2 months of growth in water media.

The roots of papyrus were soft and weak compared to those of *Miscanthidium* and the young roots could easily break off during feeding and changing the water: whereas those of *Miscanthidium* were solid and did not break. Furthermore, papyrus individual plants were easily separated whereas it was difficult with *Miscanthidium*, as these roots were strongly interlaced into each other.

The pH variation during the experimental period was not significantly different among the individual replicates. The pH of the control was also not significantly different from that of the added wastewater (Fig. 5.14). The pH of the *Miscanthidium* buckets was always higher compared to those containing papyrus ($P = 0.01$) indicating that *Miscanthidium* can grow at higher pH. This suggests that *Miscanthidium* does not necessarily prefer growing in an acid environment but the low pH observed in it's mat in the Nakivubo swamp may be a result of the growth cycle of the vegetation for instance accumulation of humic and fulvic acids in the mat from dead plants.

The average daily nutrient uptakes over the four experimental weeks are shown in Fig. 5.15. There was a rather high accumulation for both N (as ammonia) and P (reactive and total) in the plant tissue. In contrast to the experimental buckets, the control did show denitrification. Also the change in total organic phosphorus (data not shown) was very small.

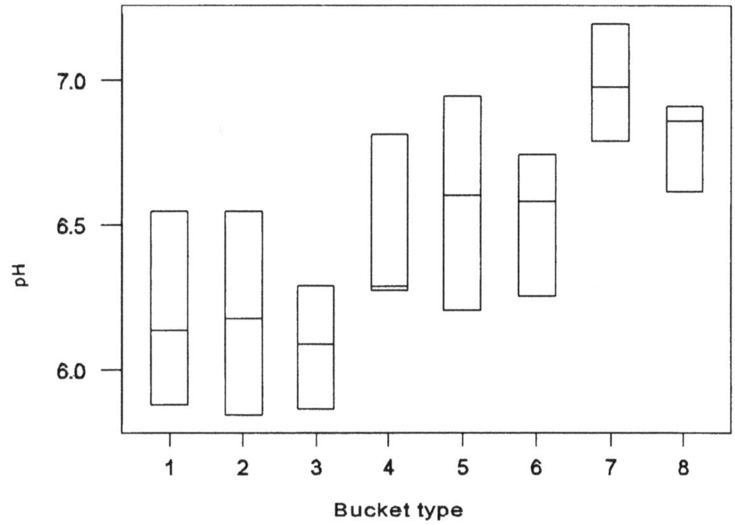

Fig. 5.14. Variation of pH in buckets planted with papyrus (1-3), *Miscanthidium* (4-6), control (7) and swamp water fed to the plants (8).

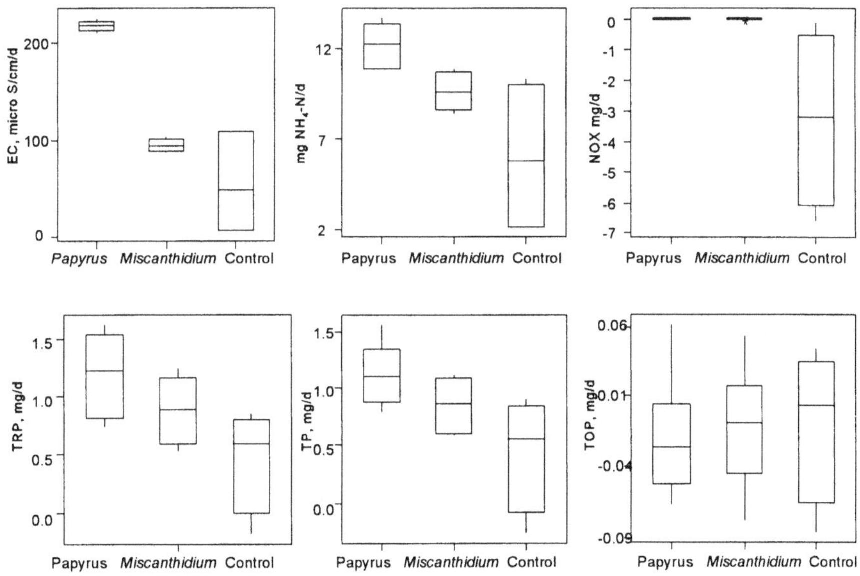

Fig. 5.15. Nutrient uptake (estimated from ambient concentration in experimental buckets) by papyrus and *Miscanthidium* over a period of one month. NOX represent NO_2 -N + NO_3-N, TRP = Total Reactive Phosphorus and TOP = Total Organic Phosphorus.

For all the weeks, there was a strong correlation between nutrient and dissolved ions (conductivity) measurements indicating similar pathways.

Fig. 5.16 shows the nutrient mass balance for both papyrus and *Miscanthidium*. The removal efficiencies were higher for papyrus. Nutrient storage was not different for both the above-ground and below-ground biomass of papyrus. For *Miscanthidium*, above-ground biomass accounted for 76% of nutrient uptake distribution. The biomass increase in papyrus was on average three times more than that of *Miscanthidium* (Table 5.2), indicating that higher nutrient uptake could have resulted in higher biomass production.

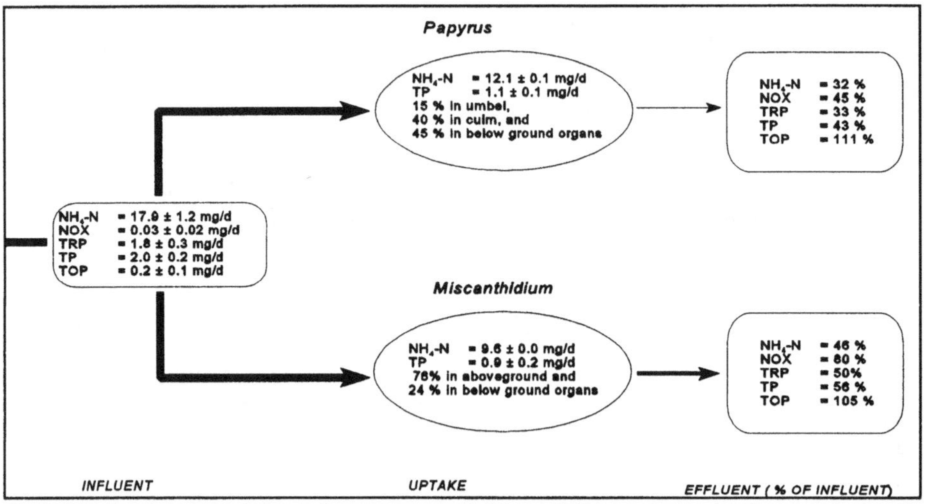

Fig.5.16. Nutrient mass balance for papyrus and *Miscanthidium* grown in buckets at a hydraulic residence time of 7 days. NOX = NO_2-N + NO_3-N; TRP = Total Reactive Phosphorus and TOP = Total Organic Phosphorus.

5.3.3 Laboratory Experiments

Leaching of nutrients under laboratory conditions

The leaching of ions from plant organs revealed that N and P leached rapidly as shown by the first day values (Figs. 5.17 and 5.18). Release rates however reduced significantly after day one. The culm and scales of papyrus showed virtually no release of nutrients after the first day.

Table 5.3. shows leached quantities of nutrients during the first day nutrient and the nutrient content of different plant organs. Considering the percentage release of nutrients per gram of plant organ, release rates appear to be higher for phosphorus than for nitrogen. In all cases, more than 80% of the released P was soluble reactive phosphorus, a form that is easily reutilized by bacteria and swamp vegetation.

Table 5.3. Nutrient release rates as a function of plant organ nutrient content over 24 h.

Plant	Portion	Leached TP% DW	Mean P content (%) ± SEM(n=4)	%P stock lost	Leached NH_4-N %DW	Mean N content (%) ± SEM(n=4)	%N stock lost
Papyrus	Umbel	0.063	0.26 ± 0.02	24	0.280	2.02 ± 0.24	14
	Culm	0.050	0.22 ± 0.03	23	0.161	0.91 ± 0.20	18
	Rhizome	0.415	0.48 ± 0.04	86	0.826	1.46 ± 0.23	57
	Peat/roots	0.029	0.22 ± 0.04	13	0.067	1.42 ± 0.09	5
Miscanthidium	Shoots	0.098	0.10 ± 0. 02	98	0.134	0.67 ± 0.11	20
	Peat/roots	0.073	0.12 ± 0.02	73	0.119	0.49 ± 0.11	24

A significant correlation between P release and plant organ P content ($r = 0.82$, $p = 0.044$) exists. Although a similar relationship is not seen for N, P and N leaching are strongly correlated (Pearson's correlation coefficient = 0.97). This is also depicted in Figs. 5.17 and 5.18. Similarly high correlations were found between leaching of ions (electrical conductivity) and nutrients (0.89 and 0.86 for P and N respectively).

Leaching enhances nutrient cycling of the water column during rainfall in the papyrus zones. For example, during the rainy season, electrical conductivity (EC) in the water rose by more

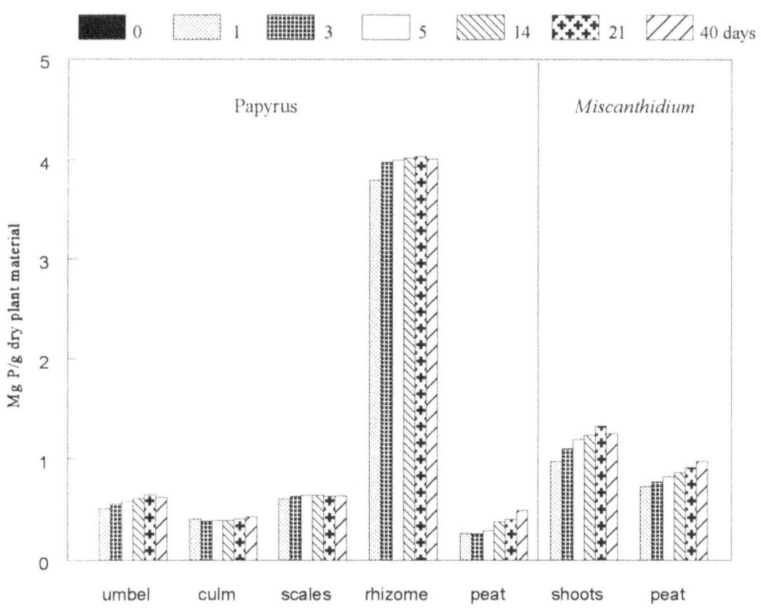

Fig. 5.17. Release of soluble reactive phosphorus from plant organs during 40 days of incubation.

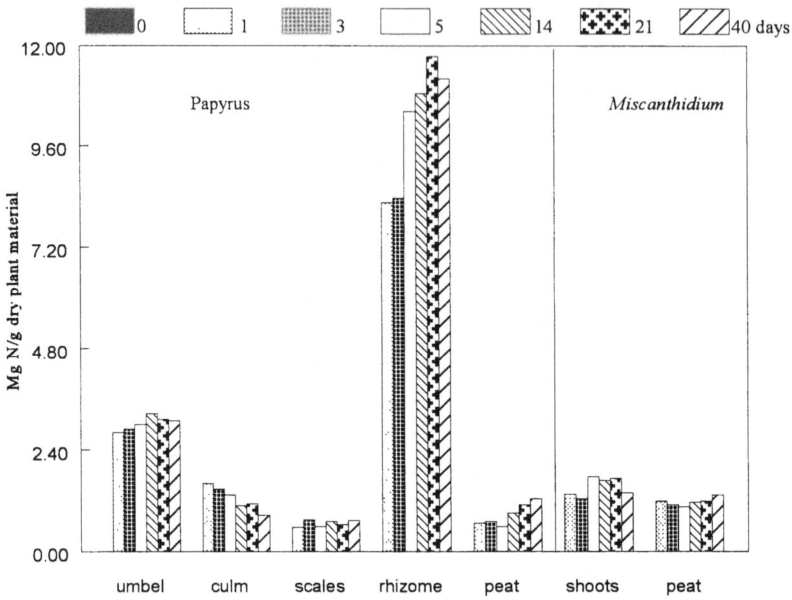

Fig. 5.18. Release of ammonium nitrogen from plant organs during 40 days of incubation.

than 50 μS/cm, instead of lowering due to dilution as would be expected (Chapter 3). Higher EC values were also observed after rainfall for surface swamp water in contact with dead papyrus plant organs. In *Miscanthidium* vegetation, EC values as low as 30 μS/cm were recorded for surface swamp water after a rainfall indicating that less ions are leached out.

5.4 Discussion

A high nutrient content was found in the above-ground biomass of papyrus plants growing in the main path of wastewater. The highest biomass per unit area was recorded at T1-359 (5844 g/m^2) which is in the middle of the wastewater flow path, and the lowest also along the same Transect at 50 m (833 g/m^2) on the sides with minimal nutrient concentration in the water. Since the western edge (Bukasa side) of the swamp at Transect 1 was not under the influence of wastewater, plants in this area presumably get their nutrients from overflows of the Nakivubo channel and rainfall run-off in addition to nutrient cycling through leaching and decomposition processes.

Since plants require nutrients for the production of new cells during growth, the availability of nutrients (like nitrogen and phosphorus), and their subsequent uptake by papyrus could explain the high growth rates and higher biomass production in areas under the influence of wastewater. Chale (1987) reported high nutrient content in plant parts and high aerial biomass (4955 g/m^2) in a papyrus swamp of Lake Naivasha, Kenya and attributed his observations to wastewater discharge into the swamp. However, no comparison was made with plants not under the influence of wastewater in the same swamp.

The observation that plants growing in the main path of wastewater were healthier (long and thick culms, long umbel and dark green shoots), whereas those not under the influence of wastewater were stunted and yellowish-green in colour, is in agreement with those of Hocking (1985), who reported that plants of *Cyperus involucratus* grown on low levels of nitrogen in a glass house had stunted growth with leaves yellowish in colour. Furthermore, healthy papyrus plants seem to out-compete other plant species for light, explaining why creepers and other plant species were less abundant in areas with high aerial biomass of papyrus in the Nakivubo swamp. Creepers were also reported to be less abundant in dense wetland vegetation of *Phragmites* (Denny, 1984).

The higher nutrients' concentration in juvenile plants is a manifestation that N and P are taken up rapidly during early stages of growth and later translocated to the storage organs when the plants mature (Gaudet, 1977; Denny, 1985; Reddy and Debusk 1987; Ndyabarema, 1991). Nutrient availability has been reported to increase plant growth rates (Koerselman and Verhoeven, 1992). However, increase in nutrients results in increased biomass only up to the carrying capacity of the system after which there is luxury nutrient uptake which does not increase the biomass (Reddy and Debusk, 1987). In other wetland ecosystems, above-ground N and P content was reported to vary in the same way as the above-ground biomass (Hocking, 1985; Koerselman and Verhoeven, 1992), suggesting that availability of these nutrients may control biomass production.

The potential of nutrient uptake by a plant is limited by it's net productivity and the nutrient content in the plant tissue. Similarly storage is dependent on plant tissue nutrient concentrations and also on the potential of biomass accumulation (i.e the maximum standing crop). Plants characterised by high biomass per unit area have a potential to store a maximum of nutrients.

Table 5.4 shows that papyrus has high potential of nutrient removal from aquatic systems compared to other emergent macrophytes like *Phragmites* and *Typha* because of it's high standing crop. As argued by Howard- Williams and Gaudet (1985), in addition to having higher biomass production compared to other emergent aquatic macrophytes, papyrus is even more productive than crop plants like sugar cane and maize. This is because papyrus can achieve high biomass production in low nutrient conditions compared to crop plants which need artificial fertilizers. However, floating aquatic macrophytes (viz., *Eichhornia* and *Lemna*) have higher uptake rates than papyrus.

Table 5.4. Nutrient (N&P) content, standing biomass and uptake rates of selected macrophytes and crops[a].

Macrophyte / crop (growth conditions)	N content g kg^{-1}	P content g kg^{-1}	Standing crop ton (DW)/ha	N uptake kg ha^{-1} yr^{-1}	P uptake kg ha^{-1} yr^{-1}	Reference
Papyrus						
Kawaga swamp Uganda (natural swamp water)	20	1	21.4	430	22	Gaudet (1976)
Nakivubo swamp (sewage)	13	2.1	37.2- 58.4	475	77	This study
Nakivubo swamp (sewage; regrowth experiment)	13	2.1	4.5	759	123	-do-
Potted experiment (sewage)	9.5	1.1	7.1	171	20	- do-
Lake Naivasha Kenya (natural swamp water)	11.2	1.6	31.2 - 37.9	491	70	Muthuri *et al.* (1989)
Miscanthidium						
Nakivubo swamp (sewage)	6.4	1.5	18.5	118	28	This study
Nakivubo swamp (sewage; regrowth experiment)	6.4	1.5	11.1	89	21	- do-
Potted experiment (sewage)	3.1	0.4	5.6	44	5	- do-
Typha (diverse)	5 - 24	0.5 - 2.4	4.3 - 22.5	40 - 193	4 -19	Reddy and DeBusk (1987)
Phragmites (diverse)	18 - 21	2 - 3	6 - 35	180 - 210	20 - 30	- do-
Eichhornia (diverse)	10 - 40	1.4 -12	20 - 240	599 - 2394	84 - 718	- do-
Lemna (sewage lagoon)	25 - 50	4.0 -15	1.3	529 - 1898	85 -569	Reddy and DeBusk (1987); Alaerts *et al.* (1996)
Sugar cane (fertilized farm land)	-	-	42			Howard - Williams and Gaudet (1985)
Maize (fertilized farm land)	-	-	40			- do-

[a] Values of Papyrus and *Miscanthidium* are for only above ground biomass except for the potted experiments.

Papyrus plants demonstrated higher nutrient uptake compared to *Miscanthidium*, both in the field and pilot experiments. In Nakivubo swamp, maximum aerial biomass observed for *Miscanthidium* was lower (1850 g/m^2) than that of papyrus (5800 g/m^2). Papyrus has been reported to have the highest biomass production compared to other macrophytes (Gaudet, 1977; Thompson *et. al.*, 1979).

Direct contact with plant roots seems necessary for nutrient uptake from the wastewater. This explains why papyrus at the interface of *Miscanthidium* (where the papyrus mat is thick and less permeable) had lower aerial biomass compared to one only a few metres away, but growing on a thin loose mat. The thick *Miscanthidium* mat impedes vertical movement of water into the mat (Chapter 4) hence slowing and reducing the contact between the plants and wastewater. This limits nutrient availability and uptake by the plants. On the other hand, the loose papyrus mat allows nutrient exchange between wastewater and plants. Contact between plants roots and wastewater was also reported to be crucial for more effective nutrient uptake by plants in constructed wetlands (Brett, 1989; Cooper, 1990; Rogers *et al.*, 1991).

In addition to nutrient availability, other factors thought to affect productivity of papyrus include water logging (Thompson, 1985; Howel *et. al.*, 1988), photosynthetic capacity of papyrus which is also limited by N and P in the umbel (Jones, 1991) and altitude (Thompson *et al.*, 1979; Denny, 1993). Higher aerial biomass has been reported in some Uganda low altitude swamps (e.g 5000 g/ m^2 for the Akika highland of Lake George) than those reported for the Busoro swamp in Rwanda (1384 g/m^2) which is at a higher altitude (Jones and Muthuri, 1985). Whereas the average biomass for papyrus in Nakivubo swamp was within the range reported in literature for East and Central African (1384 - 4955 g/ m^2, Chale, 1987; Muthuri *et al.*, 1989, Jones, 1991), locations under the influence of the wastewater had higher production values (4490 - 5800 g / m^2). This is a clear indication that nutrient availability is probably a critical factor affecting productivity in most papyrus swamps.

Since nutrients taken up by papyrus have been shown to be returned to the aquatic environment when the plants die, periodic harvesting of shoots is one of the management options for nutrients' removal from the system. For natural papyrus swamps to maintain a sustainable yield, a harvesting interval of 9 - 12 months has been recommended for Eastern Africa (Jones, 1991; Ndyabarema, 1991). Considering the area dominated by papyrus in the lower Nakivubo swamp (area =0.92 km^2) with an average above-ground papyrus biomass of 4766 g DW/ m^2, the total above-ground biomass is 4.38 x 10^9 g. Using the average papyrus nutrient content of this study (N, 1.30 % DW; P, 0.21 %DW), harvesting the above-ground biomass of Nakivubo swamp would remove 5.69 * 10^4 kg N/yr and 9.38 * 10^3 kg P/yr. Considering the total amount of nutrient (Nakivubo channel and Luzira outflow) entering the swamp (7.8 * 10^5 kg N/ yr and 6.33* 10^3 kg P/yr), harvesting of the above-ground biomass would only remove 7.3% of the N input and 14.8% of the P input of the annual total load entering the swamp.

Similarly considering the area covered by *Miscanthidium* (0.23 km^2; N 0.64 % DW, P 0.15%DW), harvesting of the above-ground biomass here, would remove 0.35% N and 1.01% P of the annual total load entering Nakivubo swamp.

Assuming a monoculture of papyrus and that all plants are inundated with wastewater (including the margins of the swamp) and have a maximum biomass production with corresponding nutrient content in plant parts in that area (as observed in this study: N 1.94% DW, P 0.33% DW), 17 % N and 35% P would be removed through annual harvesting of above-ground biomass. On the other hand annual harvesting of *Miscanthidium* monoculture, dominating the lower Nakivubo swamp removes 0.5% N and 1.4% P. So harvesting of the above-ground biomass can only remove a small portion of nutrients out of the swamp.

With an average papyrus productivity of 4766 g/ m^2 /yr and *Miscanthidium* 1847 g /m^2/yr and life per shoot is a year, the corresponding above-ground biomass nutrient release levels would be 31 gN/m^2/yr and 5 gP/m^2/yr for papyrus, and 9 gN/m^2/yr and 2 gP/m^2/yr for *Miscanthidium*. Despite the fact that the amount of nutrients leached from plants is negligible compared to the total input into the swamp, the leached nutrients plus the nutrients eluted from decomposing matter by rainwater contribute to plant growth especially in areas that are not under wastewater influence. This could possibly explain why papyrus plants at the western edge of the swamp (Fig. 4.1), exhibit poor growth during the dry season but proliferate after the rains. Leaching rates of papyrus plant organs were higher for rhizomes because during senescence, nutrients accumulated in plant tissues are translocated from shoots to the rhizomes (Wetzel, 1993b).

During the leaching experiment, the pH value of the solution in which the dried *Miscanthidium* shoots were immersed dropped from 6.4 to 5.4 whereas that of the papyrus fell from 6.6 to 5.7 suggesting that the ions leached out *Miscanthidium* organs do not immediately result in a significant drop in pH.

The pH measurements both in the field and in the pilot experiment indicate that *Miscanthidium* can grow in a wide range of pH values (4 - 7), and probably induces a low pH in its own aqueous environment. The low pH value in *Miscanthidium* dominated vegetation has been reported to limit the growth of other macrophytes (Thompson and Hamilton, 1983; Denny, 1993). The lowering of the pH by the *Miscanthidium* root mass and its tolerance to low pH values would therefore give it a competitive advantage.

Root development in the pilot experiment showed that *Miscanthidium's* main and adventitious roots have a cuspidate behaviour, interlacing into each other. Similar observations were made by Lind and Visser (1962). The interlacing nature of the roots results in trapping of dead plant material in the mat minimising their loss from the mat. Since the vertical movement of water into the *Miscanthidium* mat is restricted (Chapter 4), the decomposition products (humic and fulvic acids, compounds responsible for the low the pH in wetland ecosystems) have a higher residence time in the mat. Humic and fulvic acids are complex phenolic compounds (Buth and Voesenek, 1987; Thompson and Hamilton, 1983; Hemminga and Buth, 1991) and may stay in the system for more than 2000 years (Verhoeven, 1986). If these compounds leach out of the mat, the pH tends to fall to a lower value. A pH range of 3.5 to 5.5 has been reported for ombrotrophic wetlands and was said to be a result of exchange of metal ions for H^+ by the plants root mat, coupled with decomposition of organic matter (Kadlec and Knight, 1996).

Low dissolved oxygen concentrations accompanied by low pH values generated by the plants in Nakivubo swamp, imply that microbial decomposition of organic matter following the leaching process may be extremely slow especially in the mats of *Miscanthidium*. This would result in continuos piling of dead plant material on top of those deposited after the previous seasons growth. This results in faster organic matter accumulation than decomposition, explaining why the mat of *Miscanthidium* is five times thicker than that of papyrus.

Since papyrus and *Miscanthidium* are growing under similar environmental conditions in the Nakivubo swamp, the more modest mat build-up in the papyrus zones must be contributed to the structural composition of papyrus. It is probable that *Miscanthidium* has higher lignin content than papyrus hence is more resistant to microbial degradation. The possible low lignin content and faster decomposition rate of papyrus plants and it's detritus always sedimenting may explain why the papyrus mat is thin (Denny, 1985). This is further supported by the observations in Chapter four that the water below the *Miscanthidium* mat was always clear compared to the water under the papyrus, which had a lot of detritus. Buth and Voesenek (1987), found the rate of decomposition to be species specific and inversely proportional to the lignin content in the plants. The loose nature of a papyrus mat allows easy movement of swamp water in and out of the mat, making it easier for the oxygen penetration and washing out of decomposition products from of the mat. In *Miscanthidium* it appears that flushing is hindered by slow vertical transport of swamp water in and out of the mat.

In view of ecological adaptation, the thick mat of *Miscanthidium* helps in supporting the above-ground biomass over the swamp water, however, this situation is not favourable if plants are used for wastewater treatment as there is less contact between wastewater and plants and more flow through under the mat.

5.5 Conclusions

In this study, a high content of nutrients was found in plants growing in the main flow path of wastewater. The fact that papyrus and *Miscanthidium* plants in the Nakivubo swamp are growing under similar environmental conditions (temperature, altitude etc.), the observed differences in biomass productivity may be attributed to the wastewater distribution in the swamp. The higher nutrient uptake in papyrus and high growth rates result in higher biomass production by this macrophyte. The lower nutrient uptake in *Miscanthidium* may be due to the phenology of this plant whereby nutrient requirements will be lower. A slower decomposition process in *Miscanthidium* mat explains the high build up of the mat. Low pH in *Miscanthidium* mat is attributed to the rooting characteristics and accumulation of decomposition products which are not readily flushed out of the mat. Harvesting of above-ground biomass, if combined with other management options, may contribute to nutrient removal from the Nakivubo swamp. Harvesting of the above ground biomass in the Nakivubo swamp would only remove 7% of N in the input and 15% of P input of the annual load in the Nakivubo swamp.

5.6 References

Alaerts, G. J., Mahbubar, M. D. R. and Kelderman, P. (1996). Performance analysis of a full- scale duckweed-covered sewage lagoon. *Wat. Res.*, *30*: 843 - 852.

Breen, P. F. (1990). A mass balance method for assessing the potential of artificial wetlands in wastewater treatment. *Wat. Res*,. 24: 689-692.

Brett, D. (1989). Getting it clean with aquatic plants. *Ecos.*, *60*: 17 - 22.

Brix, H. and Schierup H. H. (1989). The use of aquatic macrophytes in water pollution control. *Ambio,* 18: 100-107.

Buth, G. J. C. and Voesenek, L. A. C. J. (1987). Decomposition of standing and fallen litter of halophytes in a Dutch salt marsh. *In*: Huiskes, A. H. L., Blom, C. W. P. M and Rozema, J. (eds.), *Vegetation between land and sea.* Dr. W. Junk Publishers, Dordrecht, pp 43 - 59Chale, F. M. M. (1987). Plant biomass and nutrient levels of a tropical macrophyte (*Cyperus papyrus* L.) receiving domestic wastewater. *Hydrobiol. Bull.*, *21*: 167 - 170.

Chambers, J. M. and McComb, J. A. (1994). Establishment of wetland plants in artificial systems. *Wat. Sci. Tech.*, 29: (4) 79 - 84.

Cooper, P. F. (1990). *European design and operations guidelines for Reed Bed Treatment Systems.* Report No. UI 17. prepared by the EC/EWPCA emergent hydrophyte for reed bed treatment systems.

Denny, P. (1984). Permanent swamp vegetation of the upper Nile. *Hydrobiologia*, *110*: 79 -

90.

Denny. P. (1985). The structure and functioning of African euhydrophyte communities. *In:* Denny, P. (ed.). *The ecology and management of African wetland vegetation*. Dr. Junk Publishers, Dordrecht, pp 125 - 151.

Denny, P. (1987). Mineral cycling by wetland plants - a review. Arch. Hydrobiol. Beih., *27*:1-25.

Denny. P. (1993). Wetlands of Africa. *In*: Whigham, D. F., Dykyjová, D. F. And Hejný, S. (eds.), *Wetlands of the world, I. Inventory, Ecology and Management*. pp 1- 46, Kluwer Academic Publishers, Dordrecht.

Gaudet, J. J. (1976). Nutrient relationships in the detritus of a tropical swamp. *Arch. Hydrobiol. 78*: 312 - 239.

Gaudet, J. J. (1977). Uptake, accumulation and loss of nutrients in tropical swamps. *J. Ecol.*, *50*: 415 - 422.

Gersberg, R. M., Elkins, B. V. Lyon, S. R. and Goldman, C. R. (1986). Role of aquatic plants in wastewater treatment by artificial wetlands. *Wat. Res.*, *20*: 363 - 368.

Hemminga, M. A and Buth, G. J. C. Buth (1991). Decomposition in salt marsh ecosystems of the S.W. Netherlands: the effects of biotic and abiotic factors. *Vegetatio., 92*: 73 - 83.

Hocking, P. J. (1985). Responses of *Cyperus involucratus Rottb.* to nitrogen and phosphorus, with reference to wastewater reclamation. *Wat. Res.*, *19* (11) : 1379 - 1386.

Howard-Williams, C. (1979). The aquatic environment. II. Chemical and physical characteristics of Lake Chilwa swamp. *In*: Kalk, M., McLachlan, A. J. And Howard-Williams (eds.), *Lake Chilwa*. Dr. W. Junk, The Hague, 81 - 90.

Howard-Williams, C. (1985). Cycling and retention of nitrogen and phosphorus in wetlands: a theoretical and applied perspective. *Freshwater Biology 15*: 391 - 431.

Howel, P., Lock, M., Cobb, S. (1988). The Jonglei canal: impact and opportunity. Cambridge University Press, Cambridge, United Kingdom.

Jones, M. B. (1991). Papyrus study. Contract No. EARO/9363/45. Uganda Wetlands Conservation and Management Programme.

Jones, M. B. and Muthuri F. M. (I985). The canopy structure and microclimate of papyrus (*Cyperus papyrus* L.) swamps. *J. Ecol. 73*: 481 - 491.

Kadlec, R. H. and Knight, R. L. (1996). *Treatment wetlands*. Lewis, Publishers, Boca, Raton.

Koerselman, W. And Verhoeven, J. T. A. (1992). Nutrient dynamics in mires of various trophic status: nutrient inputs and outputs and internal nutrient cycle. *In*: Verhoeven J. T. A. (ed.). *Fens, bogs, in The Netherlands: Vegetation, History, Nutrient Dynamics and Conservation*. Kluwer Academic Publishers, pp 397 - 432

Lind, E. M., and Viser, S. A., (1962). A study of a swamp at the north end of Lake Victoria. *J. Ecol.,. 50*: 599-613.

Muthuri. F. M., Jones, M. B. and Imbamba, S. K. (1989). Primary productivity of papyrus (*Cyperus papyrus*) in a tropical swamp; Lake Naivasha , Kenya. *Biomass, 18*: 1- 14.

Ndyabarema, R. (1991). *An investigation into sustainable use of papyrus (Cyperus papyrus) in Nakyetema swamp, Mpigi district, Central- Uganda.* M.Sc. Thesis, Makerere University, Kampala - Uganda.

Novozamsky, I., Houba, V. J. G., van Eck, R. and van vark, W. (1983). A novel digestion technique for multi-element plant analysis. *Commun. in Soil Sci. Plant analy., 14*: 39-248.

Reddy, K. R. and Debusk, W. F. (1987). Nutrient storage capabilities of aquatic wetland plants. *In:* Reddy, K. R. and Smith W. H. eds.) *Aquatic plants for wastewater treatment and resource recovery.* Magnolia Publishing Inc. Orlando - Florida, pp 337 - 357.

Richardson, C. J. and Davis, J. A. (1987). Natural and artificial wetland ecosystems: ecological opportunities and limitations. *In:* Reddy, K. R. and Smith W. H. (eds.) *Aquatic plants for wastewater treatment and resource recovery.* Magnolia Publishing Inc. Orlando - Florida, pp 819 - 854.

Rogers, K., Breen, P. F. and Chick, A. J. (1991). Nitrogen removal in experimental wetland treatment systems: evidence for role of aquatic plants. Res. J. WPCF. *63*: 934-941.

Suzuki, T. W. G., Nissanka, A. and Yashushi. K. (1989). Amplification of total dry matter, nitrogen and phosphorus removal from stands of *Phragmites australis* by harvesting and re-harvesting regenerated shoots. *In:* Hammer, D. A.(ed.). *Constructed wetlands for wastewater treatment.* Lewis Publishers, Inc. Michigan, pp 530 - 535.

Sasser, C. E., Gosselink, J. G., and Shaffer, G. P., (1991). Distribution of nitrogen and phosphorus in a Louisiana freshwater floating marsh, Aquat. Bot., *41*: 317-331.

Thompson, K. (1985). Emergent plants of permanent and seasonally-flooded wetlands. *In*: Denny, P. (ed.), *The Ecology and Management of African Wetland Vegetation.* Dr. W. Junk Publishers, Dordrecht, pp 43 - 107.

Thompson, K. and Hamilton, A. C. (1983). Peatlands and swamps of the African continent. *In*: Gore, A. J. P. (ed.), *Mires: Swamp, Bog, Fen and Moor.* Elsevier, Ecosystems of the world series, 4B, (Regional studies), Amsterdam. pp 331 - 373.

Thompson, K., Shewry, P. R. and Woolhouse, H. W. (1979). Papyrus swamp development in the Upemba basin Zaire: studies of population structure in *Cyperus papyrus* stands. *Botanical Journal of Linnean society. 78*: 299 - 316.

van Oorschot, M. M. P. (1990). The role of vegetation in nutrient removal from water. *In:* van Oorschot, M. M. P. and Verhoeven, J. T A. (eds.). *The utrecht Plant Ecology News Report,* Utrecht pp 64-85.

Verhoeven J. T. A. (1986). Nutrient dynamics in minerotrophic peat mires. *Aquat. Bot.*, *25*: 117-137.

Wetzel, R. G. (1993a). Constructed wetlands: scientific foundations are critical. *In*: Moshiri, G. A. (ed.) *Constructed wetlands for water quality improvement*, Lewis Publishers, Boca Raton, pp. 3 - 7.

Wetzel, G. R. (1993b). Humic compounds from wetlands: complexion, inactivation and reaction of surface bound and extracellular enzymes. *Verh. Internat. Verein. Limnol.*, *25*: 122 - 128.

Chapter 6

Transport and fate of faecal coliforms in the Nakivubo swamp

Kansiime, F.

6.1 Introduction

The disposal of domestic wastewater is a public health as well as an environmental problem. Disposal of untreated or partially treated wastewater results in faecal contamination of surface and ground water with pathogenic organisms. These organisms cause disease and severe public health problems and are considered to be responsible for most of the infant morbidity and mortality in developing countries (WHO, 1993). Whereas water borne diseases have been virtually eliminated in the industrialized world, outbreaks of cholera and other gastro-enteric diseases still occur with alarming frequency in developing countries (UNEP, 1995). Conventional wastewater treatment, if it includes disinfection, renders domestic wastewater more safe for disposal but has high investment, maintenance and operational costs. Most developing countries only have secondary treatment plants whose removal efficiency of pathogens is only 0 - 2 log units. Wetlands offer a low cost attractive treatment approach to these countries especially for medium and small sized communities. The role of wetlands in removal of pathogens was suggested by Gersberg *et al.* (1987) when they found that an artificial wetland planted with bulrush showed higher removal efficiency (99.1%) compared to 95.7% for the unvegetated bed at a hydraulic loading rate of 5 cm d^{-1}.

Factors that are considered to affect the survival of pathogens in wetlands include sedimentation, aggregation, inactivation by UV light, alkaline pH, exposure to biocides excreted by plants, adsorption to organic matter, grazing by protozoa and attack by lytic bacteria and viruses (Lijklema *et al.*, 1987; Gersberg *et al.*, 1989; Scheuerman *et al.*, 1989; Brettar and Höfle, 1992; Wood, 1990). Removal of indicator bacteria by wetland systems of up to 99% has been reported (Bavor *et al.*, 1987), though poor performance as low as 39 - 50% has also been recorded (Hiley, 1994). In most cases wetland systems operate at shorter hydraulic retention times than oxidation ponds. However, the latter systems have more reliable removal efficiency. Besides the modest number of studies that have examined the reduction of indicator bacteria in wetlands, there is little information on the long term survival of these organisms in wetlands.

The Nakivubo swamp has been receiving wastewater from the Bugolobi sewage works as well as storm water run-off from the Nakivubo channel and raw sewage from the Luzira prisons for over 30 years now. About 11% of the population of the city of Kampala (approx. 800,000 inhabitants) is connected to the sewerage system. The wastewater finally enters Lake Victoria at the Inner Murchison Bay. The water supply for Kampala City is abstracted from the same bay, just 4 km South East at Gaba (for detailed description see Chapter 2). Since pathogens could eventually be transported to Gaba water works, the pathogen and / or disinfection aspect of pollution is very important. There is little information on the bacteriological water quality in the Inner Murchison Bay. Previous studies only concentrated on the swamp inflow and on the Inner Murchison Bay (Kizito, 1986, Gauff/Parkman, 1983; Gauff, 1988). No study has been carried out to assess the fate of faecal coliforms as the water flows through the swamp into the bay.

Detection of all possible pathogens (viruses, bacteria, fungi, protozoa and helminths) can be costly and is a very time consuming process. Furthermore, it is impractical to attempt the routine isolation of pathogens because they are present in small numbers compared to other microorganisms. Moreover, there are many types of pathogens and each requires a unique microbial isolation technique. Methods have therefore been developed which detect organisms which are indicative of the presence of faecal pollution such as intestinal bacteria. The organisms that are currently used as indicators of faecal pollution are the coliform bacteria, particularly *Escherichia coli* and faecal coliforms. This group of bacteria, has long been used as the first choice among the indicator organisms (Bartram and Ballance, 1996). Despite their universal application, the results of *E. coli* removal may not be simply extrapolated to most other bacterial pathogens, helminths or viral pathogens. This is because the removal processes, and the behaviour in wetlands can be rather different for different pathogen types.

As the wastewater flows through the Nakivubo swamp, faecal coliforms may be transported to different swamp compartments (Fig.6.1).

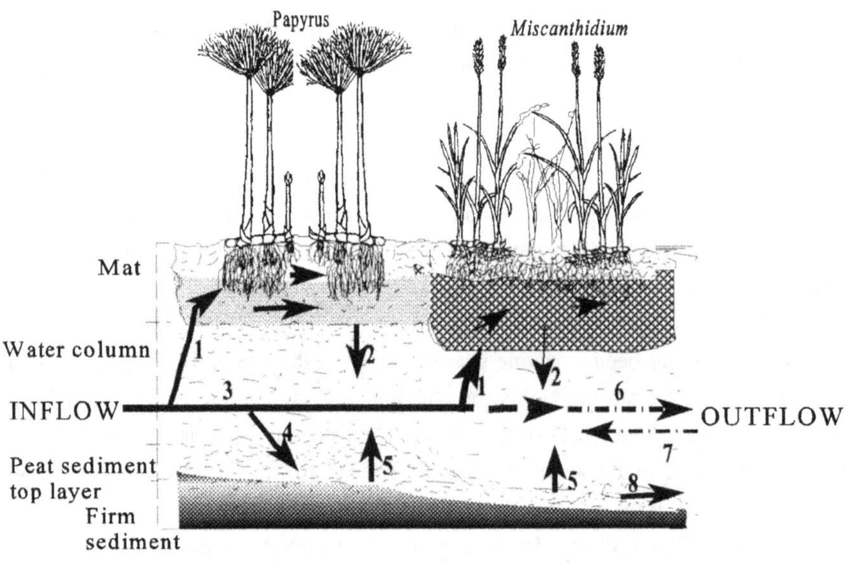

Fig. 6.1. Conceptual model of faecal coliforms transport in the Nakivubo swamp. Key: 1 = water penetration into the mat; 2 = detritus falling out of the mat with attached coliforms; 3 = flow through the water column; 4 = attachment to suspended particles in the water column; 5 = resuspension of sedimented particles; 6 = outflow of the swamp; 7 = back flow from the lake and 8 = export of peat/sedimented matter.

The coliforms may stay in the compartments for a longer or shorter time depending on the hydraulic residence time. In these compartments, several mechanisms and processes may play a part in the retention or die-off of coliforms before the wastewater reaches the Inner Murchison Bay. Part of the wastewater flows into the mat within which the faecal coliforms can get adsorbed to the plant surfaces, debris and peat. When the plant debris and peat fall out of the mat, they carry the attached microorganisms downward with them. The attached organisms would reenter the water column where they are either transported further lake-ward depending on the flow velocity and the size of particle or sediment. In the water column, the wastewater may flow into the Inner Murchison Bay and in locations where there is less contact between the wastewater and the plants, faecal coliforms retention may be minimal. The faecal coliforms can also get attached to suspended solids in the water column, which settle with attached bacteria to the swamp bottom or are transported further in the swamp. Settled particles may be resuspended into the water column. During lake seiches, coliforms may be transported back into the swamp. Settled particles may also be resuspended and exported to the Inner Murchison Bay. Overall, the retention of faecal coliforms in the Nakivubo swamp depends on the hydraulic detention time and residence time distribution of the wastewater in the swamp. High flow rates will reduce hydraulic retention time and hence reduce the likelihood of removal by both die-off time and by attachment and sedimentation. Travel time distribution in the swamp is a function of vegetation, floating nature of the mat, water depth and local velocities in the swamp.

Factors like natural die-off, predation, temperature and UV light may affect the removal of faecal coliforms in the Nakivubo swamp. Attachment to the mat, settling and sedimentation remove the coliforms from the water column and these processes may not necessarily be important especially in a short term. However, if the retention in the swamp is long enough, the coliforms may be removed due to natural die-off, predation and other mechanisms in which pH may be an important factor. Dilution by lake seiches does not necessarily result in die-off. In the Inner Murchison Bay die-off is expected to be higher due to high levels of dissolved oxygen (which might favour the growth of aerobic microbial predators) and UV. The presence of the water hyacinth in the bay can reduce the light effect and also facilitate attachment of coliforms and their sedimentation with plant debris.

Since some studies have been carried out on the bacteriological water quality in the Inner Murchison Bay and none in the Nakivubo swamp, and bearing in mind the need for adequate information on the bacteriological water quality of the Inner Murchison Bay, this study was carried out. The questions addressed were (i) does the wetland in its present condition exert a purification effect on the pathogens in the waste water discharged into it, and (ii) can we change something in the design or its management to enhance this effect. Therefore, there was need to understand the processes of pathogens removal in the swamp.

The specific aim of this study was to assess the fate of faecal coliforms as the wastewater flows through the Nakivubo swamp and to quantify the processes that may be responsible for the die-off and removal of these organisms in the swamp. In addition, the distribution of faecal coliforms in the Murchison Bay, where the water supply for Kampala is abstracted was also investigated. The objectives were to quantify:

(i) the faecal coliform loading into the swamp and evaluate the seasonal and spatial variation,

(ii) the longitudinal and transverse distribution of faecal coliforms in the swamp,

(iii) the role of the dominant aquatic macrophytes in the retention /die-off of faecal coliforms in the swamp,

(iv) the vertical distribution of faecal coliforms in different compartments of the dominant macrophytes in the swamp,

(v) the major mechanisms responsible for the removal of faecal coliforms in the swamp.

To address items (iii) and (v) bio-assays were conducted under controlled conditions.

6.2 Materials and Methods

6.2.1 Study area

This study was carried out in the lower Nakivubo swamp, hereafter referred to as the Nakivubo swamp (Fig. 4.1, Chapter 4, for a detailed description see Chapter 2). The major flows into the swamp are the Nakivubo channel and Luzira drain. Four transects (1 to 4, see Chapter 2) and the Inner Murchison Bay, were defined and arranged to be the main sampling sites. Sampling sites in the swamp are designated by Transect code and distance along the Transect as measured from the western edge (Bukasa side) of the swamp. For instance, T1-50 refers to 50 m along Transect 1.

6.2.2 Detection and estimation of faecal coliform numbers

The Multiple Tube Fermentation Technique was used to detect faecal (thermotolerant) coliforms. Ten factorial dilution series of each sample were made in a saline diluent containing bacteriological peptone (1 g/l) and sodium chloride (8.5 g/l). Lauryl sulphate (lactose) broth was used as a test media for the presumptive phase. Inoculated samples were incubated in a water bath for 24 h at 44 $\pm$ 1°C. Positive samples were subcultured into the EC media (specific for *Escherichia coli)* and incubated for another 24 h at the same temperature for a confirmatory test. The numbers of faecal coliform bacteria per unit volume of inoculum were computed using the Most Probable Number (MPN) technique according to Standard Methods (APHA, 1992). The numbers of coliforms are expressed as the Most Probable Number (MPN) per 100 ml of sample.

6.2.3 Field measurements

Spatial distribution of faecal coliforms in the Nakivubo swamp

The numbers of faecal coliforms discharged into the Nakivubo swamp were monitored at the swamp inlet (Nakivubo channel flows into swamp, under the railway bridge) and at the Luzira drain. Samples at both sites were taken just below the water surface.

Sampling holes through the vegetation of the mat were made as described in Chapter 4. Before collecting a sample from a given site, 100 - 150 ml of water was pumped through the sampling tube to rinse the residual from the previous sampling. This minimised cross contamination, as it was not possible to use individual sterile tubes for each of the sampling sites. Samples were transferred into 100 ml sterile glass bottles. After collection, samples were kept in a cold ice box before being transported to the laboratory. Samples were analysed within 4 to 6 h of sampling, which is considered an acceptable delay to minimise microbial growth.

The distribution of faecal coliforms across the swamp was monitored along T1, T2, T3 and T4. As it was not possible to carry out *in situ* measurements and analyse all samples from all the transects on a single day, monitoring along individual transects was carried out on different days. However, for a given sample replicate, measurements were done to cover all the four transects in a given week. Electrical conductivity, pH and dissolved oxygen were measured *in situ*. Water samples were taken from the water column below the mat.

In order to evaluate the diurnal variation of faecal coliforms in the Nakivubo swamp, samples were taken simultaneously at the swamp inlet and at different locations in the swamp (T2-100, T2-400, T4-200 and T4-400), for a period of 8 h between 9:00 and 17:00 h for 2 days. Samples were taken from the water column. At site T4-400, the influence of lake seiches on faecal coliform numbers was monitored by installing a data logger which recorded water level, electrical conductivity and temperature.

To assess the longitudinal transport and retention of faecal coliforms in the swamp, samples were taken from the Nakivubo channel flows (major inflow into the swamp) at different locations in the swamp at T2 (T2-100 and T2-400) and at T4 (T2-200 and T4-400). Sampling was done between 9.00 and 12.00 for 3 days during a dry season. For each sampling day, samples were taken at the same time.

To get an insight about the possible impact of discharged water into the Nakivubo swamp, on the Inner Murchison Bay - Lake Victoria, and the Gaba water works in particular (where the intake for the water supply for the city of Kampala is located), samples were taken from different locations in the Bay. Access to different locations in the swamp and the Bay is described in Chapter 4.

To determine the vertical distribution of faecal coliforms and to assess the influence of dominant macrophytes on their distribution, samples were taken in the different compartments (mat, water column and peat-sediment top layer) of the floating papyrus and *Miscanthidium* mats (Fig. 4.1). The peat-sediment layer is the top part of sediment comprising peat and decomposing plant debris (sometimes referred to by some authors as organic sludge, e.g. Gaudet, 1977), most of which originates from the mat (Chapter 5). The water column under papyrus contains considerable amounts of suspended matter, largely consisting of peat and plant debris falling out of the mat. The *Miscanthidium* mat, on the other hand, is firm and compact with little or no plant debris or peat falling off the mat, resulting in a clear water column underneath (Chapter, 5).

The sampling locations were T2-100, T4-395 (papyrus sites) T2-400 and T4-410 (*Miscanthidium* sites). The sites at T4 were close to each other (15 m apart) and at the swamp lake interface. These two sites were selected to make a comparison between papyrus and *Miscanthidium* under more or less similar conditions. Location T2-100 is a typical papyrus site far from *Miscanthidium* and with little influence from the lake. For each sampling date, samples from the locations on a given Transect were taken on the same day with an hour difference between the two locations.

6.2.4 Laboratory simulation experiments

Some of the factors influencing the faecal coliform retention/removal in the Nakivubo swamp viz. attachment to plant surfaces, natural die-off and sedimentation were assessed by simulations in the laboratory.

To determine the numbers of faecal coliforms that are attached to plant surfaces in contact with wastewater, root samples of papyrus and *Miscanthidium* were collected from the swamp at T4-395 and T4-410 respectively. In the field, whole plant units were cut out of the mat. Loose mat peat was removed by gentle shaking, and the rhizome and its roots were put in a polythene bag. In the laboratory, roots were separated and 10 g (wet) of root material was transferred to 95 ml of a saline diluent made up of 8.5 g NaCl/l and neutralised bacteriological peptone 1 g /l and this resulted in a 10^{-1} dilution. Subsequent dilutions were prepared from this dilution. Sonication was applied to detach faecal coliforms from plant roots and the peat-sediment layer. To determine the optimal sonication time, samples of mat water and roots were sonicated for 60 s in a Branson sonication bath at 40 kHz. The highest number of detached coliforms was reached after 10 - 20s of sonication (Fig. 6.2). As a result, 20 s sonication period was adopted as the standard optimum time to detach faecal coliforms from the roots and peat-sediment samples.

The natural die-off of faecal coliforms was investigated using batch reactors (conical glass vessels). Each of these was completely filled with 1 l of swamp water and incubated in the laboratory at ambient temperature (23 °C). Reactors were covered with aluminium foil to minimize light introduction and to simulate conditions under the mat. Faecal coliform numbers, electrical conductivity, pH, temperature and dissolved oxygen were monitored daily for a period of 17 days. Samples were taken with sterile pipettes at 2/3 of the total depth of the reactors.

The decay rate of bacteria of faecal coliform bacteria has been reported to follow a first order relationship described by: $\ln\frac{C_t}{C_o} = -kt$, where k is the decay constant (h^{-1}), C_o is the initial faecal coliform concentration at t = 0 and C_t is the faecal coliform concentration at time t (Chick's Law, 1908). The decay rate of faecal coliforms was determined from the plot of the logarithm of the fractional decrease in coliform concentration over time, $\ln(C_t/C_o)$ versus time. The slope of the line (fitted by linear regression) is equal to the decay constant.

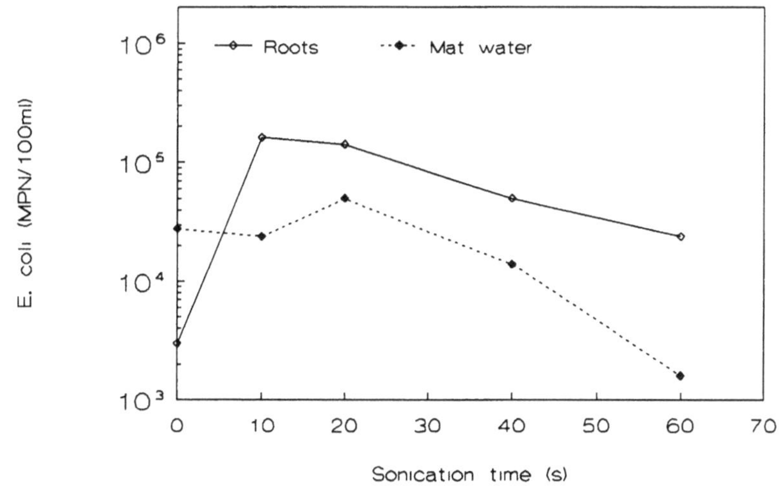

Fig. 6.2. Effect of sonication on the detachment of faecal coliforms.

Sedimentation

Two experiments were carried out to assess the role of attachment of faecal coliforms to suspended particles and eventual sedimentation from the water column. In the first experiment two batch reactors of 1 litre were compared; one filled with plain swamp water (R1) and a second in which swamp water was mixed with 250 ml of fresh peat (R2) and left to settle undisturbed for 17 days. The experiment was aimed at assessing the role of suspended material (peat from the mat in the Nakivubo swamp) on settling out of coliforms from the water column. Numbers of coliforms were determined daily. Electrical conductivity, pH, temperature and dissolved oxygen were also monitored. The second experiment was similar to the first one except that gentle stirring (manually with a stirring rod) was applied to detach coliforms from the sides of the reactors. Faecal coliforms were also monitored in the settled material. Samples were taken after 45 min, 1 h and 2 h as in the first experiment it was observed that considerable sedimentation (≥ 80%) occurred in the first two hours. All samples were sonicated for 20 sec. before inoculation.

6.3 Data analysis.

Analysis of variance (ANOVA) was used to test for differences between means of samples collected from different locations and for factors that may affect the distribution and retention of faecal coliforms in the swamp. All data were first tested for homogeneity of variances and normality using both Bartlett's and Levene's tests. Faecal coliform data were logarithmically transformed to conform to normal distribution. Following analysis of variance, differences among means were tested using Tukey's and Dunnet's multiple comparisons. The family error rate (experiment-wise error) was set at 5% and the individual error rate (compromise-wise error) was automatically adjusted by the statistical programme depending on the number of comparisons. Correlation was used to determine a relationship between a given variable and faecal coliform distribution. Linear regression was used to estimate the decay rates of coliforms with time. Statistical analysis was carried out using MINITAB software (MINITAB Release 10 for Windows).

6.4 Results

6.4.1 Field measurements

Faecal coliform loads into the Nakivubo swamp

The Nakivubo channel was found to be one of the major contributors of high faecal coliform loads into the Nakivubo swamp. On a yearly basis, the average number of coliforms for the study period 1993-1996 was not significantly different from each other ($p = 0.86$), coliforms numbers were $6.8*10^5 \pm 0.9*10^5$ MPN / 100 ml ($n = 27$). Samples taken during rainy seasons had more coliforms ($7.4*10^5 \pm 1.2*10^5$ MPN /100 ($n= 10$) than those taken during the dry seasons ($5.8*10^5 \pm 1.2*10^5$ MPN / 100 ml, $n = 17$) though the difference was not significant ($p = 0.473$). Using the average flow rate in the Nakivubo channel-swamp inflow, the load of coliforms discharged into Nakivubo swamp was estimated at $7.0*10^{14}$ no. per day.

The Luzira drain, which conveys raw sewage from the Luzira prisons, had the highest concentration of faecal coliforms, being $9.8*10^7 \pm 4.9*10^7$ MPN/ 100 ml ($n = 10$). The total load of coliforms into the swamp was estimated at $6.9*10^{15}$ no. per day.

Spatial distribution of faecal coliforms in the water column in the Nakivubo swamp

The numbers of faecal coliforms along T1, T2 and T4 are shown in Fig. 6.3.

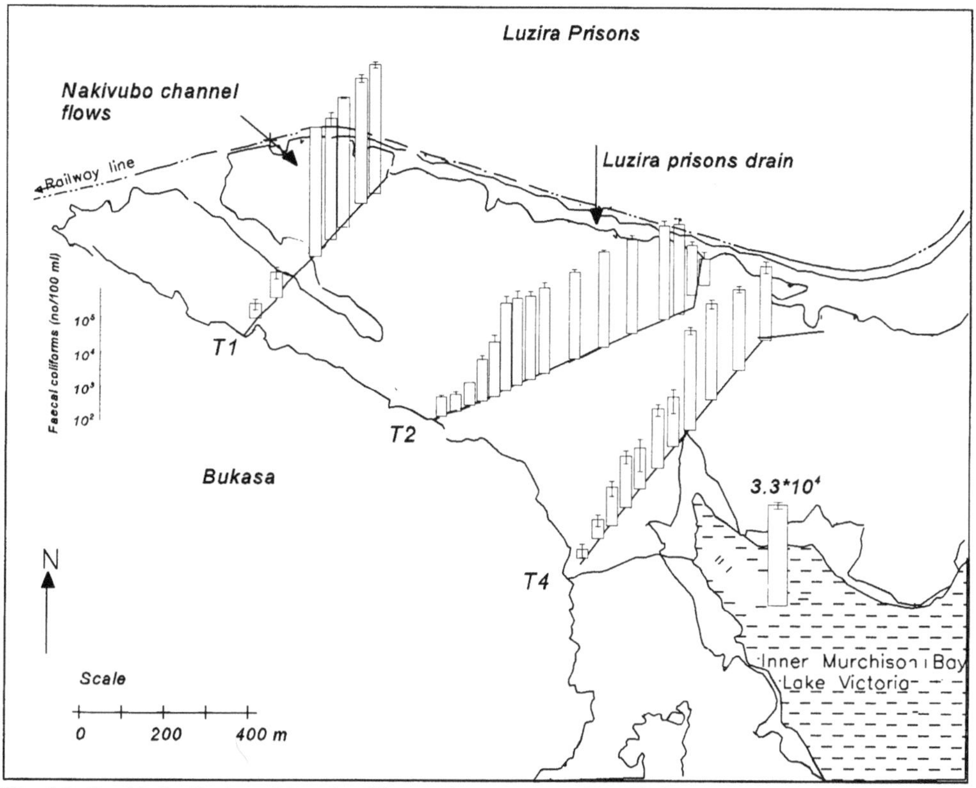

Fig. 6.3. Spatial distribution of faecal coliforms along transects T1, T2, T4 and 500 m from T4-400 into the Inner Murchison Bay.

At the T1 stretch of 259 - 459 m, which is under the direct influence of inflowing wastewater, the highest average number of faecal coliforms ($4.3*10^5 \pm 0.4*10^5$ MPN/100 ml) was measured. The water was grayish-black in colour with a foul smell of wastewater, similar to that at the swamp inlet. The first part of the Transect (distance 0 - 180 m) which is not directly under the influence of the inflowing wastewater had the lowest number of faecal coliforms ($4*10^2 \pm 1*10^2$ MPN/100 ml).

Faecal coliform numbers along T2 increased from the western edge of the swamp towards Luzira. Coliforms and surplus conductivity (with respect to the lake background value) along this Transect had a similar distribution (r^2=0.694), suggesting that the ionic species that contribute to conductivity are transported together with the coliforms. The low correlation, however, may imply that the factors affecting the two variables in the swamp are partly different. The same distribution pattern was observed for transects 3 (not shown in the figure) and 4.

Zones dominated by papyrus along T2 (distances 0 - 125 m and 650 - 750) had the lowest number of coliforms with $2.0*10^3 \pm 0.4*10^3$ MPN/100 ml for each of the two distances. The zone dominated by a mixture of papyrus and *Phragmites* vegetation, whose mat is thick and compact (1.2 m), had high numbers of faecal coliforms in the water column ($4.3*10^4 \pm 0.79*10^4$ MPN/100 ml on average). The highest numbers of coliforms were observed in the area dominated by *Miscanthidium* ($2.5*10^5 \pm 0.08*10^5$ MPN/100 ml on average) where the mat was 1.5 m thick at location T2-400.

Similarly faecal coliform numbers along T3 increased from the Bukasa, to the Luzira side of the swamp (data not shown). It was noted that the papyrus zone at distance 200 m to 300 m along T3 (a distance comparable to the one along T2), had a lower coliforms count (1 log unit less) than a similar one along T2. Since conductivity and smell of water indicated that fresh wastewater was flowing through this zone, it appears that more coliforms were retained within the papyrus vegetation zones with a thin and loose mat (and thus removed from the water) compared to those with a thick mat (papyrus and *Phragmites*) as observed at T2. In the latter zone there is less interaction between the water column and the mat. The numbers of coliforms at T3 under the *Miscanthidium* zone were comparable to those recorded under the same vegetation type at T2, suggesting insignificant removal of coliforms from the water in the *Miscanthidium* Zone.

For T4, faecal coliform numbers also increased from the Bukasa towards to the Luzira side of the swamp. The water under the pure papyrus zones had the lowest numbers of faecal coliforms, followed by the zones dominated by both papyrus and *Miscanthidium*. The areas dominated by *Miscanthidium* had the highest number of coliforms in the water column.

6.4.1.1 Diurnal variation of faecal coliforms in the swamp

Nakivubo channel-swamp inflow.

The diurnal variation of faecal coliforms and conductivity at the Nakivubo swamp inlet is presented in Fig. 6.4. High number of coliforms ($9*10^6$ MPN/100 ml) were recorded at 9:00 h in the morning, falling (by about one log unit) to a minimum level at 11.00 h. The faecal coliforms numbers rose again to a value of $1.6*10^6$ MPN/100 ml and remained more or less constant till 17 h. Assuming that ions contributing to conductivity are transported together with coliforms, it may be assumed that the maximum number of coliforms reaches the railway embankment at 4:00 o`clock in the morning as observed for conductivity in Chapter 4. This reflects the pattern of wastewater discharge into the Nakivubo channel. The modest fluctuations of the concentrations suggest that the sewage and drainage collection system is complicated and allows for ample internal mixing.

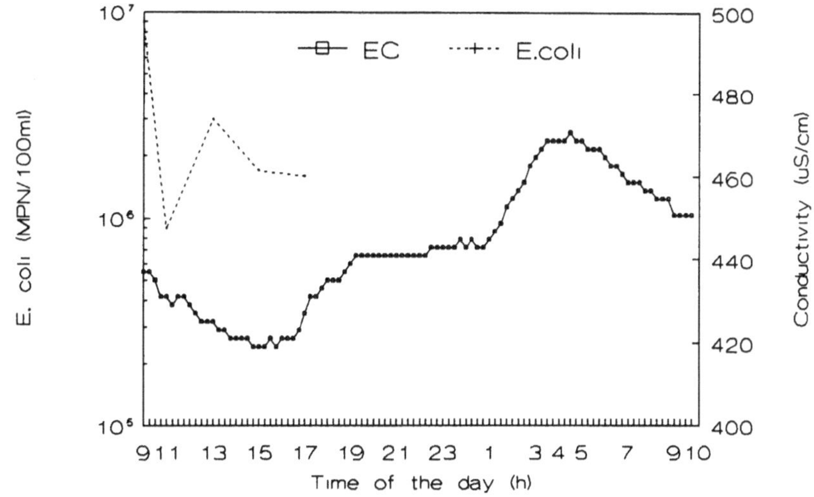

Fig. 6.4. Diurnal variation of faecal coliforms and conductivity in the Nakivubo channel at the swamp inlet.

Location T4-400 m

The diurnal variation of faecal coliforms in the swamp was strongly influenced by the lake seiches and no clear diurnal pattern was observed. Water levels, conductivity and coliforms counts for T4-410 are presented in Fig. 6.5. Rising water levels were accompanied by fall in faecal coliform numbers (r^2 = 0.67). Faecal coliform numbers and conductivity followed an identical pattern (r^2 = 0.89).

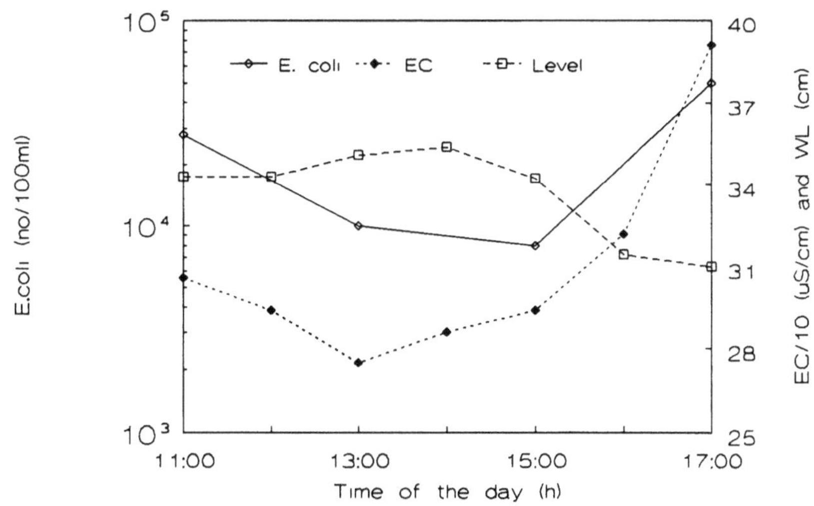

Fig. 6.5. Diurnal variation of faecal coliforms, conductivity (EC) and water level (WL) at T4-400. Changes in water level are due to lake seiches.

Inner Murchison Bay

The faecal coliform numbers significantly decreased from the swamp lake interface to the open waters of the Inner Murchison Bay (Fig. 6.6). High numbers of coliforms at distances of 500 m and 1500 m offshore indicate that water laden with coliforms reaches Inner Murchison Bay. However, low numbers of coliforms were detected at a distance of 4000 m which is close to the Gaba water intake.

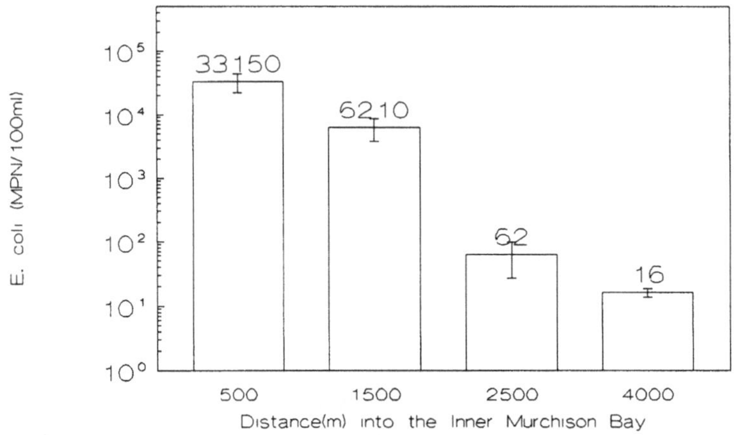

Fig. 6.6. Variation of faecal coliforms in the Inner Murchison Bay. Distance was measured from T4-400. Bars indicate standard error (n=5 except for distances 2500 and 400m where n=3). Values on top of the bar graphs are numbers of coliforms /100ml.

6.4.1.2 Vertical profiles of faecal coliforms in different swamp compartments

Location T2-100 (papyrus site)

Vertical distribution of faecal coliforms and the conductivity in swamp compartments at location T2-100 m is depicted in Fig. 6.7. The faecal coliform numbers in the water taken from the papyrus mat ($9.3*10^3 \pm 2.8*10^3$ MPN/100 ml) were not significantly different from that in the water column ($1.4*10^4 \pm 0.24*10^4$ MPN/100 ml), but significantly lower than that recorded in the peat-sediment top layer ($1.1*10^5 \pm 0.26*10^5$ MPN/100 ml). Conductivity did not differ significantly in all the compartments (range 272 - 299 μS/cm). The high numbers of faecal coliforms in the peat-sediment layer may be due to the settling of solids and plant debris from the mat to which coliforms get attached as the wastewater flows through the swamp.

Location T2-400 (Miscanthidium site)

Faecal coliform numbers were significantly lower in the *Miscanthidium* mat, ($1.5*10^2 \pm 0.72*10^2$ MPN/100 ml) compared to the water column and peat-sediment layer (Fig. 6.8). The numbers of coliforms in the water column ($4*10^5 \pm 0.15*10^5$ MPN/ 100 ml) were lower than in the peat sediment top layer ($1.0*10^6 \pm 0.85*10^6$ MPN/100 ml) though the

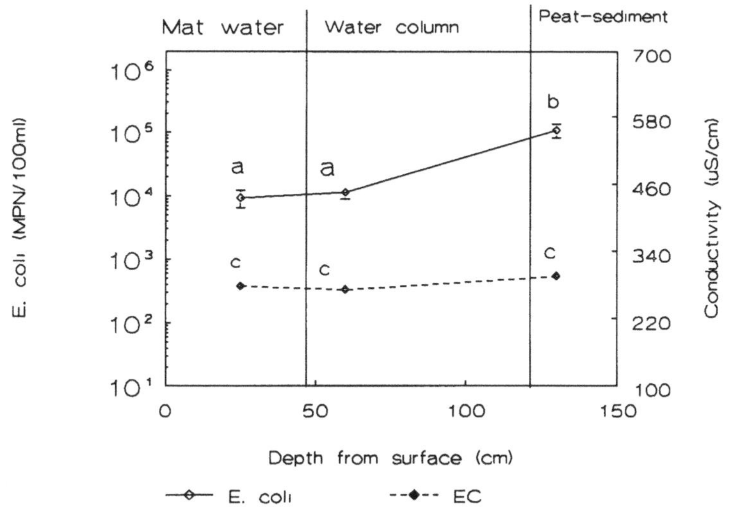

Fig. 6.7. Vertical profile of faecal coliforms and conductivity in the papyrus zone (T2 100). Bars indicate standard error of the mean (n =6). Values on the graph with the same letters are not significantly different.

difference was not significant. Electrical conductivity had the same pattern as coliforms, having the lowest value in the mat. The *Miscanthidium* mat was compact and thick (1.2 m) compared to that of papyrus which was shallow (0.4 m) and loose.

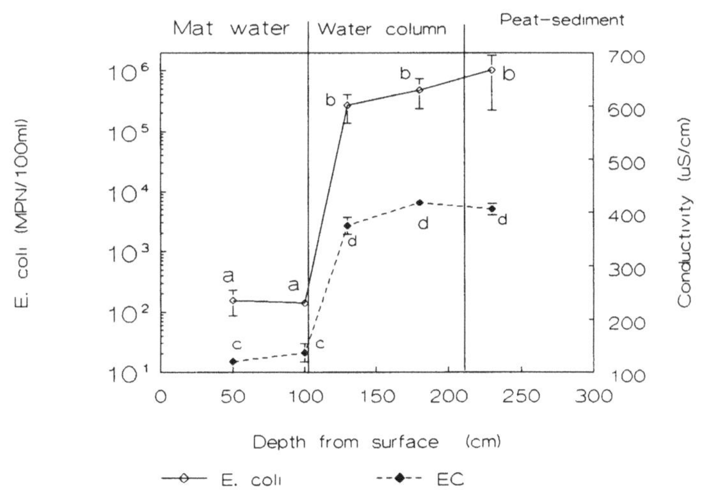

Fig. 6.8. Vertical profile of faecal coliforms and conductivity in the *Miscanthidium* zone (T2-400). Bars indicate standard error of the mean (n =6). Values on the graph with the same letters are not significantly different.

Location T4-395 (papyrus)

The vertical distribution of faecal coliforms at T4-395 is depicted in Fig.6.9. Faecal coliform numbers didn't differ significantly between papyrus compartments ($p=0.315$). Electrical conductivity had the same distribution as coliforms and was also not significantly different among all the compartments ($p = 0.40$). The distribution of coliforms and conductivity in the papyrus compartments may be attributed to better mixing of wastewater in different compartments.

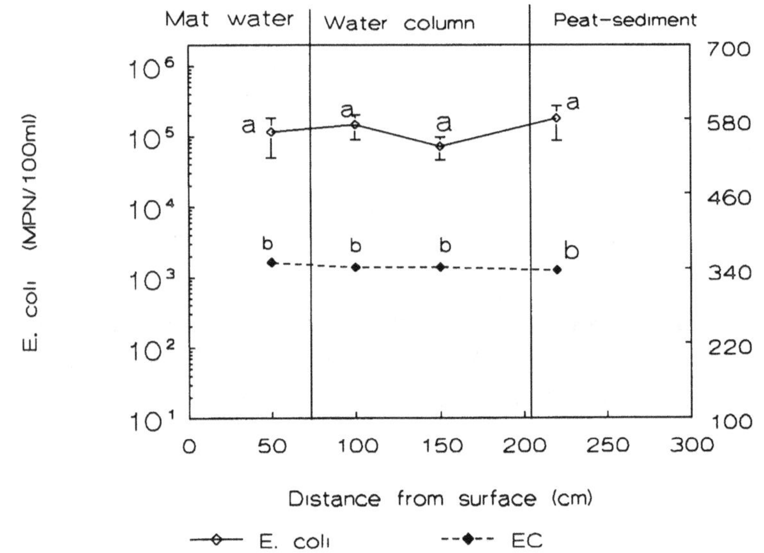

Fig. 6.9. Vertical profile of faecal coliforms and conductivity in the papyrus zone (T4 395). Bars indicate standard error of the mean ($n = 8$). Values on the graph with the same letters are not significantly different.

Location T4-410 (Miscanthidium)

Similar to the observations made along Transect 2, faecal coliform numbers were lowest in the mat of *Miscanthidium* at T4-410 m (Fig. 6.10). The coliform numbers in the water column and peat sediment top layer were not significantly different from each other. Electrical conductivity depicted the same distribution as coliforms, and was again significantly lower in the mat. At this site the mat of *Miscanthidium* was 0.8 m thick.

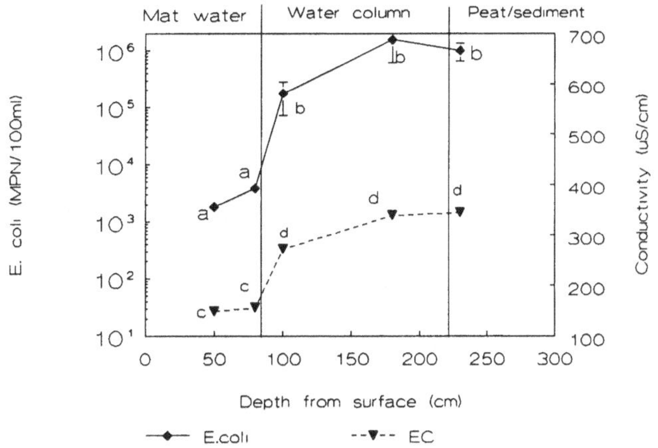

Fig. 6.10. Vertical profile of faecal coliforms and conductivity in the *Miscanthidium zone* (T4-410). Bars indicate standard error of the mean (n = 8). Values on the graph with the same letters are not significantly different.

6.4.1.3 Comparison of papyrus and *Miscanthidium* zones

Transect 2

Significantly lower number of coliforms were recorded in the *Miscanthidium* mat liquid T2-400, compared to the papyrus at T2-100 compartments (Fig. 6.11). The number of coliforms in the mat and water column of the papyrus zone (T2-100), were significantly lower than those recorded in the water column and peat-sediment layer of the *Miscanthidium* (T2-400).

Transect 4

Faecal coliforms were also significantly lower in the *Miscanthidium* mat liquid at T4-410 than other compartments (Fig. 6.12). The numbers of coliforms in the water column and peat-sediment layer of papyrus (T4-395) were significantly lower than those recorded in the corresponding *Miscanthidium* compartments. Since the sites at T4-395 and T4-410 are close to each other, low numbers of coliforms recorded at the former site may be attributed to the higher removal in the papyrus zone. This is due to a rather loose and open structure of the mat which allows easy flow and mixing of mat water with the main water body. The coliforms get attached the detritus and settle to the bottom or remain in the mat material. The mat liquid of *Miscanthidium* at this site had higher numbers of coliforms than at T2-400. No significantly different numbers of coliforms existed between the water column and peat sediment layer of the papyrus at T4-395.

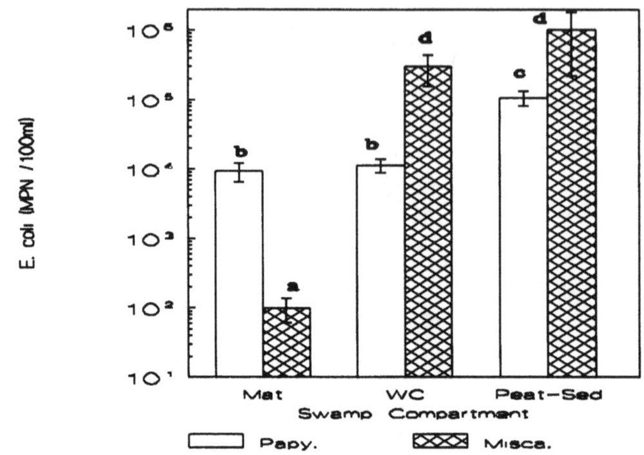

Fig. 6.11. Comparison of faecal coliform numbers in papyrus (Papy.) and *Miscanthidium* (Misca.) compartments at T2-100 and T2-400 respectively. WC = water column and Peat-sed = Peat-sediment top layer. Graphs with the same letters are not significantly different.

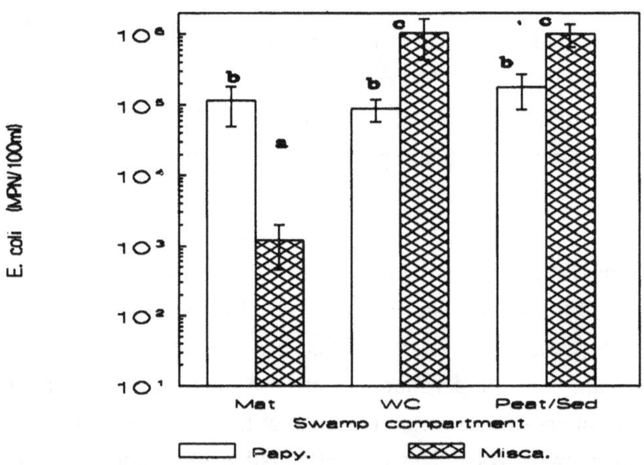

Fig. 6.12. Comparison of faecal coliform numbers in papyrus (Papy.) and *Miscanthidium* (Misca.) compartments at T3-395 and T2-410 respectively. WC = water column and Peat-sed = Peat-sediment top layer. Graphs with the same letters are not significantly different.

6.4.1.4 Profiles of physical and chemical variables

The vertical distribution of physical and chemical characteristics of water in different compartments is presented in Table 1. The samples were taken during a dry season when access to all the sites is easy. Temperature was lowest in the mat and water column at location T2-100 whereas the higher values were recorded in different compartments at other locations. The *Miscanthidium* mat liquid had the lowest pH. The pH values for other compartments were all in the same range. The mat compartments were hypoxic whereas the water column and peat sediment layer were anoxic. In general the gradients (differences) of the measured variable were higher in *Miscanthidium* than in papyrus as a consequence of differences in bulk density of the mat.

Table 6.1. Mean values ± SD of physical and chemical parameters in different swamp compartments at indicated sites. WC = water column, Peat-sed = peat - sediment top layer. Numbers in parentheses are sample replicates.

Site / Compartment	Temperature (^{O} C)	pH	DO (mg/l)
T2-100 (Papyrus) Mat	22.2 ±0.2 (10)	6.5 ± 0.0 (10)	0.1 ±0.0 (10)
Water column	22.4 ±0.1 (10)	6.4 ±0.0 (10)	0.0 ±0.0 (10)
Peat-sediment top layer	23.5 ±0.2 (5)	6.3 ±0.0 (5)	0.0 ± 0.0 (5)
T2-400 (*Miscanthidium*) Mat	23.2 ±0.2 (12)	5.0 ± 0.2 (12)	0.1 ±0.0 (12)
Water column	23.6 ±0.1 (12)	6.3 ±0.1 (12)	0.0 ±0.0 (12)
Peat-sediment top layer	24.0 ±0.1 (6)	6.4 ±0.0 (6)	0.0 ±0.0 (6)
T4-395 (Papyrus) Mat	23.3 ±0.2 (7)	6.9 ±0.1 (7)	0.1 ±0.0 (7)
Water column	23.5 ±0.1 (17)	6.8 ±0.1 (17)	0.0 ±0.0 (17)
Peat-sediment top layer	23.5 ±0.1 (10)	6.7 ±0.1 (10)	0.0 ±0.0 (10)
T4-410 (*Miscanthidium*) Mat	23.1 ± 0.1 (20)	5.7 ±0.1 (20)	0.1 ±0.0 (20)
Water column	23.7 ±0.1 (20)	6.5 ±0.0 (20)	0.0 ±0.0 (20)
Peat-sediment top layer	23.8 ±0.2 (9)	6.6 ±0.1 (9)	0.0 ±0.0 (9)

6.4.1.5 Longitudinal distribution of coliforms in the Nakivubo swamp

For the samples taken simultaneously at different locations in the Nakivubo swamp, there was in general a decrease in the ratio of faecal coliform numbers to electrical conductivity (corrected for surplus to the lake background value) in the wastewater flowing through the swamp towards the Inner Murchison Bay (Fig. 6.13). Conductivity was considered to be conservative and accounted for only dilution, so the decrease in the ratio will be due to other factors like die-off, settling etc. Considering the ratio of conductivity to faecal coliforms, 2 log units of coliforms were removed from the water under the papyrus zone (Inlet, T2-200 and T4-200), compared to 1 log unit retained in the *Miscanthidium* zone (Inlet, T2-400 and T4- 400). The removal of coliforms was higher (92%) between T2 and T4 for the papyrus zone compared to (78%) between T1 and T2. For the wastewater flowing through the *Miscanthidium* zone, the highest removal of coliforms (69%) took place between T1 and T2 whereas it was lower between T2 and T4 (54%). The zones before T2 are dominated by papyrus.

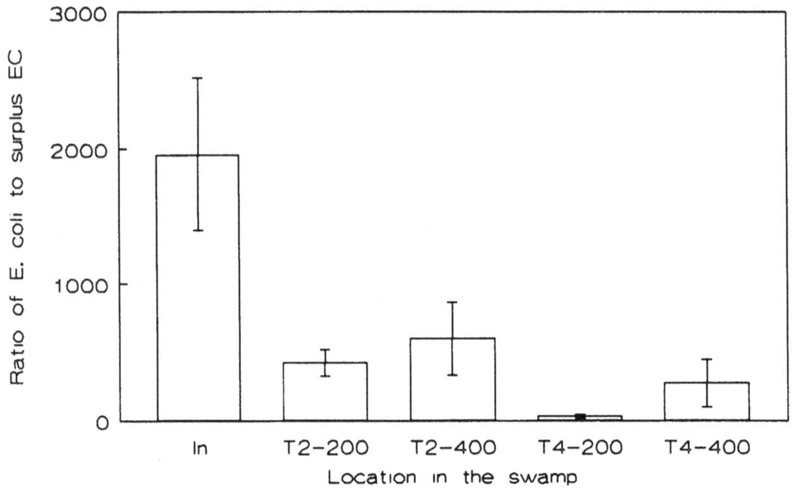

Fig. 6.13. Variation of the ratio of *E.coli* to surplus conductivity in the swamp. Locations 200 are in papyrus whereas 400 are in *Miscanthidium* vegetation.

6.4.2 Laboratory simulations

High numbers of faecal coliforms were detached from the root surfaces of papyrus ($1.8*10^4 \pm 0.4*10^4$ MPN/ g DW of root) compared to *Miscanthidium* roots ($2.2*10^2 \pm 0.95*10^2$ MPN/ g DW of root). The high density of secondary roots observed on papyrus roots (Chapter 5) could have provided a high surface area, explaining the high number of attached coliforms per unit mass. The roots of *Miscanthidium* were generally less developed and secondary roots were only observed close to the surface, whereas inside the mat, root development was poor with no noticeable secondary roots. Furthermore, the surface of *Miscanthidium* roots was smooth compared to those of papyrus.

The removal of faecal coliforms from the swamp water incubated in the laboratory is shown in Fig. 6.14. Over the experimental period (17 days), temperature varied between 23 - 25 ^oC whereas dissolved oxygen and pH remained stable. Faecal coliform numbers remained more or less the same for the first 5 days and thereafter started dropping. By day 17, the faecal coliforms had disappeared by 4 log units. The removal rate of faecal coliforms was calculated to be 0.021 h^{-1}. However, the dissolved oxygen was 2.5 mg/l whereas in the swamp oxygen is generally absent.

Addition of fresh peat material from papyrus mats to plain swamp water resulted in a more rapid removal of faecal coliforms from the water column (Fig. 6.15). Reduction in the number of coliforms was higher in the reactor to which peat was added compared to that containing plain swamp water. The removal rate constants were 0.018 h^{-1} and 0.029 h^{-1} for the respective reactors containing plain swamp water and swamp water with peat.

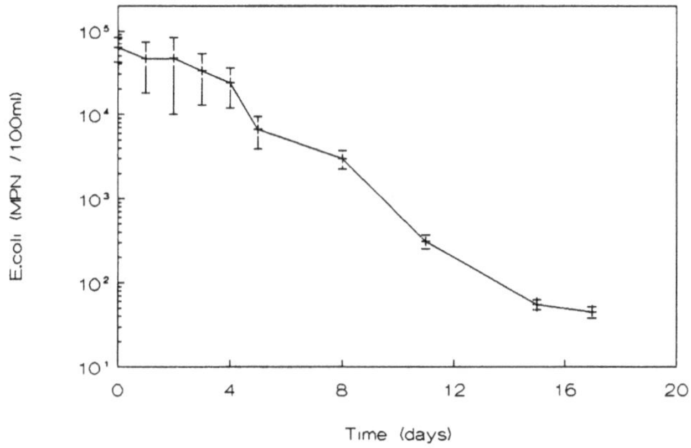

Fig. 6.14. Reduction of faecal coliform numbers in swamp water incubated under laboratory conditions. (Temp = 23^{0}C, dissolved oxygen = 2.5 ± 0.8 mg/l).

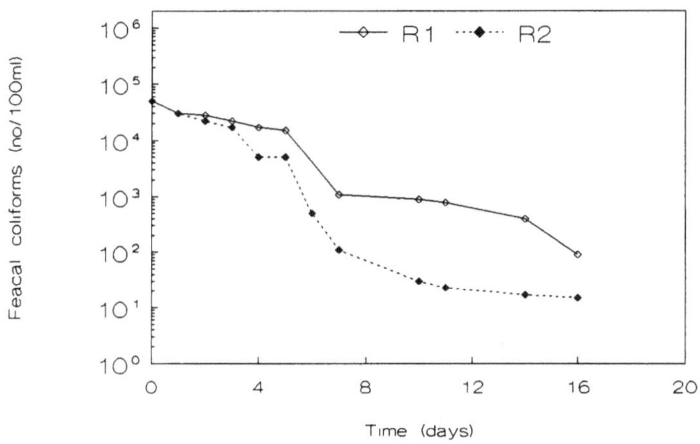

Fig. 6.15. Reduction of faecal coliforms by die-off and settling under laboratory conditions. Reactor 1 (R1) had only swamp water whereas reactor 2 (R2) contained peat and swamp water. (Temp = 23^{0}C, dissolved oxygen = 2.5 ± 0.8 mg/l).

6.5 Discussion

The coliforms in the Nakivubo channel originate from the secondary treated effluent from the Bugolobi sewage works and rainfall run-off from the city of Kampala. The channel also conveys raw sewage overflowing from old and blocked sewers in the city. Another source of coliforms to the channel are the city suburbs, most of which discharge their wastes directly into the channel (SOE, 1994). In the suburbs, pit latrines are used as well, but most overflow during the rainy season, resulting in further faecal coliform loads into the channel. The fact that the Nakivubo channel does not only carry secondary treated effluent from the sewage works, but also rainfall run-off and overflowing sewers and septic tanks, could explain why high numbers of faecal coliforms were recorded during the rainy season. Combined sewers have been reported from other parts of the world to be a major source of faecal contamination of surface waters especially during rainfall events as sewage flows are higher and resuspends sewer sediments (Pettibone and Irvine 1996, Ferguson *et al.*,1996,Wyer *et al.*,1996).

The Luzira drain is the major contributor of the faecal coliforms load entering the Nakivubo swamp. The Luzira prison does not have a proper sewage treatment system, explaining the high numbers of faecal coliforms in the drain. However, the Luzira drain flows are intermittent and this conclusion is based only on samples taken 9:00 - 12: 00 h; night flows could not be measured.

In the swamp, high faecal coliform numbers were recorded in areas where the bulk of the wastewater flows (as reported in Chapters 3 and 4). The high correlation between faecal coliforms and conductivity along Transect 2 (similar pattern for other transects) indicates that coliforms and ions contributing to the conductivity are transported together in the Nakivubo swamp. Therefore, conductivity and coliforms are complementary indicators of channelling and flow through the swamp. The low level of faecal coliforms observed at the swamp edges may be attributed to longer hydraulic detention times, allowing for significant die-off, settling, attachment to plant debris etc. Dilution from ground water was found to be minimal (Chapter 3).

Faecal coliforms concentrations in the water remained higher in the zones dominated by *Miscanthidium* compared to those dominated by papyrus. This may be explained, as observed in Chapters 4 and 5, by the restricted contact between plants and wastewater caused by the tight and compact mat of *Miscanthidium*. The thick and compact mat of *Miscanthidium* with little detritus falling out of it produces a clear water column beneath. Under these conditions, faecal coliforms remain in suspension rather than being attached to organic detritus and subsequent sedimentation to the swamp bottom. These coliforms would eventually be transported to the Inner Murchison Bay through this portion of the swamp. This results in a fast flow below the mat and hence a short hydraulic retention time. The combination of less resistance to hydraulic flow and little detrital material in the water column results in shorter retention and less die-off and settling of coliforms in this portion of the swamp. This explains why the coliforms number in the region dominated by *Miscanthidium* along T2 to T4 were more or less the same. The flow through was also confirmed during storm events, during which storm water (brown in colour) is usually seen dispersing into the Inner Murchison Bay from this part of the swamp.

The overall lower concentrations of faecal coliforms in the papyrus compartments (mat and water column) may be due to the loose nature of the papyrus mat which has high surface area detritus to which coliforms become attached as the wastewater flows through the mat. The loose nature of the mat allows moderate mixing of wastewater in the different compartments. The detritus and peat and attached coliforms eventually settle from the mat into the water column and finally to the swamp bottom. Faecal coliforms, as reported for other organisms in the microscopic range, can adhere to particles dispersed in the water and wastewater (Thoman and Mueller, 1987). The detritus and peat deposited from the mat to the swamp bottom, have been reported to be a sink of nutrients in floating papyrus swamps (Howard - Williams, 1985) and this is likely to be true also for faecal coliforms in the Nakivubo swamp and especially in the floating zones of papyrus as observed at T2-100.The possible water movement and hence transport of coliforms in the water columns of both papyrus and *Miscanthidium* is depicted in Fig. 6.16. Wastewater penetration into the *Miscanthidium* mat is restricted, whereas there is easy mixing between the water column and the mat of papyrus plus detritus settling. This further explains the high number of faecal coliforms detached from the roots of papyrus compared to those of *Miscanthidium*.

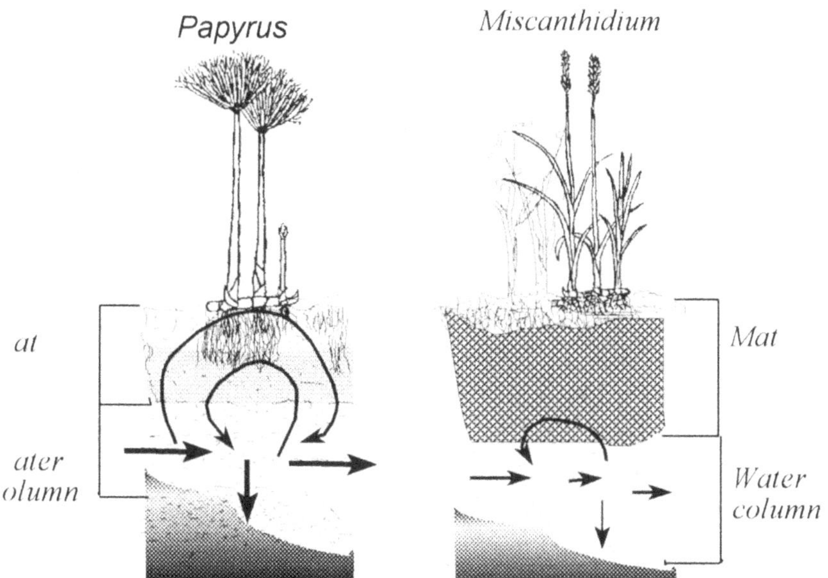

Fig. 6.16. Schematic presentation of wastewater interaction between the water column and the mats of papyrus and *Miscanthidium*.

The fact that the removal of coliforms in the Nakivubo swamp differed with the dominant macrophyte, suggests that the local environmental conditions are associated with the occurrence of specific macrophytes. Such environmental conditions in Nakivubo swamp include: (i) easier vertical penetration of wastewater into the papyrus mat, due to a loose but firm network of rhizomes, (ii) a slower rate of organic matter decomposition from *Miscanthidium* and eventual build up of the mat, and (iii) compactness of the *Miscanthidium* mat which restricts vertical penetration of the wastewater.

Aquatic macrophytes have the ability to translocate oxygen from the aerial shoots to the rhizosphere (Brix, 1994). The released oxygen might encourage the proliferation of aerobic grazers which reduce the number of faecal coliforms. This is in agreement with the observation of Cavari and Bergstein (1996), who reported a decrease of 2 log units of coliforms after 40 h for faecal coliforms incubated under aerobic conditions, whereas there was no change in cell numbers of *E. coli* when the same experiment was carried out under anaerobic conditions. This in addition to short hydraulic retention time, could explain why high numbers of faecal coliforms are observed at the lake-swamp interface of the Nakivubo swamp.

The morphological difference between papyrus and *Miscanthidium* was further manifested in differences in the vertical distribution of faecal coliforms in different swamp compartments dominated by these macrophytes. The significantly low numbers of coliforms observed in the *Miscanthidium* mat liquid at the two sites (T2-400 and T4-410) may be due to high flow through and hence less hydraulic retention time of the wastewater in addition to more restricted transport of wastewater into the compact mat of *Miscanthidium*. However, the low pH in the *Miscanthidium* mat could also result in a faster die-off of faecal coliforms in this compartment. Faecal coliforms have been reported to survive better at pH between 6 - 7 with a rapid decline above and below this range (Solic and Kortulovic, 1994; Liran *et al.*, 1994). Since some of these factors act synergetically, it was not possible in this study, to decidedly point out which of the investigated factors was responsible for the low numbers of coliforms in the *Miscanthidium* mat. For the papyrus dominated zones, the similarity in the number of coliforms in the mat and water column is attributed to easy transport of wastewater into the mat. The high numbers of coliforms in the peat-sediment layers would be due to the high flux of suspended of solids falling out of the mat (Denny, 1985), to which coliforms have been attached. The similarity in numbers of coliforms in the peat-sediment and other compartments in the papyrus site at T4 -395, is probably due to the influence of the diluting lake seiches. Also less peat/sediment accumulation was reported at this site (Chapter 8). As one progresses towards the swamp/lake interface, seiche movement becomes a major interference. The movement of water in and out of the swamp due to lake seiches, leads to resuspension of the sedimented particles and transports them into the Inner Murchison Bay. During storm events, the muddy and brown storm water is usually seen emerging out of the swamp into the Inner Murchison Bay.

Since sites T4-395 and T4-410 are close to each other, the lower coliform numbers in the papyrus zone may be explained by easy mixing between wastewater and the swamp compartments as compared to the *Miscanthidium*. This further supports the observation that easy penetration of waste water into the mat and the presence of peat and plant debris in the water column results in removal of coliforms from water in the papyrus zone. The lower number of coliforms in the *Miscanthidium* mat at T4-410 ($1*10^2$) than those recorded at T2-400 ($1*10^3$) can be attributed to the thick mat at the former site (1.3 m) compared to the latter site (0.8 m). Also the more pronounced lake influence (dilution) plus progressive die-off of the coliforms at the latter site, which is at the lake interface, may have played a role.

In addition to easy penetration of water into the mat, the high numbers of coliforms detached from papyrus roots may be attributed to the large surface area created by well-developed secondary roots of papyrus. In contrast, for *Miscanthidium*, in addition to main roots being poorly developed in the Nakivubo swamp, secondary roots were only observed on roots growing close to the surface. The root density was also lower in *Miscanthidium*. Since macrophytes in constructed wetlands have been reported to retain coliforms through adsorption and sedimentation (Watson *et al.*, 1989; Wood, 1990; Coombes and Collet, 1995), in general the higher the surface area the higher the number of coliforms that can be attached. Other factors, which may explain low attachment of faecal coliforms to root surfaces in addition to restricted vertical water transport into the mat of *Miscanthidium*, include low pH in this compartment which could have enhanced natural die-off of coliforms.

The longitudinal distribution of coliforms in Nakivubo swamp, indicates that more coliforms were removed from the water under the papyrus dominated vegetation compared to that of *Miscanthidium*. High resistance to flow in the papyrus zone results in high hydraulic retention time which allows for more time for attachment and die-off. Furthermore, easy penetration of wastewater into the papyrus mat and sedimentation of plant debris (to which coliforms may be attached), from the mat of papyrus explains the high removal coliforms from the water that flows through the papyrus zone.

The number of faecal coliforms recorded in the Inner Murchison Bay showed that wastewater with high numbers of coliforms still reaches this part of the lake. The significant reduction of coliforms by 2 log units between offshore and open water of the lake is primarily due to dilution in the lake and the exchange of water between the Inner and Outer Murchison Bay reported by Gauff, (1988). Combined effects of solar radiation, temperature, pH and the grazers acting synergetically as reported by Solic and Kortulovic (1994), to reduce the pathogens' survival in open waters, will further be responsible for the significant reduction of coliforms in the Inner Murchison Bay. The values recorded for the open waters of the Inner Murchison Bay, are in the range reported by WHO (1970), Gauff and Parkman (1983), and (Kizito, 1986). The importance of dilution in such a bay was reported by Pereira and

Machado (1987), who observed that the inner zones of the Rio Formosa Estuary in Portugal, which were not flushed during tidal changes, exhibited high bacterial counts originating from crude domestic wastewater disposal

The higher removal rate of coliforms (0.029 h^{-1}) in swamp water to which peat was added compared to plain swamp water (0.018 h^{-1}), demonstrates that the presence of settling material acted as seed material for sedimentation. This was confirmed by the presence of more coliforms in the settled material than in the water column in the reactor to which peat was added. This may explain why high numbers of coliforms were recorded in the peat sediment top layer compared to the water column of papyrus dominated areas especially at T2-100. Faecal coliforms have been reported to accumulate in the sediment of various aquatic systems where their survival is prolonged (Lijklema *et al.*, 1987; Scheuerman *et al.*, 1989). According to Brettar and Höfle (1992), survival may also be due to reduced grazing pressure in the sediment and even regrowth. Finally settled particles to which coliforms are attached may prolong the survival of coliforms.

The decay rates of faecal coliforms in the swamp water incubated in the laboratory (0.02 h^{-1}), was in the range reported by Gersberg *et al.* (1987) of 0.019 - 0.022 h^{-1}, for samples incubated in situ in dialysis bags placed in depressions below the gravel surface of the wetlands. However, our value was lower than that reported by Scheuerman *et al.* (1989), for coliforms monitored in experimental corridors (retention time of 4 days), of a Florida Cypress wetland (0.03 h^{-1}). The fact that numbers of coliforms were reduced by 4 log units after 17 days, under laboratory conditions, suggests a possibility of longer survival of coliforms in Nakivubo swamp where there is restricted contact between the wastewater and plants. This means that faecal coliforms can be transported into the Inner Murchison Bay especially under *Miscanthidium* where the residence time is thought to be much shorter than 17 days. For instance, during storm events the hydraulic retention time in the *Miscanthidium* route is 16 - 24 h. Furthermore, anoxic conditions in Nakivubo swamp and limited sunlight will prolong the survival of faecal coliforms and lower the actual decay rate.

6.6 Conclusions

The Nakivubo channel and Luzira drain discharge high loads of faecal coliforms into the Nakivubo swamp. In the Nakivubo swamp, the retention of faecal coliforms in areas dominated by papyrus was higher. Less resistance to flow and minimal interaction between wastewater and mat (where plants are rooted) may be responsible for the low removal of faecal coliforms in areas dominated by *Miscanthidium*. Peat and plant debris in addition to attachment to plant surfaces were responsible for the high retention of faecal coliforms in the papyrus dominated zones. Overall, attachment, sedimentation and natural die-off were found to be important mechanisms responsible for the retention of coliforms in the Nakivubo swamp. However, still high numbers enter Inner Murchison Bay- Lake Victoria. As high as 1.3 $*10^{14}$ faecal coliforms /d enter the Inner Murchison Bay compared to the input into the swamp of 1.4 $*10^{15}$ no./d. Dilution in the bay in addition to factors like sunlight, grazing, pH and temperature are responsible for a rapid reduction of faecal coliforms in the Bay with distance. Very low numbers of faecal coliforms were recorded at a site close to Gaba Water Works.

6.7 References

APHA, (1985). *Standard methods for examination of water and wastewater*. American Public Health Association, Washington. D C.

Bavor, H. J., Roser, D. J. and Mckersie, S. (1987). Nutrient removal using shallow lagoon -solid matrix machrophyte systems. *In:* Reddy, K. R. and Smith W. H. (eds.). *Aquatic plants for wastewater treatment and resource recovery* . Magnolia Publishing Inc. Orlando - Florida, pp 227 - 235.

Batram, J. and Ballance, R. (1996). *Water quality monitoring: a practical guide to the design and implementation of fresh water quality studies and monitoring programmes.* UNEP/WHO, E&FN SPON, London.

Brettar, I. and Höfle, M. (1992). Influence of ecosystematic factors on survival of *Escherichia coli* after large scale release into lake water mesocosms. *Appl. Environ. Microbiol. 58,* 2201 - 2210.

Brix, H. (1994). Functions of macrophytes in constructed wetlands. *Wat. Sci. Tech. 29*, 71-78.

Cavari, B. and Bergstein, T. (1996). Factors affecting survival of pathogens and indicators of pollution in fresh waters. *Israel Oceanographic & Limnological Res.* Tiberis, Israel. pp, 412-415.

Chick, H. (1908). Investigations of the law of disinfection. *J. Hyg., 8:* 92

Coombes, C. and Collett, P. J. (1995). Use of constructed wetlands to protect bathing water quality. *Wat. Sci. Tech.. 32*, 149 - 158.

Denny, P. (1985). The structure and functioning of African hydrophyte communities. The floating leaved and submergent vegetation. *In*: Denny, P (ed.). *The ecology and management of African Wetland vegetation*. Dr. W. Junk Publishers, Dordrecht. pp 125-152.

Ferguson, C. M., Coote, B. G.., Ashbott, N. J. and Stevenson, I. M. (1996). Relationship between indicators, pathogens and water quality in an estuarine system. *Wat. Res. 30*, (9) 2045 - 2054.

Gaudet, J. J. (1977). Uptake, accumulation and loss of nutrients in tropical swamps. *J. Ecol. 50*, 415 - 422.

Gauff/Parkman, (1983). Water quality surveys. Gaba, Entebbe and Jinja. Update of the WHO/ Resource studies report C/D. Lake Investigations studies report `B'. Water quality. Report prepared for the Ministry of Water and Mineral Development and National Water and Sewerage Corporation; Kampala - Uganda.

Gauff , H. P (1988). Seven town water supply and sewerage expansion project. Water quality studies. Report prepared for the Ministry of Water and mineral Development and National water and Sewerage corporation; Kampala - Uganda.

Gersberg, R. M., Brenner, R., Lyon, S. R. and Elkins, B. V. (1987). Survival of bacteria and viruses in municipal wastewater applied to wetlands. *In:* Reddy, K. R. and Smith W. H. (eds.). *Aquatic plants for wastewater treatment and resource recovery*. Magnolia Publishing Inc. Orlando - Florida pp 237 - 245.

Gersberg, R. M., Lyon, S. R., Brenner, R. and Elkins, B. V. (1989). Integrated wastewater treatment using artificial wetlands: a gravel case study. *In:* Hammer, D. A. (ed.). *Constructed wetlands for wastewater treatment.* Lewis Publishers, Inc. Michigan, pp 477-483.

Hiley, P. (1994). The reality of sewage treatment using wetlands. *In: Proceedings of the 4 th International conference on wetland systems for water pollution control.* Guangzhou, Peoples Republic of China. pp 68-83.

Howard-Williams, C. (1985). Cycling and retention of nitrogen and phosphorus in wetlands: A theoretical and applied perspective. *Freshwater Biology 15*, 391-431.

Kizito, S. (1986). *The evaluation of pollution levels in Nakivubo Channel, Kampala- Uganda.* M.Sc. Thesis, Makerere University, Uganda.

Lijklema, L., Habekotte, B., Hooijmans. C. M., Aalderink, R. H. and Havelaar, A. H. (1987). Survival of indicator organisms in a detention pond receiving combined sewer overflow. *Wat. Sci. Tech. 19,* 547 - 555.

Liran, A., Juanico, M. and Shelef, G. (1994). Coliform removal in stabilisation reservoir for wastewater irrigation in Israel. *Wat. Res. 28*, 1305 - 1314.

Pereira, M. A. S. and Machado, L. (1987) Environmental impact assessment of the discharge of the effluent from the Faro waste stabilisation ponds in the Ria Farmosa, Algawe, Portugal. *Wat. Sci. Tech. 19,* 337 - 339.

Pettibone, G. W. and Irvine, K. N. (1996). Levels and sources of indicator bacteria associated with the Buffalo river area of concern, Buffallo, New York. *Journal of Great Lakes Research 22*, (4) 896 - 905

Scheuerman, P., Bitton, G. and Farrah, S. (1989). Fate of microbial indicators and viruses in a forested wetland. *In*: Hammer, A. D. (ed.) *Constructed wetlands for wastewater treatment*. Lewis Publishers Chelsea, Michigan. pp 657-663.

SOE, (1994). *State of Environment Report of Uganda*. Ministry of Natural Resources. Kampala.

Solic,M.and Kortulovic, N. (1992).Separate and combined effects of solar radiation, temperature, salinity and pH on the survival of faecal coliforms in sea water. *Marine Pollution Bulletin. 24*, 411 - 416.

Thoman, R. V. and Mueller, J. A. (1987). Principles of surface water quality modelling and pollution control. Harper and Row Publishers, Inc. New York.

UNEP, (1995). *Water Quality of World River Basins*. UNEP Environment Library No 14. Nairobi.

Watson, J.T., Reed, S.C., Kadlec, R.H. and Knight, R.L.(1989). Performance expectations and loading rates for constructed wetlands. *In:* Hammer, D. A. (ed.). *Constructed wetlands for wastewater treatment.* Lewis Publishers, Inc. Michigan, pp 319 - 351.

WHO, (1970). Master plans for Water Supply and Sewerage for greater Kampala and Jinja Areas (1970 - 2000). Special studies Reports ``C'' and ``D'' Lake investigations. Special Studies report ``B'' Water quality.

WHO, (1993). *Guidelines for drinking-water quality,* Volume 1, recommendations. Second edition, World Health Organization, Geneva.

Wood, A. (1990). Constructed wetlands for wastewater treatment, Engineering and design considerations. *In*: Cooper P. F. and Findlater B. C. (eds.) *Constructed wetlands for water pollution control*. Pergamon Press, Oxford.

Wyer, M. D., Kay, D., Dawson, H. M., Jackson, G. F., Jones, F. Jannette. Y. and Whittle, J. (1996). Delivery of microbial indicator organism to coastal waters from contaminated sources. *Wat. Sci. Tech. 33*, 37 - 50.

Chapter 7

Nitrification in the Nakivubo swamp

Kansiime, F.

7.1 Introduction

Reduced nitrogen compounds are among the principal constituents of concern in wastewater because of their role in eutrophication, their oxygen demand on the receiving water (through nitrification) and their toxicity to aquatic organisms. High-rate removal of nitrogen and other nutrients from wastewater in tertiary treatment is beyond the reach of most developing countries. As a result, nitrogen removal from wastewater using wetlands is considered to be an economically and ecologically feasible alternative for many communities (Hsieh and Coultas, 1989; Hammer and Knight, 1994). Wetlands remove nitrogen compounds notably through nutrient uptake by macrophytes that can be harvested, and by microbiological denitrification of nitrate.

The major pathways of nitrogen transformation in wetlands are shown in Fig. 7.1. Nitrogen in wastewater can be present both in organic and inorganic forms. Organic nitrogen is mineralised to ammonium by ammonifying bacteria under aerobic and anaerobic conditions.

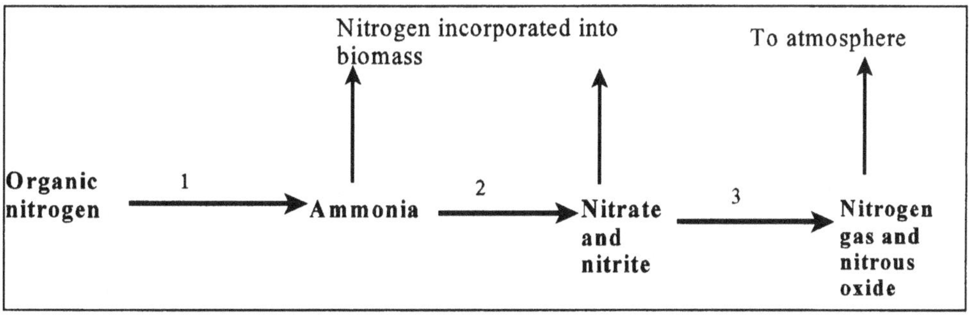

Fig. 7.1. The main pathways in which nitrogen is lost from fresh water wetlands.
1= ammonification (aerobic and anaerobic conditions), 2= nitrification (aerobic conditions), 3= denitrification (anaerobic conditions).

Some of the ammonium is incorporated into biomass and part of it may be oxidised and converted to nitrite through a process of nitrification which is mainly carried out by different genera of chemolithotrophic ammonia oxidising bacteria. These include *Nitrosomonas*, *Nitrosospira*, *Nitrosolobus*, *Nitrosovibrio and Nitrosocystis* (Watson *et al.*, 1981). Nitrite can further be oxidised into nitrate. The nitrate formed is either taken up by plants or diffuses into adjacent anaerobic zones where it is denitrified. Oxidation of nitrite to nitrate is carried out by nitrifying bacteria of the genera *Nitrobacter*, *Nitrococcus*, and *Nitrospina* (Randal and Ingraham, 1981; Kuenen and Robertson, 1988). Denitrification, the final step in nitrogen transformation, involves the conversion of nitrate into nitrogen gas. Nitrous oxide (N_2O) may also be formed. This process is mainly carried out by denitrifying bacteria and takes place under anaerobic conditions. Other heterotrophic bacteria, actinomycetes and fungi are also capable of oxidising organic nitrogen compounds to nitrite and nitrate. However, heterotrophic nitrification is considered to be of quantitatively minor importance compared to autotrophic nitrification (Laanbroek and Woldendorp, 1995).

Low concentrations of oxygen limit nitrification in wetlands (Hammer and Knight, 1994; Wittgren and Tobiason, 1995; Sikora *et al.*,1995). Despite the fact that nitrification has been reported to take place in anaerobic sediments (Watson *et al.*, 1981), it depends on the availability of oxygen and therefore has been reported to be predominantly a process that takes place in the upper zones of wetlands (Randal and Conway, 1989) and the rhizosphere in flooded soils (Tiedje, 1988; Paul and Clark, 1989).

Storage of nitrogen in wetlands is only a temporary solution, whether the nitrogen is incorporated into the biomass or adsorbed (fixed by soil) because sooner or later it will reach the system capacity (Howard-Williams, 1985; Brix, 1997). Since the pH of fresh water wetlands is usually acidic (Kadlec and Knight, 1996), ammonia volatilisation is not likely to be a significant pathway of nitrogen removal. As a result, nitrification-denitrification seems to be the only long term process for nitrogen removal from wetlands where biomass is not harvested (Reddy *et al.*, 1989; Patrick, 1990).

Nitrification can occur at low concentrations of dissolved oxygen (DO) but the reaction rate is relatively slow at concentrations less than 2 mg/l (Sikora *et al.*, 1995). Although the low concentrations of DO in the bulk of wetlands may inhibit nitrification, the reaction may occur at faster rates in the rhizosphere where DO concentrations may be higher due to localised oxygen release by rooted aquatic plants (Brix, 1993; Reddy and D'Angelo, 1997). Oxygen transport through air spaces (aerenchyma tissue) of stems and roots of some aquatic macrophytes (like papyrus) into the root zone promotes nitrification of ammonia, with the nitrate formed diffusing into the adjacent anaerobic zone in which it undergoes denitrification (Fig. 7.2). Oxygen leakage from the rooting mat of the plants creates oxidised conditions in the otherwise anoxic environment and this is believed to stimulate both anaerobic decomposition of organic matter and growth of nitrifying bacteria (Brix, 1997).

In addition to low concentrations of oxygen and ammonia other factors like temperature, soil moisture and pH plus their interplay limit nitrification or reduce its rate (Belser, 1979; Underhill, 1990). Low numbers of nitrifiers may also limit nitrification (Both *et al.*, 1992). At temperatures below 5 °C and above 40 °C the nitrification rate is very low whereas the optimal temperature range is 30 - 35 °C. The nitrification rate declines markedly below pH 6.0, and is negligible at pH > 8.0 (Belser, 1979; Underhill, 1990; Hammer and Knight, 1994; Kadlec and Knight, 1996).

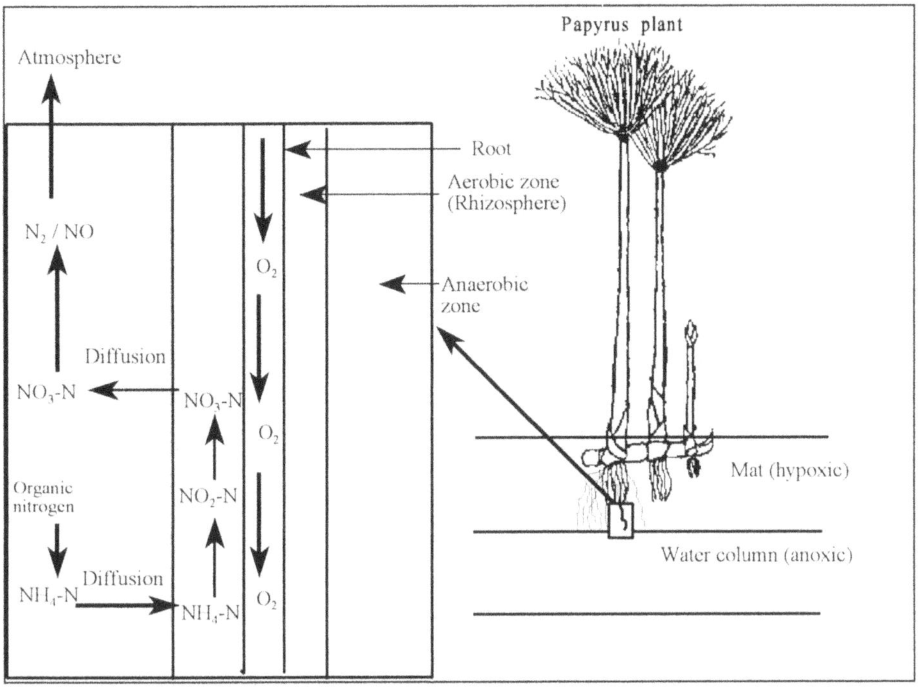

Fig. 7.2. Schematic representation of nitrogen transformation in the rhizosphere (enlarged) of papyrus.

Utilisation of atmospheric nitrogen gas as source of nitrogen is called nitrogen fixation and represents an import of N into the aquatic system. Nitrogen fixation might not typically be important in wetlands unless leguminous plants, cyanobacteria and other nitrogen fixing bacteria are present (De Boer *et al.*, 1995), and is not observed in nitrogen-rich ecosystems (Kadlec and Knight, 1996). All nitrogen fixation activities are repressed as long as there is sufficient ammonium (e.g. in environments rich in ammonia) or nitrate (van Oostrom and Russell 1994; Kadlec and Knight, 1996). Nitrogen fixation uses stored energy from either autotrophic or heterotrophic sources and this process may not be advantageous when nitrogen is otherwise available for growth. This has been referred to as an 'ammonia switch off' mechanism (Brock *et al.*, 1994).

The Nakivubo swamp has been receiving secondary treated effluent from Kampala city and raw sewage from Luzira for more than 30 years. The wastewater is heavily laden with nitrogen compounds of which $NH4^{+}$ is the major component (Chapter 4) with a typical average concentration of 18 mg/l and total load of 1000 kg N/d. In the swamp part of the nitrogen is taken up by the aquatic macrophytes (Chapter 5). Another mechanism for the removal of nitrogen from the Nakivubo swamp, is its loss through biological nitrification and denitrification (Fig. 7.1). However, there is little or no information on the nitrification/denitrification processes in the Nakivubo swamp, and the role of plants in these transformations (Fig. 7.2). Therefore, and as nitrification is the limiting step in nitrification and subsequent denitrification, this study was carried out to get more insight in the role of nitrification in nitrogen removal from the Nakivubo swamp. The role of the dominant macrophytes in the nitrification process was also investigated.

It was hypothesised that biological nitrogen removal through the processes of nitrification and denitrification in the Nakivubo swamp primarily depends on the numbers of the nitrifiers and/or the nitrification activity (the rate of conversion of $NH4^+$ by nitrifiers over a given time). In turn, the distribution and activity of nitrifying organisms were thought to be substantially influenced by the availability of oxygen and the aquatic macrophytes in the swamp.

The objectives of the study were to assess:

(i) the factors that affect the distribution and activity of nitrifiers in the Nakivubo swamp.

(ii) the vertical and longitudinal distribution of nitrifiers in different compartments in the swamp.

(iii) the role of the nitrification process in the removal of nitrogenous compounds from the swamp, using bioassays under controlled conditions.

7.2 Materials and methods

7.2.1 Sampling locations

This study was carried out in the Nakivubo swamp, a predominantly floating swamp dominated by papyrus and *Miscanthidium* (Chapter 2). Three sampling sites in the swamp were selected (one at Transect 1 (T1) and two at Transect 4 (T4)) Fig. 4., Chapter 4.

At T1, site T1-259 was chosen. T1-259 refers to the location 259 m along T1 measured from the western edge of the swamp. For T4 two sites were selected: one in the papyrus (T4-395) and another in the *Miscanthidium* (T4-410). The two transects (T1 and T4), are located at the swamp inlet and the swamp lake interface respectively and were selected to reflect a rough spatial distribution of nitrifiers along the longitudinal axis of the swamp. In addition, the sites at T4 were selected to assess the influence of the two dominant macrophytes on the distribution and activity of nitrifying bacteria. At each location *in situ* measurements of physical and chemical variables were done. Samples of roots, peat and water were also collected for enumeration of nitrifiers and determination of the potential nitrification capacity. Peat and root samples were taken from the mat 10 - 50 cm deep whereas water samples were taken from the water column.

To get an insight in the factors that might affect the distribution and activity of nitrifiers, the physical and chemical variables (pH, temperature, dissolved oxygen (DO), ammonium, nitrite and nitrate), were also measured at the sampling sites. *In situ* temporal variation of dissolved oxygen in the Nakivubo swamp was also monitored. Measurement and analyses were carried out as described in Chapter 4.

7.2.2 Oxygen release by plants

The release of oxygen by aquatic macrophytes was first monitored in the field by continuous measurement of DO in the mat/root zone of papyrus and *Miscanthidium* for several days. Oxygen was measured with an oximeter (with a stirrer) placed in the mat and connected to a data logger.

Secondly, whether oxygen release by plant roots occurred was estimated by monitoring DO with an oximeter (connected to a data logger) placed in the buckets containing papyrus plants. The observed DO will be a resultant of several processes, of which release is one. The potted plants were established in an enclosed compound at Makerere University, Institute of Environment and Natural Resources, Kampala-Uganda. Triplicates of papyrus were acclimatised for one month in plastic buckets (10 l) before monitoring the oxygen release. The plants were fed with fresh wastewater collected from Nakivubo swamp influent (Fig.4.1) and a hydraulic retention time of 7 days was used. Oxygen was also monitored in a control with only wastewater. The buckets were covered with aluminum foil to minimise algal growth. The ambient temperature varied between 23-25 ^{0}C.

7.2.3 Effect of seiches on oxygen transport into the swamp

Oxygen transported into the swamp with the back-flow of water due to the lake seiches was monitored both in the mat and the water column at T4-400. Continuous measurement of oxygen was made with an oximeter (connected to a data logger) placed either in the mat or water column. A temperature probe was also connected to the data logger.

7.2.4 Enumeration of nitrifying bacteria

Nitrifying organisms were enumerated using the Most Probable Number (MPN) technique (Mutalewich *et al.*, 1975). A modified mineral solution of Schmidt and Belser (1982) was used and contained (g/l): $(NH_4)_2SO_4$, 0.5; $CaCl_2.2H_2O$, 0.04; $MgSO_4.7H_2O$, 0.1; Phenol red, 0.005; K_2HPO_4, 0.2; $FeSO_4.7H_2O$, 0.0005; Na_2EDTA, 0.0005 as a mineral solution. The trace elements solution contained (g/100 ml): $NaMoO_4.2H_2O$, 0.01; $MnCl_2$, 0.02; $CoCl_2.6H_2O$, 0.0002; $ZnSO_4.7H_2O$, 0.01; $CuSO_4 5H_2O$, 0.002. One ml of trace element solution was added to 1 litre of mineral solution. Growth of ammonia oxidisers is accompanied by a drop in pH which was detected by including phenol red in the media which changes from red to yellow as the pH value drops below 6.5.

To prepare a 10^{-1} dilution of peat and root samples, 10 g of the sample was added to 95 ml of sterile physiological saline diluent. The mixture was shaken on a rotary dual shaker for 2 minutes at 120 rpm. Thereafter, the samples were sonified in a Braun water bath at 40 KHz for 20 sec to desorb attached microbial cells. The 10^{-1} dilution was thereafter shaken vigorously and subsequently, 10 ml of this solution was transferred to 90 ml of dilution water to make a 10^{-2} dilution. The working dilution water was made up of 1 mM phosphate buffer. The pH of the media was adjusted to 7.8 with 1 M NaOH. One ml of serial 10-fold dilutions of the inoculum was transferred to 5 replicate tubes per dilution containing 4 ml of autoclaved (121 °C for 15 min) test medium. Thereafter the samples were mixed and incubated at room temperature (23 - 25 °C) in a dark box (covered with aluminium foil) to eliminate the light effect. After inoculation, roots and peat were oven dried at 103 °C for dry weight determination. Incubation was carried out for a period of 2 months.

As a control three uninoculated test tubes containing only the test medium were also incubated. A beaker containing distilled water was also put in the incubation box to reduce evaporation loss from the media.

7.2.5 Nitrification activity

Nitrification activity of ammonia oxidisers was carried out on root, peat (soil surrounding the roots) and water samples. Before sonication, root samples were gently washed in distilled water to remove the attached peat. The samples were collected from T1-259 and T4-395 m and T4- 410 m. Due to limited facilities and to ensure replication, the activity on water was only carried out on samples taken from the water column at T4-395 m. Peat and rhizome/root samples were collected in polythene bags and transported to the laboratory. In the laboratory, roots were separated from rhizomes before treatment.

The method of Schmidt and Belser (1982) was used in determining the potential nitrifying activity of the samples. Six replicate samples of peat and of roots, 10 g each, were incubated in a test media composed of 1 g of neutralised bacteriological peptone (Oxoid), 8.5 g NaCl, 0.095g of $(NH_4)_2SO_4$, (as a test substrate) and 1.06 g $NaClO_3$ per liter. The chlorate inhibits the conversion of nitrite to nitrate (Belser and Mays, 1982). Under these conditions, the rate of ammonia oxidation was assumed to be equal to the rate of nitrite accumulation. Aerobic incubation of the samples was performed in 500 ml Erlenmeyer flasks containing 300 ml of the test media. Samples were shaken on a dual shaker at 120 rpm. For water samples, 100 ml of the sample was added to 200 ml of test media. Similarly a blank containing only the test media was incubated. Samples were filtered before analysis to remove suspended solids. Nitrite production was monitored after every two hours for a period of 6 - 8 hours, a period during which microbial growth or death was thought to be negligible. Regression analysis was used to estimate the rate of nitrite production. The pH of the samples was measured after incubation and varied between 6 and 6.5 and DO was 5.7 - 6.3 mg/l.

7.2.6 Statistical analyses

Statistical analysis was carried out using MINITAB software (MINITAB Release 10 for Windows). Data of activity tests were analysed with ANOVA to test for differences among sample means. Data were first tested for normality and homogeneity of variances. They were $\log_{10}$ transformed, if these requirements were not met. Following analysis of variance, differences between means were tested with Tukey's and Dunnet's multiple comparisons.

7.3 Results

7.3.1 Physical and chemical variables

The physical and chemical variables of water samples taken from the mat and the water column from the sampling sites are presented in Table 7.1. Temperature varied between 23.3 - 23.7 °C and was generally higher in the water column than in the mat. The pH range was slightly acidic (6.4 - 6.6) and in the range favourable for growth of nitrifiers (6 - 8, Underhill, 1990), except in the *Miscanthidium* mat where pH was more acidic (5.0). The ammonium concentration was highest in the water column at T1-259 but decreased towards the lake swamp interface. The *Miscanthidium* mat had the lowest concentration of ammonium (3.7 mg/l). The ratio of ammonium-nitrogen to conductivity generally increased from the mat to the water column, suggesting more transformation of ammonia in the mat system. It was assumed that conductivity is a conservative variable. The DO concentration was low (≤ 0.1); the highest value (0.2 mg/l) was recorded in the papyrus mat at T1-259 whilst the water column was anoxic at all sites.

Table 7.1. Physical and chemical variables at sampling sites.

	Site					
	T1 259 m		T4-395		T4-410	
Variable	mat	water column	mat	water column	mat	water column
Temp (°C)	23.5 ± 0.1	23.4 ± 0.1	23.3 ±0.1	23.6 ± 0.1	23.4 ± 0.1	23.7 ± 7
pH	6.6 ± 0.01	6.6 ± 0.02	6.6 ± 0.04	6.6 ± 0.04	5.0 ± 0.1	6.4 ± 0.1
NH_4-N (mg/l)	17.5 ± 0.9	20 ± 1	12.4 ±0.7	15.1 ±0.6	3.7 ± 0.5	17.8 ± 1.0
NH_4-N : EC	0.039	0.045	0.044	0.004	0.024	0.115
NO_3-N (mg/l)	0.01 ± 0	0.03 ± 0.01	0.05 ±0.01	0.02 ± 0.002	0.03±0.004	0.03 ±0.004
NO_3-N : EC	0.0002	0.0001	0.0001	0.0001	0.0002	0.0002
NO_2-N (mg/l)	ND	ND	0.004 ± 0.001	0.007± 0.004	0.001 ±.0	ND

7.3.2 Oxygen release by plants

In the field, there were no detectable diurnal changes in dissolved oxygen in the root zone or mat, neither in the papyrus nor the *Miscanthidium* vegetation. Most likely, the oxygen translocated from the aerial shoots to the rhizosphere, if any, was immediately consumed in the conversion of organic matter (by heterotrophs) and other oxygen consuming processes close to the root surface.

The DO concentration in the planted buckets and the blank (which contained only wastewater) varied with temperature (Fig. 7.3). Oxygen and temperature rose in parallel, but temperature experienced a lag. The time lag shift to the right in the temperature curve can be explained by slow rise in water temperature following the onset of solar radiation. On the other hand, photosynthesis starts immediately as soon as the photosynthetic solar radiation becomes available. Maximum values for both temperature and dissolved oxygen were observed at around 14.00 h, probably when photosynthesis was at maximum. The oxygen released by papyrus rooting mat is attributed to photosynthesis.

To estimate the oxygen release by plants the DO concentration was corrected for the blank to compensate for surface aeration or any other oxygen input processes like oxygen release from primary production by algae. The net change in DO (resulting from different processes) in the water in which papyrus shoots (8 culms) were growing was on average 1.6 mg/l. Considering the water volume of 5 l, the total oxygen released was $8*10^{-3}$ g. However, the difference between DO in the planted bucket and the blank is not a direct measure of the plant root release, but a qualitative indicator of the occurrence of this release. A greater difference between the planted buckets and the blank corresponds to a higher release rate, but its magnitude can only be assessed if other (cell) processes rates are quantified. In the blank there are two active processes viz., BOD exertion and re-aeration and the resulting DO follows from their dynamic balance. On the other hand, in the planted bucket, there is also oxygen transport from aerial shoots and the re-aeration will be different as most of the surface area of the bucket is covered by plants. However, under our experimental conditions (sensitivity of the measuring equipment), the estimate made gives an indication of the oxygen release by papyrus plants and its diurnal variation.

Oxygen produced in aerial shoots during photosynthesis translocated to other parts of the plant like the roots, where it is used in the active uptake of nutrients and for respiration. The excess oxygen then diffuses into the surrounding water. The increase in DO can be attributed to photosynthesis because of the correlation with solar radiation and dissolved oxygen in the water containing plants. The buckets were covered with aluminium foil to minimise algal growth.

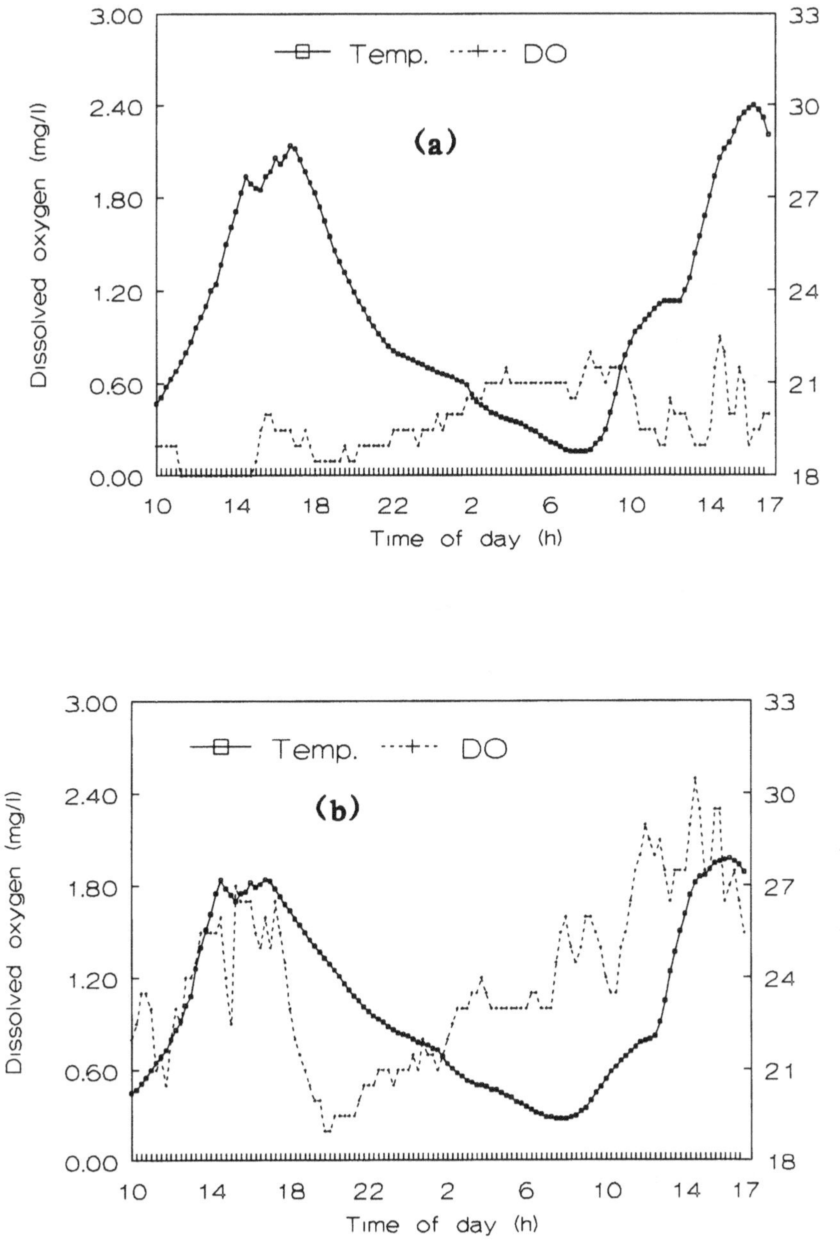

Fig. 7.3. Variation of dissolved oxygen (DO) and temperature (Temp) in the blank (a) and experimental buckets (b) containing papyrus plants in an enclosed compound. The blank contained only the wastewater.

7.3.3 Movement of oxygenated lake water into the swamp

The variation of DO oxygen in the water column at T4-400 is shown in Fig. 7.4.

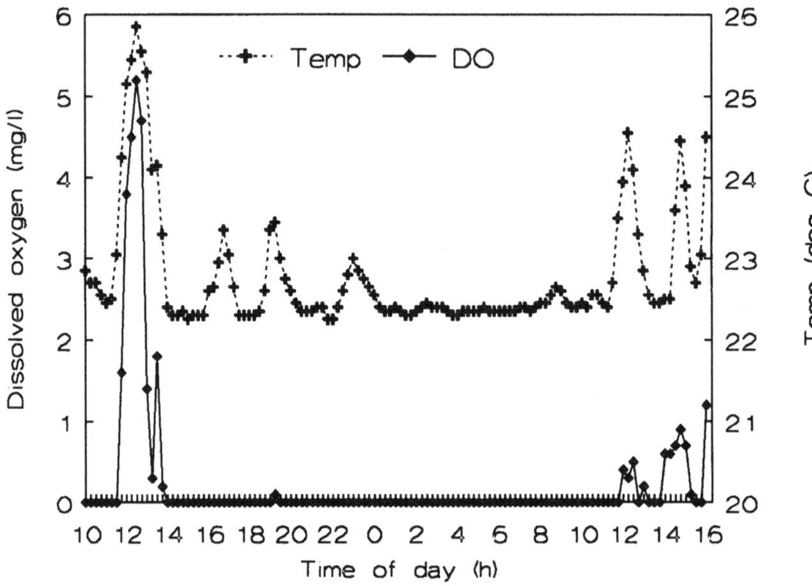

Fig. 7. 4. Diurnal variation of dissolved oxygen and temperature in the water column at T4-410.The peaks in both dissolved oxygen and temperature are attributed to back-flow of water due to lake seiches.

Apart from the back-flow of oxygenated lake water into the swamp due to lake seiches, in most cases the water column was anoxic. The DO of 5.2 mg/l in the back-flow was accompanied by rise in temperature of the swamp water (23- 26 °C), which was attributed to the warmer lake water flowing into the swamp.

7.3.4 Distribution of nitrifying bacteria

The spatial distribution of nitrifying bacteria enumerated from roots and peat samples is presented in Fig. 7.5. Overall, the numbers of nitrifying bacteria were 3 orders of magnitude higher in papyrus root samples than those in *Miscanthidium*. The highest numbers of nitrifiers ($5.2 * 10^4$ MPN /g DW) were recorded in the peat at T4-395 (papyrus).

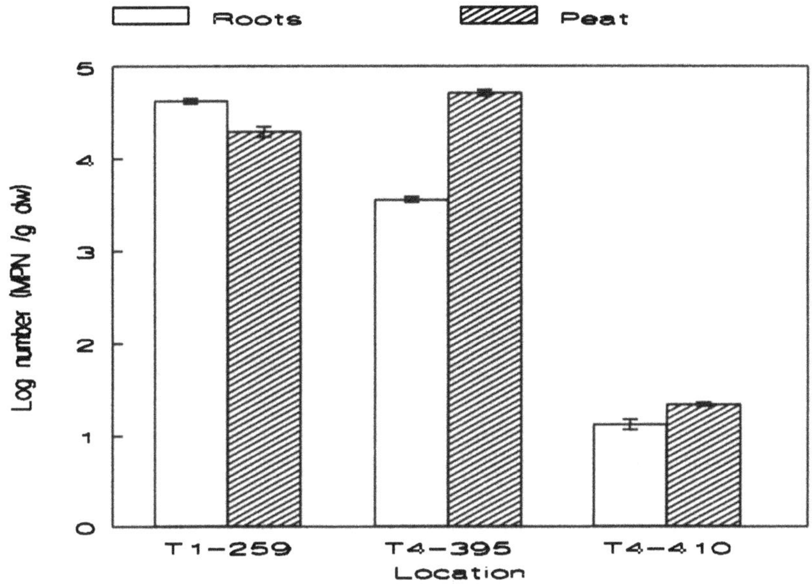

Fig. 7.5. Average spatial distribution of nitrifiers in the peat and on root samples from different locations in Nakivubo swamp. T1-259 and T4-395 are papyrus sites and T4-410 is a *Miscanthidium* site. Bars indicate standard error (n= 2).

Site T4-410 which is located in the *Miscanthidium* zone had the lowest number of nitrifiers (< 21 MPN /g DW). High numbers of nitrifiers were attached to the roots of papyrus at T1-259 ($2.8*10^5$ MPN/g DW), compared to the numbers attached to the roots at T4-395 ($3.6*10^3$ MPN/g DW). *Miscanthidium* roots had the lowest numbers of nitrifiers (< 16 MPN/g DW), suggesting a different degree of influence by the vegetation or by conditions in the mat.

The distribution of nitrifiers in the water column along the longitudinal axis of the swamp is depicted in Fig. 7.6. There were high numbers of nitrifiers in the water column at T1-259 ($1.7*10^4$ MPN/100 ml) compared to the water column at T4-395 ($2.2*10^3$ MPN/100 ml) which were in the same range as those in the water column below the *Miscanthidium* mat ($1.8*10^3$ MPN/100 ml). The general decrease in the number of nitrifiers from the swamp inlet to the lake interface may be explained by die-off and lack of growth and possibly the increasing attachment to roots or to settling of the nitrifiers. The nitrifiers may have died or become dormant due to lack of oxygen during transport from T1 to T4. The high concentration of nitrifiers at the inlet may have originated from the trickling filters at Bugolobi sewage works and may be their subsequent growth in the Nakivubo channel as result of better aeration and high concentrations of ammonia.

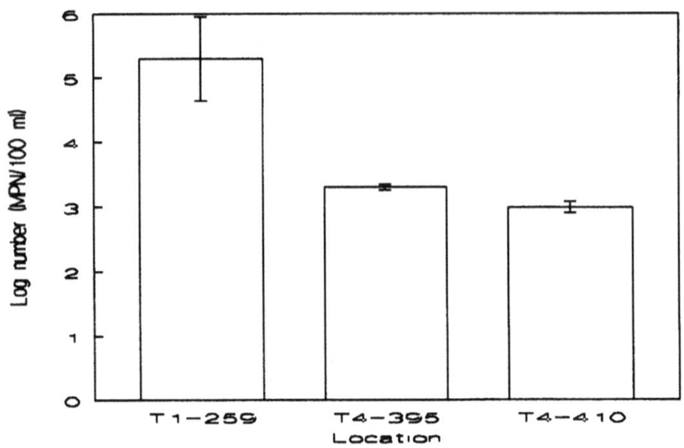

Fig. 7.6. Average spatial distribution of nitrifiers in the water column at different locations in Nakivubo swamp. Bars indicate standard error (n=2).

7.3.5 Nitrifying activity

Generally there was an immediate conversion of ammonium to nitrite (within 2 hours of sample incubation) by papyrus root samples. The nitrite production activity of *Miscanthidium* roots exhibited a lag phase of up to 4 hours. The activity of nitrifiers attached to papyrus roots at T1-259 and T4-359 was not significantly different ($p > 0.05$), but it was significantly higher than that recorded for *Miscanthidium* roots at T4-410 (Fig. 7.7). The peat sample at T1-259 had a significantly higher activity than that at T4-395 and the latter was not

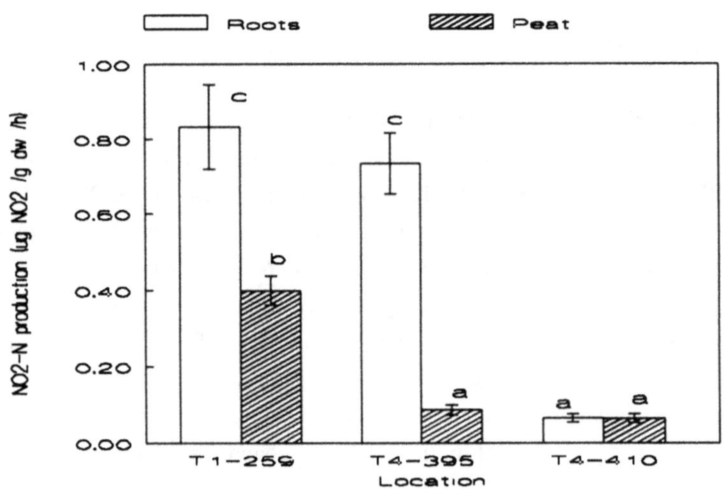

Fig. 7.7. Average nitrifying activity of peat and on root samples from different locations in Nakivubo swamp. Bars indicate standard error (n=6). Bar with the same letters are not significantly different from each other at p= 0.05.

significantly different from that of the *Miscanthidium* peat. There was no correlation between the numbers of nitrifiers and the nitrification activity. The nitrification activity of the water column was 0.0038 mg NH_4-N/l/d.

7.4 Discussion

The distribution of nitrifying bacteria and nitrification activity in most ecosystems depends largely on the abundance of the $NH4^+$, dissolved oxygen and other environmental variables (Underhill, 1990). The pH range for the survival of nitrifiers has been reported to be 6 - 8, and the temperature range of 23 - 40 °C (Belser, 1979; Underhill, 1990; Kadlec and Knight, 1996). In this study, the temperature range was 23 - 24 °C, and the pH was slightly acidic (6.4 - 6.6), except in the *Miscanthidium* mat where it was as low as 5.0. High numbers of nitrifiers are often isolated in areas where oxygen concentration is usually high (Belser, 1979). Most of the physical and chemical variables in the Nakivubo swamp are in the range that would favour the survival of nitrifying bacteria, except DO, which is mostly so low that either hypoxic or anoxic conditions exist in most compartments. Possibly also pH limits the growth of nitrifiers.

The low pH in the *Miscanthidium* mat, is likely to be unfavourable for nitrification as indicated by the low numbers and activities of nitrifiers observed in this vegetation zone. Activity of nitrifiers is reportedly low at pH < 6, though nitrification at pH values as low as 3 has been reported (De Boer *et al.*, 1995). However, at this low pH, the activity stopped when the pH was not adjusted. Even the nitrite oxidising bacteria that have been isolated at pH 4 are acid sensitive (Hakinson and Schmidt, 1988; Kuenen and Robertson, 1988; Laanbroek and Woldendorp, 1995). Under such conditions of low pH, the activity of nitrifiers is supposedly suppressed and the microbial cells eventually lose their activity.

Ammonia may not be limiting in most papyrus dominated parts of the Nakivubo swamp as its concentration was high (usually above 10 mg/l), except in the *Miscanthidium* mat where it was 3.7 mg/l, leaving oxygen as the major limiting factor for nitrification. Oxygen availability has been reported to limit nitrification in wetlands (Wittgren and Tobiason, 1995; Kadlec and Knight, 1996). In the Nakivubo swamp the mat was always hypoxic or anoxic. The water column was in most cases anoxic except when there was back-flow of oxygen-rich lake water due to lake seiches. A second source of oxygen is translocation from the aerial plant shoots of macrophytes to the rhizosphere (Armstrong and Armstrong, 1990; Brix, 1993; Meirong and Jones, 1995). In this study no diurnal changes in oxygen concentration in the mat/root zone in the swamp were observed. This could be due to the fact that translocated oxygen to the rhizosphere by the aquatic macrophytes is immediately consumed by the oxygen demanding processes, notably organic matter decomposition and nitrification. Furthermore, the oximeter used was not sensitive enough to detect small changes in oxygen concentration. Oxygen translocation and release by papyrus plants was evident in the potted experiment in an enclosed compound (Fig. 7.3). The increase in DO in the experimental buckets may be attributed to photosynthesis because of the correlation with solar radiation and dissolved oxygen in the water containing plants as the highest DO coincided with maximum solar radiation.

Oxygen release by aquatic plants promotes the nitrification of ammonia in wetlands like the Nakivubo swamp. Considering the oxygen release of 8 * 10^{-3} g /culm and the average plant density of papyrus in the Nakivubo swamp is 17 culms/m^2, the amount of oxygen released by papyrus into the Nakivubo swamp can be estimated to be 1.7*10^{-2} g/m^2/d. Assuming the total area of the Nakivubo swamp (1,150,000 m^2) is dominated by papyrus, the total amount of oxygen released into the swamp would be 2.0*10^4 g/day. From the stoichiometric conversion of NH_4^+ to nitrate through nitrification, 4.6 g of oxygen is required for every gram of NH_4-N converted. Considering the total amount of oxygen released into the swamp and assuming that all the oxygen released is used in nitrification, 4.39*10^3 g of NH_4-N would be converted in a day. This is only 0.028 % of the total NH_4-N load into the swamp. Brix (1993) reported an oxygen release rate by *Phragmites australis* of 0.02 g/m^2/d, whereas Armstrong and Armstrong (1990) reported a higher range of 0.5 - 5.2 g O_2/m^2/d for the same plant. The variability in the oxygen reported by several authors has been attributed to the different methodologies. However, as pointed out by Brix (1997), the non-homogeneity of the oxygen release pattern for wetland plant roots makes it difficult or impossible to extrapolate from the results obtained for oxygen from a micro-electrode technique in a controlled experiment to *in situ* release. On the other hand as pointed out by Focht and Verstraete (1977), nitrifiers are poor competitors for oxygen compared to heterotrophic bacteria, since heterotrophs have a much higher affinity for oxygen which suggests that in wetlands with high organic matter like the Nakivubo swamp, the little available oxygen would be fully used by heterotrophs. This explains why no oxygen was detected in the mat of papyrus in the Nakivubo swamp.

The input of oxygen into the swamp due to lake seiches was estimated by using an average water level rise of 0.1 m which results in 837,300 m^3 of lake water flowing back into the swamp (Chapter 3). Such a seiche was commonly found to occur at 13.00 h though in most cases (especially when the water hyacinth was resident in the bay) the lake water was not oxygenated. Using the maximum oxygen concentration observed in the water column at T4-400 (5.2 mg/l), the oxygen brought with the lake water amounts to 4.35*10^6 g/day. Assuming active nitrification (and negligible heterotrophic oxygen consumption, though this is unlikely to occur the Nakivubo swamp which is rich in organic matter) in the swamp, this oxygen input would convert 9.47 *10^5 g NH_4-N /day. This corresponds to 49% of the total daily ammonia load into the swamp.

The distribution of nitrifiers, revealed that there were more nitrifiers at T1-359, a site close to the swamp inlet, than at other sites. This could be attributed to the fact that nitrifiers originate from Bugolobi sewage works (trickling filters whose effluent already contains nitrifiers) and are carried with the Nakivubo channel water into the swamp. Matulewich and Finstein (1978), reported a number of nitrifiers of 1.4* 10^4 MPN/100ml, in a river receiving a discharge of domestic wastewater from an activated sludge treatment plant. The Nakivubo channel is fast flowing and shallow in most places and always has some residual oxygen (0.2 - 0.5 mg/l), probably from surface aeration. It appears that nitrification begins upstream in the channel and that the nitrifiers are transported downstream into the swamp. Availability of oxygen and high concentrations of NH_4-N in the channel could have encouraged the growth of nitrifiers, probably explaining why samples taken from T1 had substantially higher numbers of nitrifiers than those from T4, despite that at T4 DO is much higher.

The enumerated nitrifiers were higher on the roots of papyrus and the surrounding peat than in the *Miscanthidium*, suggesting that the vegetation of the Nakivubo swamp and the conditions in their mats may affect the distribution of nitrifiers and their activity. The low numbers in the *Miscanthidium* root/mat complex may be due primarily to low pH and restricted mixing of the water in the water column and that in the mat. This is further reinforced by the fact that the number of nitrifiers in the water column under the mats of *Miscanthidium* and papyrus were in the same range. The low numbers of nitrifiers in the water column at T4 compared to those in the mat/root zone (assuming that the density of swamp water is 1 g/cm^3) also suggest that these organisms grow and survive easily in the root/mat zone. However, it is difficult to make an accurate comparison of water and mat samples as the units of measurement were different. Fewer numbers in the water column may also be due to attachment of these organisms onto the root/mat compartment or settling from the water column with suspended solids. Nitrifiers have a long doubling time and can only survive from being washed out of the swamp by attaching/growing on surfaces like roots. Lack of surface for the attachment of nitrifying bacteria has been reported to limit nitrification (Wittgren and Tobiason, 1995). Ottavá *et al.* (1997) reported that nitrifying bacteria were absent in the wastewater in a constructed wetland, though high numbers were found attached to the roots and rhizomes of the 'treatment' grass *Glyceria maxima*. Also Both *et al.* (1992) found more nitrifiers in the root zone of *Glyceria maxima* compared to the surrounding reduced bulk soil. Ammonia oxidising bacteria have also been found to be mainly associated with roots in the gravel bed system (May *et al.*, 1990). Mutalewich and Finstien (1978) reported the following order of numbers of nitrifying bacteria within habitats of a polluted river: rooted aquatic plants > algae ≈ rocks > sediment >> water.

Higher numbers of nitrifiers in the papyrus zones than in *Miscanthidium* may be attributed to the favourable pH in the mat of the former vegetation. Furthermore, high concentrations of ammonia and probably a higher supply of oxygen from the plants and lake seiches may encourage the growth of nitrifiers. As reported in Chapter 5, there is easy mixing of waters in the swamp water column and in the mat of papyrus which flushes out the decomposition products like humic and fulvic acid from the mat. These compounds cause the acidic pH and appear to accumulate in the mat of *Miscanthidium* where there is restricted mixing. This results in low pH value (as low as 5) which may restrict the growth and survival of nitrifiers. This restricted mixing also implies that oxygen transported into the swamp with lake seiches does not enter the thick mat, leaving nitrifiers to rely only on oxygen translocated by the aerial shoots to the rhizosphere. However, since the pH in the water column and in the mat of papyrus are similar, it implies that it is not only pH, but also other factors associated with plants that encourage the growth of nitrifiers in the root mat zone of papyrus.

The activity of nitrifiers was not correlated to their number in the Nakivubo swamp. The discrepancy between MPN numbers and nitrification activity in soil samples has also been observed elsewhere (Belser and Mays, 1982) and was explained by the low counting efficiency of the MPN method applied. However, Verhagen *et al.* (1992) attributed this discrepancy to the fact that the bacteria could have lost their activity but survived as viable cells during the period when the conditions were not favourable. Since the incubation period for ammonifiers using the MPN method is long (Mutalewich *et al.*, 1975), the dormant cells could have grown but were not able to regain their activity during the short period of activity measurements. As pointed out by Dolfing and Bloemen (1985), activity tests measure the activity of active cells since cell growth is minimal during this short incubation period. De Boer et al., (1995) reported high numbers of nitrifiers which indicated that active nitrification took place in the recent past after which the cells became dormant. In this study the population detected with the MPN enumeration might have been dormant but because of the long incubation period (60 days), the cells were able to grow again. During the activity test, which was carried out for a period of only 8 hours, most cells might not have regained their activity. Furthermore, this period could have been too short to allow the full adaptation of the cells to the new conditions of the experiment (oxic and high substrate concentrations). Some samples especially those obtained from the *Miscanthidium* zone, had a long lag phase (up to 4 h) probably because low pH inactivated the cells.

The fact that the activity on the roots of papyrus was higher compared to that in the peat may be attributed to the periodic translocation of oxygen from the shoots to the roots during photosynthesis (Fig. 7.7).This could have resulted in cells getting accustomed to oxic conditions and remaining viable and active. High numbers of nitrifiers in the peat could have been viable but not active. This may explain the long lag phase on samples obtained from *Miscanthidium* and papyrus peat zones at T4-395 (data not shown). Higher activity was observed for biofilms attached to roots of *Phragmites australis* and gravel, than in the water column (Williams *et al.*, 1994).

Two scenarios have been suggested (Shaher *et al.*, 1993) when the dormant facultative cell is suddenly exposed to aerobic conditions. One is that the cell will reactivate or rebuild its nitrifying capacity and start nitrifying within a short lapse of time by utilising the available oxygen at that time. The advantage of this strategy is certainly the ability to utilise temporary aerobic conditions. However, the price the cell pays can be very high. It is possible that the energy demanded to adapt to temporarily aerobic conditions and back to anaerobic will be higher than the energetic gain obtained from a short period of nitrification (especially due to the relatively low energy yield of nitrification). The second strategy is to delay the change in order to get maximum assurance that the new aerobic conditions are stable and reliable. The second strategy is hypothesised to be the one employed by nitrifying bacteria which do not have systematic oxygen supply. Then, nitrification takes place only following a significant lag period after the anaerobic conditions commenced.

Considering the average activity in the mat (roots and peat) of papyrus of 1.23 μg NH_4-N /g/d and the mat bulk density of 52 g/cm^3, and assuming that active nitrification takes place in the top 10 cm of the mat, ammonia removal was estimated at 0.064 g N/m^2/d. Assuming that the whole swamp is dominated by papyrus, this would relate to <1% of the NH_4-N inflow. However, if a whole mat thickness of 50 cm is considered, the ammonia removal would amount to 20%. Ammonification rates of 3 to 35 mg N/m^2/d have been reported for a natural wetland in central Minnesota (Kadlec and Knight, 1996). Higher nitrogen mineralisation rates (41 to 125 mg/m^2/d), have been reported for organic soils of Florida (Reddy, 1989).

7.5 Conclusions

Nitrification and subsequent denitrification appeared to be one of the two major processes that remove nitrogen from the Nakivubo swamp. Oxygen was one of the factors that limited the distribution of nitrifiers and their activity in the swamp. Low pH values in the *Miscanthidium* mat limited the viability and the activity of the nitrifiers in this zone. The translocated oxygen from the aerial shoots to the rhizosphere resulted in prolific growth of nitrifying bacteria on the roots and peat surrounding the roots, probably leading to high nitrifying potential. The main source of oxygen in the swamp appears to be the oxygenated lake water which is transported to the swamp by the lake seiches.

Laboratory measurements of the specific activity and MPN enumeration of nitrifiers may have little relation to the actual situation in the field. These activities represent a nitrifying potential which will be achieved in the Nakivubo swamp only when conditions are optimal for nitrification, which occurs when O_2 is brought in by the seiches. In addition, activity measurements can only measure activity of active cells, not for dormant ones. The scarcity of oxygen, and the high concentration of organic biodegradable matter throughout the swamp suggests that the oxygen will be preferentially consumed by the heteretrophs. Nitrification in the papyrus mat of the Nakivubo swamp can remove as much as 20% of the ammonia load in the wastewater into the swamp. Oxygen release by papyrus plants was insignificant as it was estimated that if used in nitrification only 0.03% of the ammonia in the input would be converted. The input of oxygen into the swamp by seiches as the major source of oxygen could convert 59% of the ammonia load in the input at active nitrification as an upper limit.

7.6 References

Armstrong, J. and Armstrong, W. (1990). Pathways and mechanisms of oxygen transport in *Phragmites australis*. *In:* Cooper P. F. and Findlater B. C. (eds.). *Constructed wetlands for water pollution control.* Pergamon Press, Oxford, pp 529 - 533.

Belser, L. W. (1979). Population ecology of nitrifying bacteria. *Ann. Rev. Microboil.*, *33*: 309 - 333.

Belser, L. W. and Mays, L. (1982). Use of nitrifier activity measurements to estimate the efficiency of viable nitrifier counts in soils and sediments. *Appl. Environ. Microbiol.*, *43*: (4) 945 - 948.

Both, G. J., Gerads, S. and Laanbroek, H. J., (1992). The occurrence of chemolithoautotrophic nitrifiers in water-saturated grasslands soils. *Microb. Ecol.*, *23*: 15 - 26.

Brix, H. (1993). Macrophyte-mediated oxygen transfer in wetlands: transport mechanisms and rates. *In.* Moshiri, G. A. (ed.). *Constructed wetlands for water quality improvement.* CRC Press, Inc. Boca Raton.

Brix, H. (1997). Do macrophytes play a role in constructed treatment wetlands? *Wat. Sci. Tech.*, *35* : 11 17.

Brock, T. D., Madigan, M. T., Martinko, J. M. and Parker, J. (1994). *Biology of microorganisms.* Prentice-Hall International, Inc.

De Boer, W., Gunnewieck, P. J. A. and Laanbroek, H. J. (1995). Nitrification at low pH by a bacterium belonging to the genus *Nitrosospira.* *Soil. Biochem.*, *27*: 127 - 132.

Dolfing, J. and Bloemen, W. G. B. M. (1985). Activity measurements as a tool to characterize the microbial composition of methanogenic environments. *J . Microbiol. Methods. 4*: 1-12.

Focht, D. D. and Verstraete, W. (1977). Biochemical ecology of nitrification and denitrification. *Adv. Microb. Ecol.*, *1*: 135 - 214.

Hakinson, T. R. and Schmidt, E. L. (1988). An acidophilic and neutrophilic *Nitrobacter* strain isolated from the numerical predominant nitrite-oxidizing population in an acidic forest soil. *Appl. Environ. Microbiol.*, *54*: 1536 - 1540.

Hammer, D. A. and Knight, R. L. (1994). Designing constructed wetlands for nitrogen removal. *Wat. Sci. Tech.* *29* : 15 - 27.

Howard-Williams, C. (1985). Cycling and retention of nitrogen and phosphorus in wetlands: theoretical and applied perspective. *Freshwater Biology 15*: 391 - 431.

Hsieh, Y. P. and Coultas, C. L. (1989) Nitrogen removal from freshwater wetlands: nitrification- denitrification coupling potential *In:* Hammer, D. A. (ed.). *Constructed wetlands for wastewater treatment.* Lewis Publishers, Inc. Michigan, pp 493 - 500

Kadlec, R. H. and Knight, R. L. (1996). *Treatment wetlands.* Lewis, Publishers, Boca, Raton.

Kuenen, J. G. and Robertson, L. A. (1988). Ecology of nitrification and denitrification. *In:* Cole, J. A. and Ferguson, S. J. (eds.) *The nitrogen and sulphur cycles*, Cambridge University Press. Cambridge, England, pp 181- 218.

Laanbroek, H. J. and Woldendorp, J. W. (1995). Activity of chemolithotrophic nitrifying bacteria under stress in natural soils. *Adv. Microb. Ecol., 14*: 275 - 304.

Meirong, L. And Jones, M. B. (1995). CO_2 and O_2 transport in the aerenchyma of *Cyperus papyrus* L. *Aquat. Bot. 52*: 93-106.

May, E. Bulter, J. E., Ford, M. G., Ashworth, R. F., Williams, J. and Baghat, M. M. M. (1990). Chemical and microbiological processes in Gravel Bed Hydroponic (GBH) systems for sewage treatment. *In:* Cooper P. F. and Findlater, B. C. (eds.). *Constructed wetlands for water pollution control.* Pergamon Press, Oxford, pp 33 - 40.

Mutalewich, V. A., Strom, P. F., and Finstein, M. S. (1975). Length of incubation for enumerating nitrifying bacteria present in various environments. *Appl. Microbiol. 29*: 265 - 268.

Mutalewich, V. A. and Finstein, M. S. (1978). Distribution of autotrophic nitrifying bacteria in a polluted river (the Passaic). *Appl. Environ. Microbiol. 35*: 67 - 71.

Ottavá, V., Balcarová, J. and Vymazal, J. (1997). Microbial characteristics of constructed wetlands. *Wat. Sci. Tech. 35* (5): 117-123.

Patrick , W. H. (1990). Microbial reactions of nitrogen and phosphorus in wetlands. *In:* van Oorschot, M. M. P. and Verhoeven, J. T. A. (eds.) *The Utrecht Plant Ecology News Report,* Utrecht, pp 52 -63.

Paul, E. A. and Clark, FE (1989). *Soil microbiology and biochemistry.* Academic press, Inc., San Diego.

Randal L. D. and Conway, T. E. (1989). Nitrification and denitrification at the Iselin marsh / pond / meadow facility. *In:* Hammer, D. A. (ed.). *Constructed wetlands for wastewater treatment.* Lewis Publishers, Inc. Michigan, pp 477-483.

Randal, M. J. and Ingraham, J. L. (1981). The denitrifying prokaryote. *In:* Mortimer, S. P; Stolp, H; Trüper, H. G. Balows, A. and Schlegel, H. G. *The Prokaryotes*, Vol I Springer-Verlag, Berlin. pp 913 - 925.

Reddy, K. R., Patrick, W. H . and Lindau, C. W. (1989). Nitrification-denitrification at the plant root-sediment interface in wetlands. *Limnol. Oceanogr. 34*: 1004 - 1013.

Reddy . K. R. and D`Angelo, M. E. (1997). Biogeochemical indicators to evaluate pollutant removal efficiency in constructed wetlands. *Wat. Sci. Tech., 35* : 1 - 10

Schmidt, E. L. and Belser, L. W. (1982). Nitrifying bacteria. *Methods of soil analysis*, Part 2: Chemical and microbiological properties- Agronomy Monograph no. 9, pp 1027 - 1042.

Shaher, D. Kochba, M. and Avnimelech, Y. (1993). Nitrification pattern in a fluctuating anaerobic-aerobic pond environment. *Wat. Res., 27* : 1469 - 1475.

Sikora, F. J., Tong, Z., Behrends, L. L., Steinberg, S. L. and Conrod, H. S. (1995). Ammonium removal in constructed in constructed wetlands with recirculating subsurface flow: removal rates and mechanisms. *Wat. Sci. Tech. 32* (3) : 193 - 202.

Tiedje, O. M. (1988). Ecology of denitrification and dissimilatory nitrate reduction to ammonia. *In:* Zehnder, A. J. B. (ed.). *Biology of anaerobic microorganisms.* John, Wiley and sons, Inc. New York, pp 179 - 242.

Underhill, S. E. (1990). Techniques for studying the microbial ecology of nitrification. *Adv. Microbiol. 22*: 417 - 445.

Van Oostrom, A. J and Russell, J. M. (1994). Denitrification in constructed wastewater wetlands receiving high concentrations of nitrate. *Wat. Sci. Tech. 29:* 7 - 14.

Verhagen, F. J. M., Duyts, H. and Laanbroek, H. J. (1992). Competition for ammonium between nitrifying and heterotrophic bacteria in a continuously percolated soil column. *Appl. Environ. Microbiol. 58*: 3303 - 3311.

Watson, S. W., Valois, W. F. and Waterbury, J. B. (1981). The Family Nitrobacteriacae. *In:* Mortimer, S. P., Stop, H; Trüper, H. G., Balows, A. and Schlegel, H. G. (eds.). *The Prokaryotes*, Vol I Springer-Verlag, Berlin. pp. 997 - 1004.

Williams, J. B., May, E., Ford, M. G. and Butler. J. E. (1994). Nitrogen transformation in gravel bed hydroponic beds used as tertiary treatment stage for sewage effluents. *Wat. Sci. Tech., 29* : 29 - 36.

Wittgren, H. B. and Tobiason, S. (1995). Nitrogen removal from pretreated wastewater in surface flow wetlands. *Wat. Sci. Tech.* 32: 69 - 78.

Chapter 8

Phosphorus retention in the sediments of the Nakivubo swamp

Nalubega, M.

8.1 Introduction

The role of sediments as the main sink or as a source of nutrients, notably phosphorus (P), and their importance in water quality management and eutrophication control have been extensively discussed by several authors (*i.a.*, Lijklema, 1980b; Håkanson and Jansson, 1983; Kelderman, 1985; Thomann and Müller, 1987). Sediments can be important P sources even in systems that have long ceased to receive external loads. Decreasing the external load can cause release from sediments due to re-equilibration to a lower solid phase concentration (Ku *et al.*, 1979). Indeed, the contribution of sediment release to the total phosphorus (TP) load in the aquatic system may exceed that of external sources (Auer *et al.*, 1993). In systems like the Nakivubo swamp that are under continuous nutrient loading, the role of sediments in nutrient retention or enrichment must be quantified before ample measures can be taken to protect adjacent water bodies; these measures must be based on a proper insight of nutrient fluxes and their control.

The most important physical characteristics of sediments that affect the internal P load in aquatic systems are grain size and its distribution, compaction and porosity, density of particles, vertical and horizontal inhomogeneities, and sediment mixing depth (Lijklema *et al.*, 1986), as well as organic matter, oxygenation state and the P content. In addition, nutrient release is a result of complex physical (advection, pore water diffusion, transport and mixing of solids by resuspension, sedimentation, gas ebullition and bioturbation), chemical (sorption, dissolution, co-precipitation and inclusion) and biological (mineralization) sediment-water interaction processes (Håkanson and Jansson, 1983; Lijklema, 1993).

Phosphorus can be exchanged from the sediment into the water phase or *vice versa*. The solubility of particulate P under prevailing pH, redox and ionic strength conditions is controlled by the chemical speciation of P present in the sediment and its interactions with other minerals or amorphous material in its vicinity (Hieltjes and Lijklema, 1980).

The sediment P retention function depends on the redox potential, solubility equilibria and the pH of the overlying water. Lijklema (1980a) showed that for freshly precipitated amorphous Fe(III) hydroxides, P adsorption reduced with pH and that this phenomenon is not completely reversible. The pH effect was interpreted as a competition between OH^- and $H_2PO_4^-$ for positive adsorption sites. The sediment concentration of potential adsorbents (containing Al, Fe, Mn, Mg and Ca oxides and their compounds) also affects P retention and for long term changes in P retention, the ratio of P loading to that of potential adsorbents is even more important than the P loading itself (Lijklema, 1986).

Phosphorus fixation onto sediments within a water body by adsorption and/or the formation of insoluble compounds is the only significant sink in aquatic systems for the conservative P.

Phosphate anions are taken up by for example kaolinite ($Al_4Si_4O_{10}(OH)_8$), montmorillonite ($Al_4Si_8O_{20}(OH)_4$) and by (especially) freshly deposited ferric and aluminium (hydr)oxides. The latter have a strong tendency towards chemical bonding between phosphate ions and the metal complexes.

Literature information on nutrient dynamics of sediments in **floating wetlands** is scanty (Gaudet, 1980; Cooke, 1992). For P, besides plant uptake, interaction with the solid phase components and precipitation are the major removal processes. Formation of refractory organic material results into a P sink. Microbial pools are reported to play a minor role in P storage but bacteria can play a major role in triggering P release from wetlands (Patrick, 1990).

Efficient P removal in wetlands is rather problematic (Kadlec and Knight, 1996). Biomass increases (without harvesting) should not be counted as part of the long-term sustainable P removal capacity of wetlands since virtually everything is recycled back into the system. In many natural, unharvested wetlands therefore, sediment accretion including some refractory P containing material may be the only permanent P sink.

Sediment sources in wastewater treatment wetlands are both allochthonous and autochthonous. Biodegradation and stabilization of settling matter (debris, litter, inorganic precipitates, etc.) yield new sediment material comprising of organic material of humic nature, and inorganic clays, with Fe and Ca as the leading constituents (Patrick, 1990). The resulting humic material does not contain P as a structural element but may adsorb P via interaction with Fe(III) or Al complexes. Humic substances (which are also typically chemical reductants) are in competition with P for Fe and therefore their presence may reduce P adsorption by Fe, thus altering the redox effects on P dynamics.

For the Nakivubo wetland that has received waste water for over 30 years, and with no foreseeable prospects for alteration of this state, it is necessary that the capacity of its sediments to receive and deplete nutrients will be quantified. This will help in defining optimum control of eutrophication in the Murchison Bay.

Objective of the study

The major objective of this study were to determine the contribution of the Nakivubo swamp sediments in the removal of phosphorus from the wastewater.

It was presumed that the continuous flow of nutrient-rich waste water along the major flow path in the swamp could lead to local sediment P saturation, while sediments in the rest of the swamp might have low P concentrations and high retention capacity. It was therefore hypothesized that the P content and exchange potential of sediments outside the major flow path will be substantially different from that in the major channel. Furthermore, it was expected that the back-flow of nutrient-poor lake water into Nakivubo swamp at irregular intervals (see Chapter 3) is likely to favour the enhancement of P release, and hence would cause a decreasing gradient of sediment P concentrations from the inlet towards the lake. Due to the high productivity of tropical swamps, it was hypothesised that there is a high rate of peat formation with subsequent burial, and that this is a P major sink.

To test the above hypotheses, the autochthonous and allochthonous sediment loads into the water system were estimated and the kinetics of the sediment-P equilibria simulated in the laboratory. The different P species were also quantified to estimate the release potentials. Selected sediment characteristics (nutrient content and release rates, mineral content, fine to coarse ratio, vertical sedimentation flux) at different locations in the swamp were compared to assess their role in, or response to differential hydraulic and P loading characteristics across and along the swamp, as well as to the different vegetation types.

8.2 Materials and methods

8.2.1 Study area and sampling sites

Sediment dynamics were studied for the lower Nakivubo swamp, hereafter called the Nakivubo swamp (for locations, see Fig.8.1). In this study, sediments refer to the material within 40 cm above the transition between loose peat material (typically 5 - 40 cm thick) and firm sediment, and 30 cm below this level. The zero level is assigned to the transition. Layers above are assigned a positive value while negative values are given to the layers below.

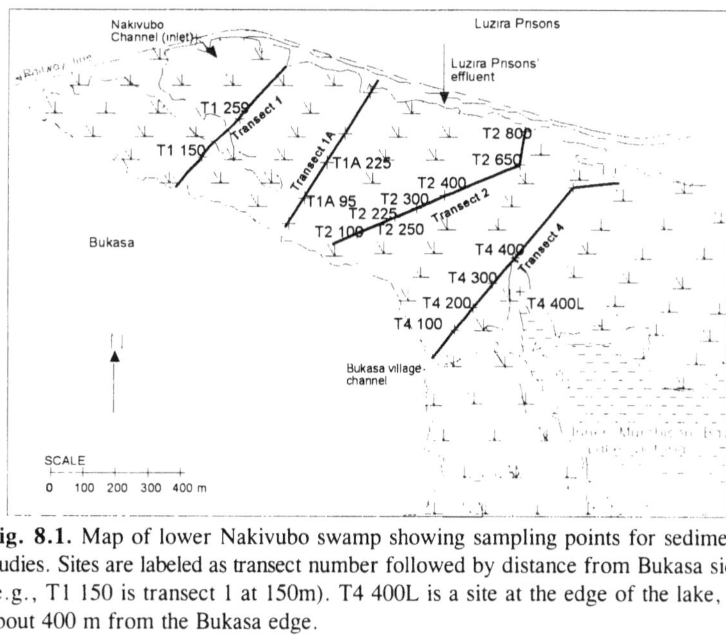

Fig. 8.1. Map of lower Nakivubo swamp showing sampling points for sediment studies. Sites are labeled as transect number followed by distance from Bukasa side (e.g., T1 150 is transect 1 at 150m). T4 400L is a site at the edge of the lake, at about 400 m from the Bukasa edge.

A detailed account of sediment profiling was presented in Chapter 3. The sediment depths and thickness of the mat/peat layer had an impact on the flow rate of water in the swamp and in turn on the sedimentation/resuspension fluxes.

8.2.2 Sediment load into the Nakivubo swamp

i) Allochthonous loads

The settleable solids (SS) load into the Nakivubo swamp was measured from Nakivubo Channel inlet by collecting 20 l of water during rainy seasons and dry seasons and allowing it to settle for 24 hours before decanting and drying the residue in the oven at 103 ± 2 °C for dry weight determination. On some occasions, the SS load was determined direct by filtration of 500 ml, and determining the change in weight of the filter papers (Schleicher and Schuëll, acetate fibre nr. 595) after oven-drying for 24 hours at 103 ± 2 °C.

ii) Autochthonous load: sediment traps

The sediment load generated by the continuous decay of the floating mat, overlying vegetation and the transported material was measured using sediment traps (see Fig. 8.2). The traps were 4 cm in diameter and 30 cm long (aspect ratio = 7.5, in accordance with the accepted value of >4; Håkanson and Jansson (1983). Each trap set consisted of 4 cylinders (for one tier traps) or 8 cylinders (for two tiers as in Fig. 8.2). The traps were attached to a galvanised iron pipe of adjustable length. The traps were kept vertical by driving the pipes deep into the firm clay

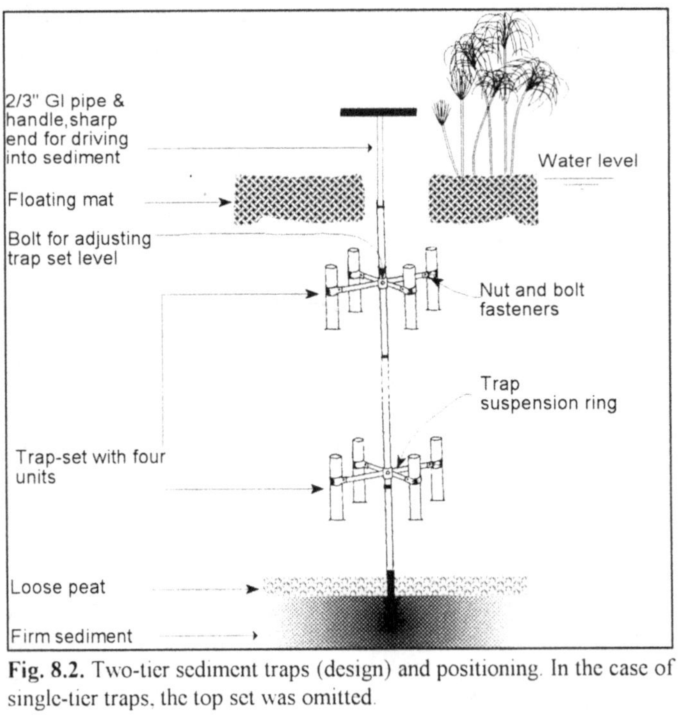

Fig. 8.2. Two-tier sediment traps (design) and positioning. In the case of single-tier traps, the top set was omitted.

sediment. Quadriplicates were used per trap tier to: (a) enable collection of an adequate quantity of material for analysis without excessively increasing the diameter and hence length of the tubes; and (b) investigate standard deviations in the quantity of settling material per trap tier.

Single-tier traps were initially located at the inlet, T1 150, T2 100, T2 400, T4 200, T4 400 and at the swamp lake interface near T4 400, designated as T4 400L (Fig. 8.1). Due to vandalism at the inlet and uncertain results at T2 100 (due to significant mat disturbance during sample collection), the two locations were abandoned and double-tier traps instead introduced at T2 250 and T4 400. It was expected that the top layer would collect local contributions from the mat above while the bottom layer would collect combined local and transient particulate matter in the travelling water. It is noted that if there was adequate vertical mixing in the system, then there would be no difference in the material collected at different tiers. The lower tier was maintained as for single-tier traps while the upper tier was placed just beneath the mat. Trap placement details for single-tier traps are given in Table 8.1.

Overlying water in sediment traps was collected in acid-washed plastic bottles and transported to the laboratory for TRP (Total Reactive Phosphorus or orthophosphate) and NH_4-N analysis. Analysis was done with the HACH DR 2000 spectrophotometer according to APHA (1992) as discussed in Chapter 4.

Table 8.1 Profile characteristics of trap locations for single-tier traps.

Location of traps	Mat thickness (m)	Trap tier above sediment (m)	Total water depth to sediment (m)
T1 150	0.50	0.45	0.96
T2 100	0.46	0.57	1.12
T2 250	1.40	0.33	2.53
T2 400	1.50	0.67	2.52
T4 200	0.53	0.47	1.75
T4 400	1.06	0.37	3.00
T4 400L	_	0.55	3.50

Between December 1995 and February 1996, traps were emptied at weekly intervals and at about monthly intervals between March and August 1995. Samples of supernatant water in the traps were analysed for pH, ortho-P and ammonia. The wet sediment volume was determined after which the trapped material settled in the dark for 24 hours at 22 - 24°C. The supernatant water was then siphoned off and the material was oven-dried at 103±2°C to constant weight, yielding the sedimentation flux ($g/m^2/d$). Sub-samples were ignited in a muffle furnace at 550 ±2°C for the determination of the organic matter content while 0.3 g was digested at 330°C on a TECTOR aluminium block using a sulfuric-salicylic acid mixture with selenium as catalyst according to the procedure of Novozamsky *et al.* (1983). Sodium and potassium were determined using flame photometry while Mg, Ca and Fe were read using atomic absorption spectrometry in the Soil Science Department of the Faculty of Agriculture-Makerere University, Kampala.

8.2.3 Sediment profiles and characteristics

Sediments were obtained using an Eijkelkamp multisampler (1 m long; 0.04 m internal diameter) which could extract up to 1 m depth (depending on the stiffness of the material) of sediment and overlying water. The sampler enabled extrusion of undisturbed and intact sediment cores. Depending on the required test, the sediment was immediately sliced using a stainless steel kitchen knife and stored in polythene bags (to maintain the moisture content), before transportation to the laboratory for subsequent tests. Any sediment that was not being analysed immediately was kept in the refrigerator at 4 °C.

Triplicate sediment cores were obtained from 2 locations along T1 (T1 150 and T1 259), 5 locations along T2 (T2 100, T2 250, T2 400, T2 650 and T2 800) and 5 locations along T4 (T4 100, T4 200, T4 300, T4 400 and T4 400L; Fig. 8.1). The cores were sliced into 5 to 20 cm thick layers, with the zero level assigned to the transition zone between the loose peat material and the consolidated material (firm sediment) as discussed in Section 8.2.1. Typical layers (wherever possible) were 40 to 20, 20 to 0, 0 to -5, -5 to -10, -10 to -20, and -20 to -30 cm. Analyses included the following:

i) *Moisture content, MC*

$= \frac{wet\ weight\ -\ dry\ weight}{dry\ weight} * 100\ (\%)$, after oven-drying to constant weight at 103±2 °C.

ii) *Loss on ignition, LOI*

$= \frac{dry\ weight\ -\ ignited\ weight}{dry\ weight} * 100\ (\%)$, after igniting the oven-dried samples at 550±2 °C for 4 hours, on pre-ignited glass dishes. The strong correlation between LOI and particulate carbon content (C) for sediments found elsewhere (Peck *et al.*, 1980; Håkanson and Jansson, 1983; Psenner and Pucsko, 1988) was used to estimate the carbon content of the sediments here. From their results and from literature, Håkanson and Jansson (1983) concluded that C can be estimated as LOI/2 if LOI is greater than 10%.

iii) *Digestion and analysis of samples for N, P and metallic content*:

About 0.3 g of the oven-dried samples was digested using a sulphuric acid-selenium mixture according to Novozamsky *et al.*(1983). The digested samples were analysed for N, P, Na, K, Mg, Ca and Fe as discussed before.

iv) *Particle size analysis:*

Apart from the organic debris from decaying plant parts, and the sandy edge samples (Chapters 2 and 3), the swamp sediment particle size was always smaller than 2 mm. The percentage of fine material passing through the 75 μm mesh sieve was determined for a known weight of sediment sub-sample by wet-sieving (US Dept. of Interior, 1969; Klute, 1986). The residue was oven dried at 103$\pm$2 °C. By considering the moisture content of a parallel sample, the weight of the portion finer than 75 μm was deduced. This gave an indication of soil permeability and adsorbent capacities (the smaller the particle size, the higher the surface area) of the soil.

Particle size distribution of fine materials (less than 75 μm) for selected samples was determined by the **hydrometer method**, according to ASTM (1986) and Klute (1986). The hydrometer type used was 152H. The method is suitable for clays and fine silt, which constituted the bulk of the Nakivubo swamp sediments. The mean particle diameter, ***d*** [mm], was calculated from:

$$d = \sqrt{\frac{300 \ x \ \eta \ x \ h}{980 \ x \ (G_s - G) \ x \ t}} \tag{8.1}$$

where:
- η = coefficient of viscosity of water [N s/m^2]
- h = effective depth [cm] deduced from hydrometer reading (ASTM, 1986)
- G_s, G = specific gravity of particles and water, respectively [-]
- t = time of reading after start of sedimentation [s]

The percentage of soil remaining in suspension, ***P*** (% passing a corresponding ***d***) was calculated from

$$P = (Ra/W) \ x \ 100 \tag{8.2}$$

where:
- R = hydrometer reading with composite correction applied [g]
- a = correction factor for the specific gravity of the sediment [-]
- W = oven-dry mass of soil [g]

The clay fraction $P_{2\mu m}$ (%) was estimated from the particle size distibution curves as the fractions corresponding to $\leq$ 2 μm.

8.2.4 Determination of sediment P sorption kinetics and isotherms

Sediment samples were collected as described in 8.2.3 from sites in the major waste water flow path (T1 259, T2 400, and T4 400) and from the less affected areas (T2 100, and T4 100). The latter 2 sites and T1 259 are papyrus vegetated zones with a thicker peaty material layer while T2 400 and T4 400 are *Miscanthidium* vegetated zones with a very thin peat layer (Chapter 3).

To determine the optimum experimental conditions for assessment of the sorption rates and equilibria in the ranges representative for field conditions, about 50 samples containing 0.5 g of dry sediment and different concentrations of P (in 136 ml glass vials) were studied over time. The concentration of ortho-P in the initial solution ranged from 0 to 2.0 mg P/l in 0.01 M $CaCl_2$. Sediment samples from T4 100 (0-5 cm depth) were used in duplicate. A second set of experiments was conducted with 0.25 g of sediment from the same location, with initial P concentrations ranging from 0 to 5 mg P/l. All sample vials were tightly closed and incubated (T=22 ± 2 °C) in the dark.

The sediment samples from the 5 locations were subdivided into slices from 0-5 and 5-10 (cm) depth, were air-dried, pulverized and homogenized. Glass vials (136 ml) with a rubber cap and an aluminium seal were used in all experiments. To enhance settling of the fine clay particles, 0.01 M $CaCl_2$ (Reddy *et al.*, 1980) in distilled water was used. A one minute per day vigorous manual shaking (for intensive contact between the particels and P in the solution) was applied throughout all the experiments. More frequent or prolonged shaking is reported not to affect the results (Stirling and Wormald, 1977). Loosely bound P was determined in consecutive extractions. Triplicate samples were shaken for one minute and left to settle for at least 6 hours. 100 ml of the supernatant was taken for analysis of ortho-P and replaced with distilled water. This procedure was repeated until no more P could be extracted from the sediment.

Adsorption isotherms were determined in duplicate for all locations using 0.25 g of air-dried sediment material with initial P concentrations ranging from 0 to 10 mg/l in 136 ml of distilled water plus 0.01 M $CaCl_2$ (initial pH adjusted to 6.8 - 7.0).

8.2.5 Nutrient release from sediments

i) Anoxic Batch release experiments

Two sets of batch release experiments (Table 8.2) were carried out underthe given conditions, with sediments from different locations. Phosphorus and nitrogen release rates were studied on sediment columns from 5 sites in the swamp. Rsults will be presented for two locations along the major water flow path in the swamp (T2 400 and T4 400) a well as for T4 400.

Table 8.2 Experimental framework for batch nutrient release experiments. EC unit is μS/cm.

Sample sources (no. of replicates)	Overlying water	Sampling schedule	Test hypothesis
I. T4 400 Papyrus (6) and T4 400 *Miscanthidium* (6)	3 of each topped with swamp water (EC ≈ 360, pH ≈ 7.5), 3 with diluted swamp water or lake water (EC ≈ 85, pH ≈ 7.5).	Daily for 8 days	Difference between nutrient release from papyrus and *Miscanthidium* sediments at the same location (T4 400) with swamp or lake water.
II. T2 250 (6) T2 400 (6)	3 with diluted swamp water (EC ≈ 188, pH ≈ 7.5), 3 filled with 'lake' (swamp diluted with tap) water (EC ≈ 115, pH ≈ 7.5).	6 hourly for 3 days, and after 5 days	Difference between release rates for inner swamp sediments receiving both waste water and lake water

Intact, undisturbed sediment columns were collected and transported to the laboratory where the overlying water was replaced. The columns (Fig. 8.3) were closed at the top and bottom after inserting a tube for sampling and a stirrer, and stored at room temperature. The overlying water was gently mixed some 3 minutes prior to sampling, leaving the sediment-water interface intact. The volume removed for sampling was replaced with the type of water originally added to the column. Samples were analysed for total P and NH_4-N. No artificial aeration was done.

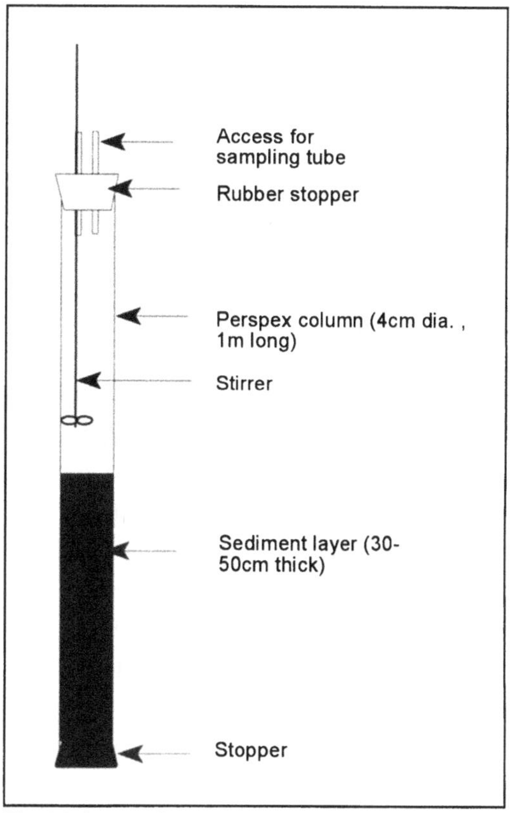

Fig. 8.3. Set-up of column experiments for batch nutrient release from sediments.

ii) Continuous anoxic P release experiments

A continuous flow system was prepared to simulate the effect of continuous water flow conditions on P release. 200 g of sediment material from T1 259, T1A 95, T1A 225, T2 70, T2 450, T4 100 and T4 400 was put in 1 l glass bottles to which water was added. A continuous flow setup

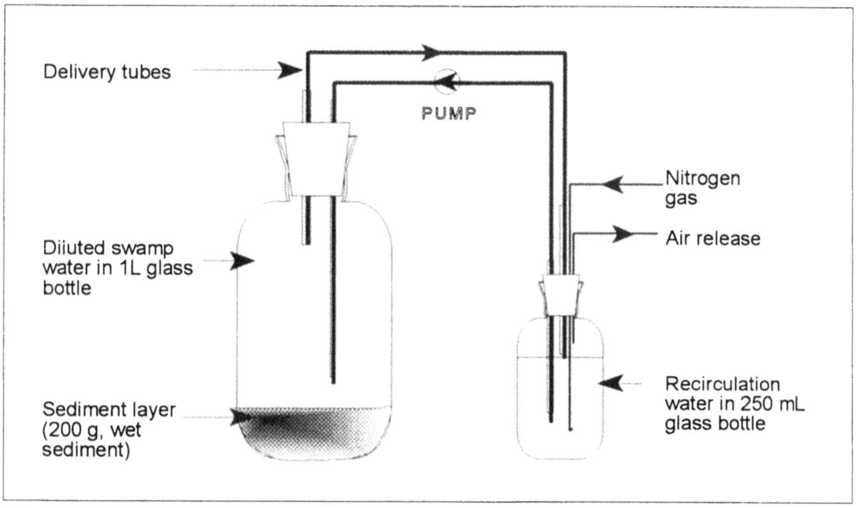

Fig. 8.4. Set-up of anoxic P release experiments with continuous flow.

in which nitrogen gas was bubbled continuously to expel air (Fig. 8.4) was arranged. After 2, 4.5, 17, 26 and 42 hours, water samples were taken and the P content and redox potential of the overlying water was determined.

The flow rate in the system was 156 ml/h giving a retention time of about 6.5 hours. Fifty ml water samples were drawn from the sediment bottles and replaced with equivalent amounts of nitrogen flushed initial water.

8.2.6 Phosphorus fractionation

In order to assess the different P binding forms of the sediments in the swamp, fractionation experiments were performed. Pooled sediment samples were taken from T1 259, T1A 95, T1A 225, T2 70, T2 450, T4 100 and T4 400, by extracting 5 cores from each location, putting all the top 10 cm sediment portions together, followed by homogenization. Prior to analysis, the wet sediment samples were stored in a refrigerator at 4°C. Samples were analysed in triplicate, following an adaptation of the Furumai and Ohgaki (1982) procedure shown in Fig. 8.5.

The method facilitated the determination of organic P as the difference between TP and TRP in the main extraction and follows a parallel procedure for total P determination. Scheme A

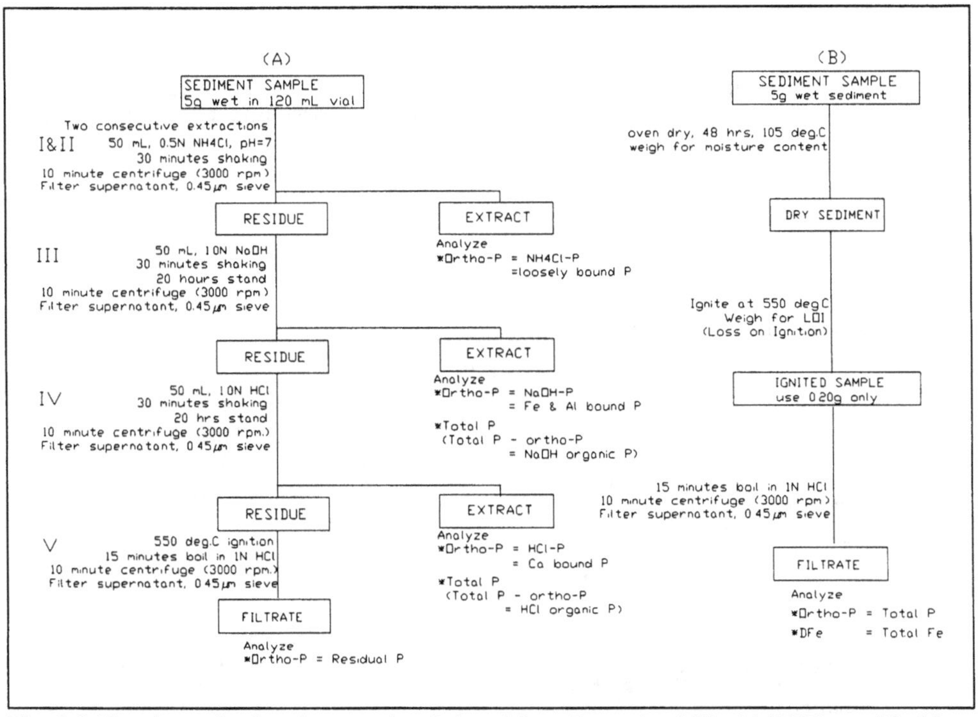

Fig. 8.5. Phosphorus fractionation procedure [adapted from Furumai and Ohgaki (1982), modified from Chang and Jackson (1957), Williams (1967) and Hieltjes and Lijklema (1980)] -Scheme A. Scheme B is for the determination of dry weight, organic matter content and total P and Fe of parallel samples.

allows to fractionate loosely-bound P, Fe and Al-bound P and Ca-bound P. Scheme B is the parallel scheme for TP determination where a known weight of the ignited sample was boiled in 30 ml of 1N HCl for 15 minutes, filtered and the filtrate was analysed for TP and Fe.

8.2.7 Sediment resuspension

Resuspension experiments were done using a model 'micro flume' of dimensions 60cm x 14.5cm x 11cm (Fig. 8.6). The inlet baffle was designed to ensure nearly laminar flow; this was confirmed by a fluorescent dye tracer test.

Characteristic sediment samples were collected from T2 100 (papyrus zone) and T2 400 (*Miscanthidium* zone). About 2 cm thick samples were placed in the flume and left to settle. When the overlying water was clear again (at least after 24 hours), tap water was allowed to flow into the system, initially at a very low but steady rate. The outflow volume was recorded and the SS in the effluent determined. The results were analysed for minimum (critical) resuspension velocity. Similar measurements were subsequently made with increasingly higher

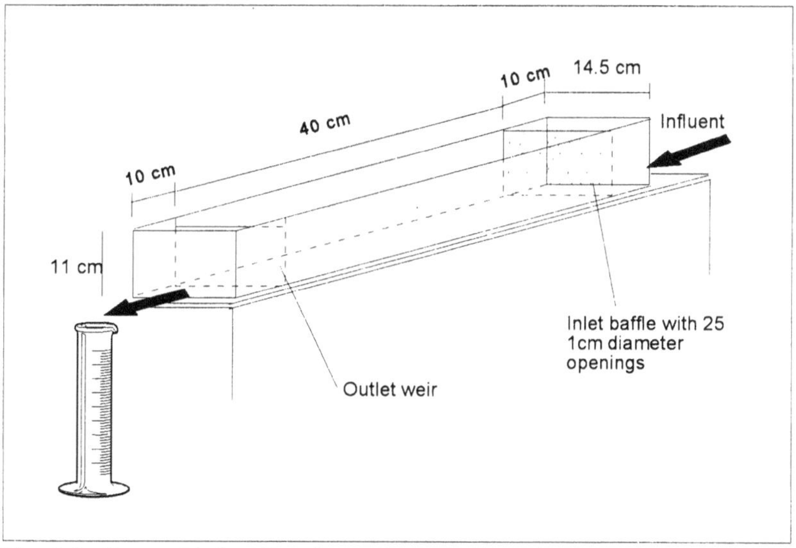

Fig. 8.6. Setup of 'micro-flume' experiment for determination of critical sediment resuspension velocity.

velocities.

8.2.8 Statistical data analysis

All data were analysed using MINITAB statistical package (Release 10.1, 1994). Differences between variable means within and among different locations were tested using ANOVA. Data was pre-tested for normality and homogeneity of variances. Non-linear regression was done using SPSS statistical package. Unless otherwise stated, means are presented $\pm$ 1 Standard Error of the mean [SE].

8.3 Results

8.3.1 Sediment distribution and particle size

The spatial variation in swamp sediment characteristics depends on both the overlying vegetation type (Chapter 2) and the water flow characteristics (Chapter 3). These, in turn, depend on the location relative to the swamp edge, the lake and the major waste water sources into the swamp. From observations made during the study, four sediment types are identified:

(i) the rich loam soils (up to 50 cm thick) overlying stiff grey clays, near the main waste water inlets namely after location T1 150 and the Luzira effluent diffuse point (Figure 8.1). Beyond T1 250 and near the Luzira effluent, these soils are overlain by *Enhydra fluctuans*, a very fast-growing weed that appeared to thrive in nutrient-rich water and soil. Occupancy is about 2% of the study area;

(ii) the sandy clays within 50 m of the eastern and western swamp edge. These are the areas from which clays and sands are mined for building purposes;

(iii) grey shiny clays underlying up to 50 cm of loamy ferruginous, quite well decomposed peat in areas vegetated by papyrus. This is about 70% of the study site;

(iv) grey shiny clays underlying very thin strips of poorly decomposed plant debris in areas vegetated by *Miscanthidium* and its associates. This is about 20% of the study site.

The peaty material under both Miscanthidium and papyrus vegetation zones is unconsolidated and susceptible to erosion. Water flow velocities therefore may play a big role in the accumulation and/or transport of this material.

Characteristics of the peat layer are also related to the thickness and type of the overlying mat. Areas vegetated by papyrus have generally thinner mats whose decaying plant debris freely falls through the rhizomes to form a very peaty water column beneath. In *Miscanthidium*, dead plant material is retained at the surface of the mat since it cannot fall through the thick mat tuft (Chapter 5). The material that falls to the bottom of the swamp is the more decomposed debris that is slowly eroded from the mat bottom. This results in a very thin peat layer of different composition and a much clearer water column. A schematic presentation of the sediment zones that play the leading role in the waste water nutrient dynamics in the Nakivubo swamp is shown in Fig.8.7. The figure shows the flow path of particulate material and P along a gradient from the swamp inlet to the lake.

Further discussion on the spatial variation of sediment depth and peat thickness has been presented in Chapter 3 (Section 3.3.1 and Figures 3.3 and 3.7).

Swamp "firm" sediments are silty clays with 90% of all sediment material finer than 75 μm (except for the edge cores at T2 650 where it is >75%). The particle size distribution for the

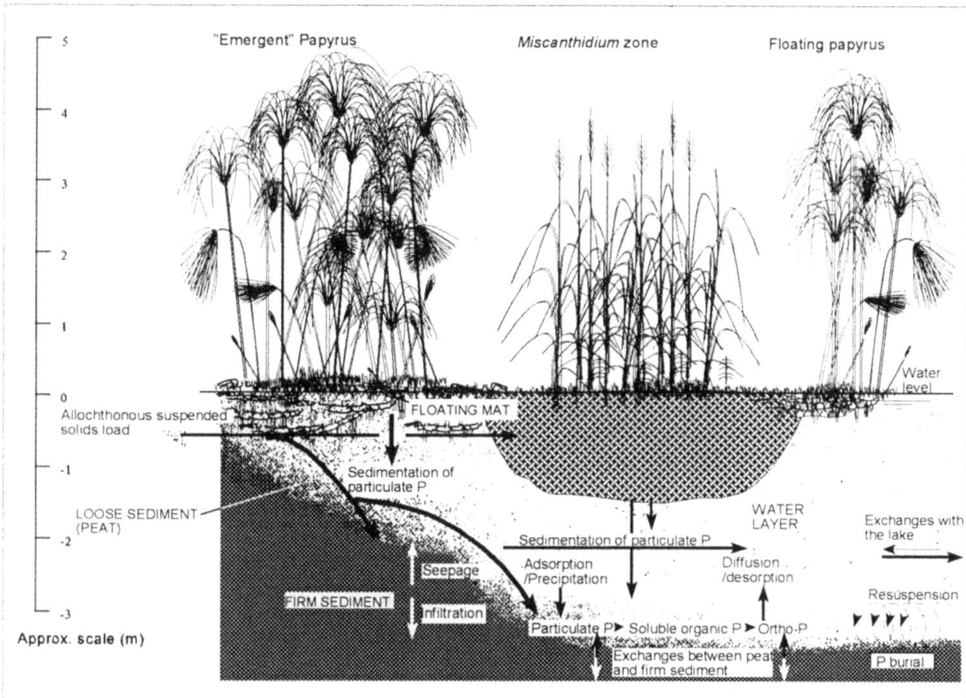

Fig. 8.7. Vertical profile of the lower Nakivubo swamp showing the allochthonous and autochthonous sediment (and P) pathways along a longitudinal profile from the railway bridge to the bay (left to right). Note the proportions of the different compartments (mat, water, peat and firm sediment) for the edge papyrus, floating papyrus and *Miscanthidium* vegetated zones. Not drawn to horizontal scale.

materials analysed for T2 and T4 is shown in Figures 8.8.a and b. The effective size (d_{10} corresponding to 10% smaller) for all locations but T4 300 is < 1 μm. At T4 300, d_{10} is about 3 μm. Particle specific densities ranged between 1830 kg/m^3 (in the swamp interior) and 2450 kg/m^3 (mostly along T4). The tendency was for larger particles to be closer to the edges (because of the sandy material) and closer to the lake (possibly due to the washout of finer material by higher flows/velocities). This implies that the adsorbed P content of these sediments (as a function of adsorption sites on the basis of surface area) would also be lower.

8.3.2 Sediment loading into the swamp

i) Allochthonous loads into the lower Nakivubo swamp

The average concentration of settleable suspended solids (SS) at the swamp inlet (measured at the railway culvert) amounted to 19 $\pm$ 2 mg/l (n = 28; range 6 - 44 mg/l). For a mean daily flow via the Nakivubo channel of 103,575 m^3/d (Chapter 3), the allochthonous load into the swamp thus averages 1970 kg/d (720 tons/year). Due to the large flows involved and the direct flow of its water through the centre of the swamp, the Nakivubo Channel is the largest source of allochthonous SS load into the Nakivubo swamp. By virtue of their location and because of the emergent vegetation at the edges of the swamp (Chapter 2), the contribution from Luzira

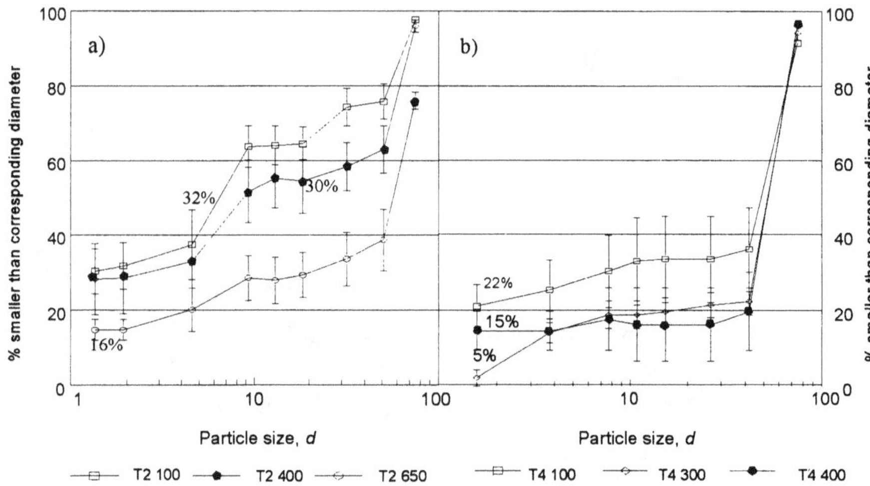

Fig. 8.8. a. and b. Sediment particle size (d in μm) distribution for samples T2 (a) and T4 (b) (T2 and T4, respectively). The vertical axis represents the proportion of particles that are smaller than a corresponding particle size. For the exact locations, refer to Figure 8.1. The clay fractions are marked as percentages on the corresponding lines. Vertical bars indicate standard errors (n=3).

and the storm water runoff of Bukasa and Luzira hills are effectively retained by the ground cover of the watershed and the thick vegetation at the swamp edge. It was therefore assumed that this water does not join the major waste water path in the swamp water column.

ii) Autochthonous loading rates

(a) Sediment fluxes

The mean material deposition rates measured by sedimentation traps between December 1995 and August 1996 are shown in Table 8.3. Site T2 100 has a papyrus mat with typically loose peaty material and a very turbid 'swamp' water column. T4 400 (the lower trap) occasionally contained much decaying organic matter, presumably brought into the swamp from the water hyacinth-laden lake by waves and seiches leading to high standard errors. Apart from these extreme conditions, the material flux ranged between 31 and 88 $g/m^2/d$.

The lowest flux for all substances was observed at T2 400, a site with a thick floating mat of *Miscanthidium* (Fig. 8.1) where the local flow rate is moderate with respect to both the inflows and the lake seiches effects compared with either the inlet or T4 400 near the lake. Where the peat in and beneath the papyrus mat is more fixed (that is, not easily disturbed by movement around the system like at T1 150 and T4 200), the sedimentation rates in the *Miscanthidium* vegetated zones are lower than in the papyrus zones. The values for T2 400 (31 $g/m^2/d$) can be taken as the typical deposition rates for *Miscanthidium* zones while the average of T1 150

Table 8.3 The mean material deposition rates (vertical material flux) in the Lower Nakivubo swamp. Values are: means ± 1 Standard Error of the mean, SE.

Location	Material flux (g DW/m^2/d)
T1 150	55 ± 12 (n=6)
T2 100	435 ± 161 (n=5)
T2 225 upper	65 ± 9 (n=5)
T2 225 lower	74 ± 10 (n=9)
T2 400	31 ± 7 (n=10)
T4 200	55 ± 12 (n=10)
T4 400 upper	70 ± 27 (n=4)
T4 400 lower	174 ± 82 (n=8)

and T4 400 (55 g/m^2/d) will be taken as the undisturbed sedimentation rates for papyrus zones.

Considering the above material flux range for the whole study area (≈ 1,150,000 m^2, Chapter 2), the "autochthonous" material deposition in the swamp would average 7,100 kg/d for the *Miscanthidium* zone and 50,600 kg/d otherwise. This quantity, that either sinks to the bottom of the swamp or is washed away by flowing water, is thus about 30 times larger than the allochthonous load. The swamp mat and vegetation are therefore the greater sources of swamp sediment.

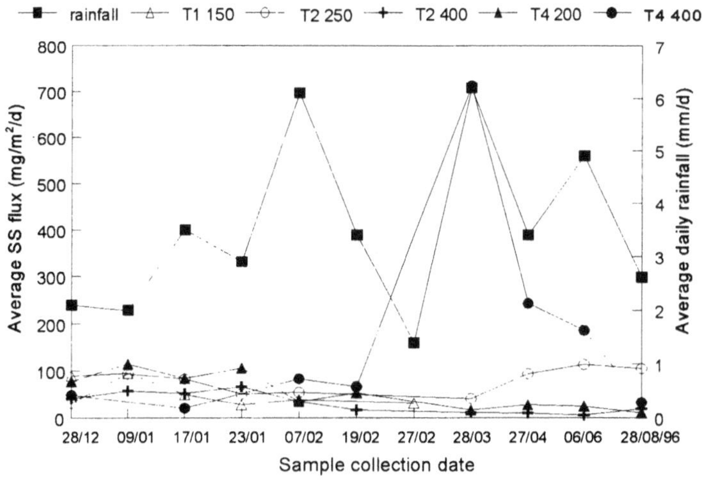

Fig. 8.9. Sediment fluxes into lower traps in the Nakivubo swamp in relation to average period rainfall (December 1995 - August 1996).

(b) *Seasonal variation of sediment fluxes*

Figure 8.9 shows the relationship between the sediment flux and rainfall at different locations within the swamp. The seasonal variation in deposited SS loads for sites within and outside the major waste water flow path showed negative correlation with rainfall for all sites except T4 400 ($r=0.69$, $n=11$). Papyrus vegetated sites had strong negative correlations while T2 400 and T2 250 had poor negative correlations (-0.47 and -0.10, respectively). The negative correlation could imply a higher dilution of the suspended solids in the water column as opposed to bottom scouring in the papyrus areas while for the area nearer to the lake (T4 400), higher water transport definitely increases the resuspended matter (hence the positive correlation).

(c) *Vertical distribution of SS in the water column*

A comparison of the sediment fluxes into the upper and lower-tier traps when the samples were measured simultaneously ($n=4$), gave 65 ± 9 and 88 ± 16 gDW /m^2/d, respectively, at T2 250. T4 400 values were 70 ± 27 and 248 ± 122 gDW /m^2/d, respectively. The upper trap load at T2 250 was therefore 74% of the lower trap load while at T4 400, it was only 28% (note high SE). The higher material flux for the lower traps could be attributed to scouring and transport of bottom material due to the higher flows. The large variation in sediment fluxes close to the lake (T4 400) may be due to flow differences as a result of seiches and the water hyacinth. T2 250 showed less variations in sediment fluxes because it is further from the lake and off the centre of the major flow path. It is therefore less affected by the water hyacinth and seiches.

iii) Sediment accretion rates

An estimate of the sediment accretion and decomposition/erosion rates in the swamp was made by considering that the measured mean material deposition rates gave the gross flux. From paragraph 8.3.2.2 (Table 8.3), it was shown that the average sedimentation rate was 55 g/m^2/d for papyrus areas (T1 150 and T4 200). Similarly, for the *Miscanthidium* location T2 400, 31 g/m^2/d was found. These prototype values for locations least disturbed by sampling or the water hyacinth in the lake were used.The estimated bulk density (dry weight/24 hours settled volume) of the sediment trapped under the mat was respectively 64 kg/m^3 and 42 kg/m^3. These values give gross sediment accumulation rates of typically 0.31 m/yr and 0.27 m/yr under papyrus and *Miscanthidium,* respectively. So if no transport / decomposition takes place, the swamp would be silted up in about 6 years.

Sediment core observations however showed average peat thickness layers of only about 0.5 m for papyrus zones and 0.05-0.10 m for *Miscanthidium.* These low values suggest that the annual net sediment accretion rate is much less than the above gross figures. Considering the depth profile of P at T1 150, about 60 cm of consolidated sediment and peat has accumulated since the diversion of the channel (about 16 years ago). This gives a sediment accretion rate

Table 8.4 Average 2-year water quality variables for the lower Nakivubo swamp at sediment source sites. Values are: mean ± 1 standard error (number of replicates). Note: TRP (water) and sediment P (sections 8.3.3.2) are poorly correlated (r=0.082).

Location	TRP (mg/l)	EC (μS/cm)	Dissolved Oxygen (mg/l)	Temperature (°C)	pH
Inlet	1.56±0.05 (50)	437±4(36)	0.22±0.04(15)	23.1±0.1(32)	6.8±0.0(20)
T1 150	0.17±0.05 (4)	220±34(5)	0.18±0.03(3)	21.1±0.2(3)	-
T1 259	1.55±0.10 (21)	425±11(14)	0.64±0.17(5)	23.1±0.1(7)	6.8±0.0(2)
T2 100	0.20±0.03 (4)	219±0(15)	0.17±0.01(7)	22.6±0.0(8)	6.2±0.0(3)
T2 225	1.52±0.06 (41)	411±4(11)	0.13±0.05(3)	22.9±0.1(11)	6.5±0.0(9)
T2 400	1.49±0.05 (55)	409±3(16)	0.00±0.00(8)	23.3±0.0(10)	6.7±0.0(9)
T2 600	1.49±0.05 (16)	409±6(9)	0.07±0.01(3)	22.9±0.2(4)	6.6±0.1(3)
T4 100	0.57±0.08 (8)	266±6(12)	0.08±0.02(5)	21.4±0.2(5)	5.8±0.0(5)
T4 200	0.71±0.04 (43)	320±6(12)	0.02±0.02(8)	22.3±0.2(5)	6.3±0.1(8)
T4 400	1.10±0.05 (66)	352±15(15)	0.3±0.2(11)	22.9±0.2(11)	6.5±0(11)

of about 0.04 m/year. The discrepancy can be attributed to incidental high erosion rates during severe storms.

8.3.3 Chemical composition of the Nakivubo swamp sediments

8.3.3.1 Influence of water quality

The overlying water quality is expected to affect the P sorption characteristics. Detailed water quality analysis in the lower Nakivubo swamp was discussed in Chapter 4. A summary of average 2-year physical characteristics and TRP in the water column at different locations is given in Table 8.4. Very low DO concentration and a near neutral pH of the water prevail throughout the swamp while high nutrient concentrations and EC are found along the central waste water flow path.

The pH of water strongly influences P adsorption especially as it controls the solubility of P bound to Fe, Al and Ca. In a low-calcium alkaline medium (as in shallow productive lakes where pH may rise to more than 9 during photosynthesis), P adsorbed earlier onto Fe and Al hydroxides will be substituted by hydroxyl ions and returned into solution. Formation of the hydroxyl-apatite P mineral may lead to phosphate sorption by calcium (Lijklema, 1980; Håkanson and Jansson, 1983; Stumm and Morgan, 1996). In the Nakivubo swamp, where the daily pH changes and ranges are limited (probably because of the vegetative cover that inhibits algal growth), pH mediated release may not occur. The pH range of the water (5.8 - 6.8) favours P deposition on Fe(III) and Al(III) (hydr)oxides. At these conditions, the $H_2PO_4^-$ is likely to be the dominant ligand (Stumm and Morgan, 1970) in the Nakivubo swamp. There is also likely to be much organic P. The highly anaerobic swamp sediments and the low redox potential of the water over the whole swamp (typically -200 to -456 mV along T2 and T4), however, favour the release of Fe bound P due to Fe reduction (Stumm and Morgan, 1970). Temperature is almost constant throughout the year and its variations are considered not to affect the dynamics of P sorption.

8.3.3.2 Chemical content of bottom sediments

i) Spatial variation of chemical content

The major ions of interest were the P-adsorbing metals Fe, Ca and Mg, and the nutrients N, P and C. Al was not determined due to analytical limitations. The Ca and Mg content of sediments was very low (less than 0.06 mg/g dry weight except at T4 200 and T4 400 where the average Ca was 0.1 and 0.6 mg/g respectively) and are not presented further. The spatial variations in the chemical characteristics of sediment material, and the classification of sediments whose means are not statistically different from each other within each graph, are shown in Fig. 8.10. Sites with similar levels are denoted by letters a, b, c and d.

No clear trend in N and C concentrations either in the major waste water flow path (from the inlet to the lake) or across the transects was seen, although concentrations tended to be lowest near the edges and under the *Miscanthidium* vegetated zone. However, a positive correlation between N and C ($r=0.653$, $n=135$) suggests similar sources and pathways. Both N and C are fairly uniformly distributed in the longitudinal axis but have a strange pattern in the cross-section.

Iron depicted a definite negative gradient (-0.012 gFe/m) from the swamp inlet towards the lake ($p<0.001$). Phosphorus was higher near the inlet and in the centre of the swamp. Lowest values were found close to the lake (T4 400). Across the transects, concentrations remained fairly constant. The higher Fe contents close to the inlet may be a result of allochthonous contributions while the lower values nearer the lake may be partly due to the constant flushing and dilution of the solubilised Fe^{2+} by lake seiches. This also may be responsible for the lower P content closer to the lake. The higher P content of the sediments at T4 100 compared to, e.g. T4 400 are coupled with much lower N and C content as seen from the C:N:P ratios in Table 8.6, and not to the Fe content since this is the same.

ii) Nutrient variation over depth

Results for nutrient and chemical characteristics over the depth showed a significant reduction ($p \leq 0.001$, $n=112$) in LOI (or carbon content) from 47 $\pm$ 6% in the top 10 to 20 cm layer to 21$\pm$2% in the -20 to -30 cm layer. The organic matter content of the peat layer was significantly different from that of the firm sediment ($p< 0.001$, $DF=112$). Similar observations were made for N (peat: 9.9 $\pm$ 1.4 mg/g; sediment: 5.0 $\pm$ 0.5 mg/g) and for Fe. However, there was no significant difference in particle size, P and Na among the layers (at $p<0.05$).

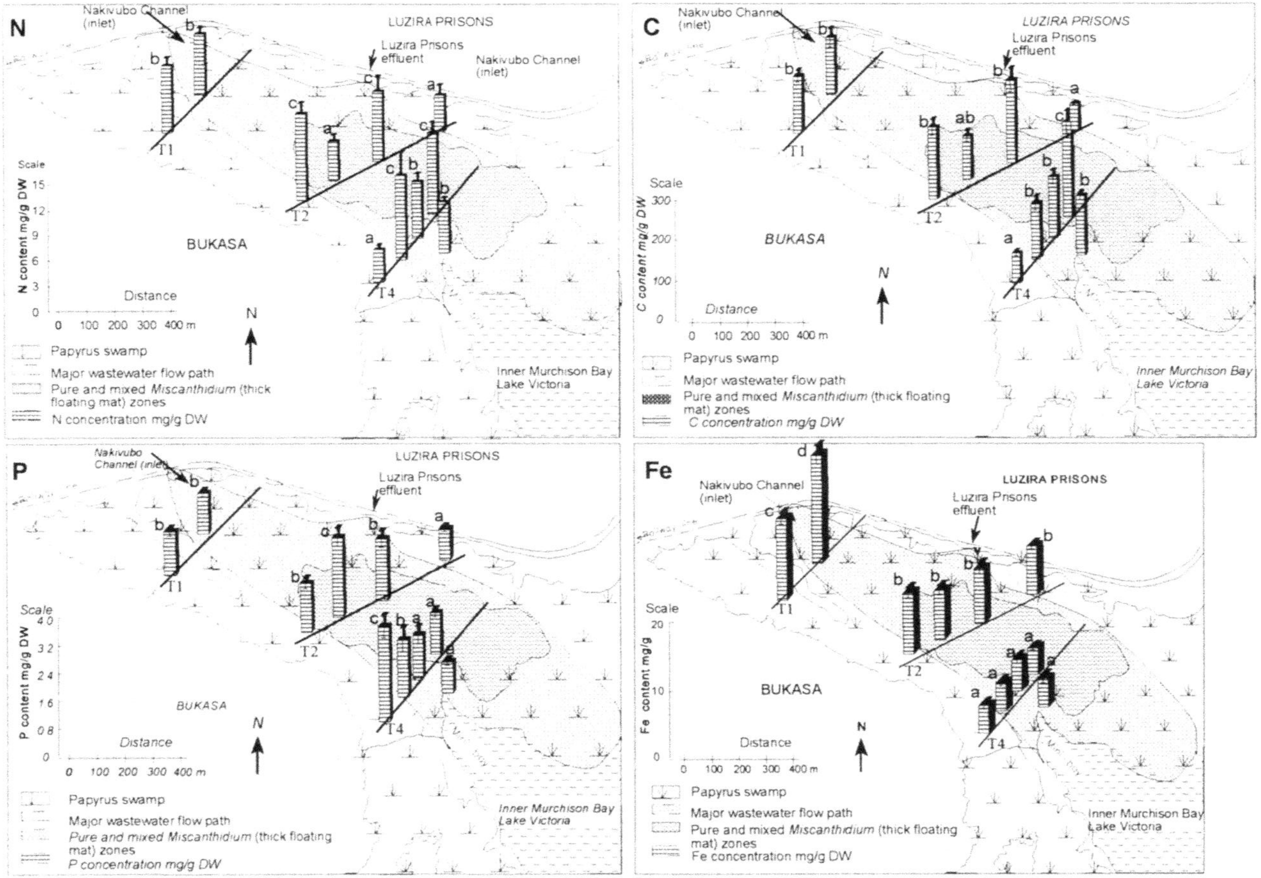

Fig. 8.10. Nutrient and Fe distribution over the swamp sediments showing sites with statistically similar nutrient contents: For N, F = 16.9; p<0.001; There is generally no significant difference in carbon content C, of sediments from most sites except T2 600 (edge) and T4 100 (edge). F=15.23, p<0.001 The lowest P content was found near the lake; F = 38.6; p<0.001. A clear negative gradient towards lake for Fe; F=76.75 p<0.001. T1, T2 and T4 are the transect locations.

A closer look at the depth profile of nutrients within the sites is shown in Fig.8.11. The figure

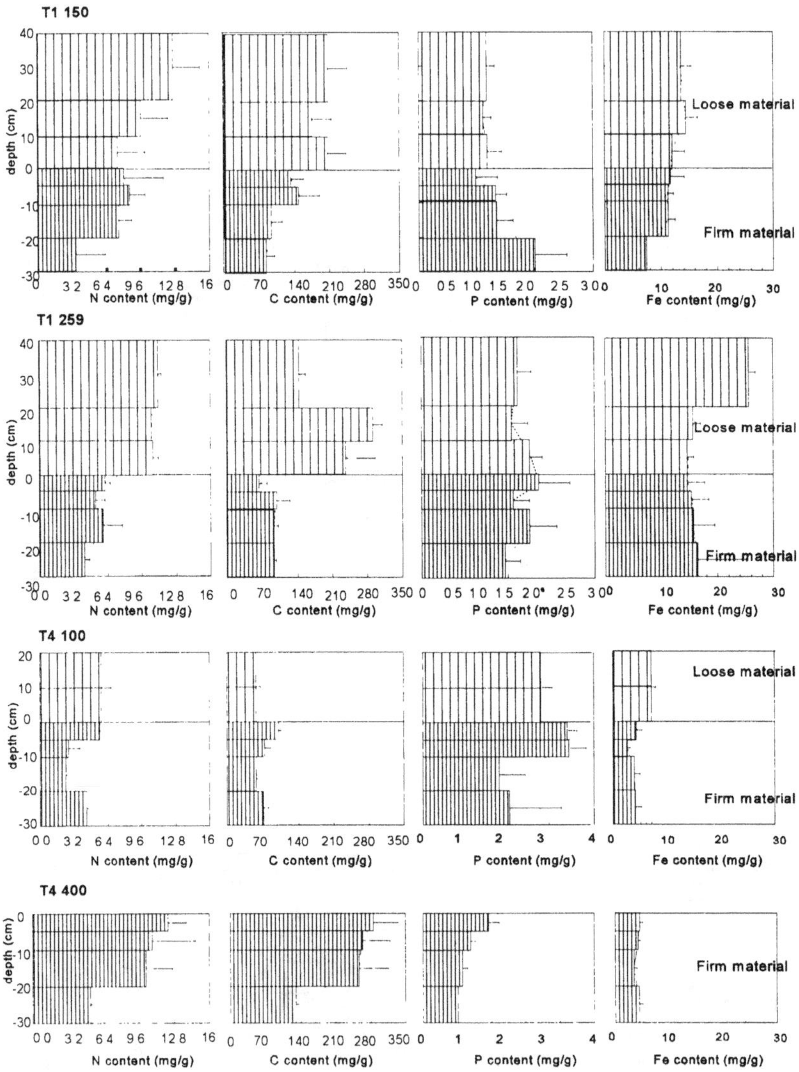

Figure 8.11 Depth profiles of N, C, P and Fe for sediments from T1 (closer to the swamp inlet) and T4 (near the lake). A comparison of sediments from major waste water flow path (T1 259 and T4 400) and from the less influenced areas of the swamp (T1 150 and T4 100). The < 5 cm thick debris layer at T4 400 is part of the 0 to -10 layer.

shows depth profiles of N, C, P and Fe for 4 locations in the swamp, *i.e.*, namely the papyrus zone close to the inlet (T1 259); papyrus zone in the old route of the channel but no longer influenced by waste water (T1 150); papyrus zone near the lake (less waste water impact - T4 100); and *Miscanthidium* zone near the lake (T4 400). The following was noted:

a. there was a generally negative gradient with depth for all elements, except for P at T1 150. This may suggest different states of decomposition and diagenesis; and ultimate burial (sink) of material;
b. the nutrient content of T1 150 is very comparable to that of T1 259 except P where the buried sediment has a higher content. This is related to the loading history/old route of Nakivubo Channel (before its diversion during the wars of the 1980s (Chapter 2). At T1 150, the reduction in P content in the last nearly 20 years after the diversion was within about 20 cm of the firm sediment depth, suggesting a sediment accretion rate of about 1 cm/year and a lower allochthonous load in the recent past; and
c. the P content at T4 100 was very high while other nutrients were very low (low C:N:P). This may be one reason why the productivity at this point is the lowest along T4 (Chapter 5).

iii) *Nutrient ratios*

Nutrient and metal ratios are important in P sorption. Table 8.6 shows the variation of nutrient and Fe content of the sediment near the edge portions of the swamp and along the major flow path respectively. It is seen that:

a. the Fe:P (molar) ratio decreases from the inlet towards the lake offering no explanation for the reduction in P content nearer to the lake;
b. the C:P (molar) ratio is less than the critical value for the cessation of P release (about 200 - Patrick, 1990) except T4 400 (closest to the open lake). Therefore, the P content in the sediment is liable to be released;
c. the C:N:P ratio for areas with high P content is very low (T4 100, T4 200, T2 225 and T2 400). These regions also have a lower Fe and C content;
d. depth variations of the C:P and N:P ratios showed a general reduction in both ratios for all locations except T4 100 (Fig. 8.11). This may again be due to higher mineralization rates for N and C compared to P.

8.3.3.3 Chemical content of trapped sediments

(i) *Average nutrient content*

The average vertical nutrient fluxes for all sites are shown in Table 8.5. Despite the variation in material fluxes, the P flux is more or less uniform over the swamp at about 0.10 $gP/m^2/d$. Excluding extreme values, the N flux is about 0.6-0.8 $gN/m^2/d$ for *Miscanthidium* and 1.1 $gN/m^2/d$ for papyrus zones. Nutrient values at T2 100 and T4 400 (lower trap) are related to the high material flux which resulted from a loose mat and seiches effects on the decaying hyacinth, repectively.

(ii) *Seasonal variation in nutrient fluxes*

Figure 8.12 shows the average N, P, Fe and C contents in the trapped material and the seasonal SS fluxes for 2 sites in the swamp namely: T1 150 (papyrus, firm mats) and T4 400 (*Miscanthidium*, near the lake). The following was observed:

Table 8.5 Mean material and nutrient sedimentation in the Lower Nakivubo swamp. All vertical fluxes are measured relative to the dry weight of material.

Location	Material flux ($g/m^2/d$)	P flux ($gP/m^2/d$)	N flux ($gN/m^2/d$)	C flux ($g/m^2/d$)	n
T1 150	55 ± 12	0.10 ± 0.02	0.98 ± 0.21	16.1 ± 5.2	6
T2 100	435 ± 161	0.57 ± 0.21	5.57 ± 2.09	69.1 ± 27.0	5
T2 225 upper	65 ± 9	0.12 ± 0.01	1.31 ± 0.21	20.8 ± 4.4	5
T2 225 lower	74 ± 10	0.09 ± 0.01	0.93 ± 0.13	23.0 ± 3.2	9
T2 400	31 ± 7	0.09 ± 0.02	0.59 ± 0.08	7.2 ± 2.2	10
T4 200	55 ± 12	0.11 ± 0.02	1.26 ± 0.15	14.9 ± 5.3	10
T4 400 upper	70 ± 27	0.10 ± 0.05	0.85 ± 0.30	19.1 ± 6.9	4
T4 400 lower	174 ± 82	0.49 ± 0.08	1.71 ± 0.38	55 ± 24.3	8

a. the P flux of trapped material was almost constant and showed little correlation with rainfall (r=-0.146 at T1 150 to r=0.261 at T4 400);
b. nitrogen flux was negatively and weakly correlated with rainfall (r=-0.251 at T1 150 to r=-0.124 at T4 400);
c. the carbon flux was strongly correlated with the seasonal SS load (r=-0.96 and 0.61, for the two sites respectively). This may imply dilution for T1 150 and increase in scouring of material due to higher flows for T4 400;
d. the impact of flows/rainfall was most visible at T4 400 especially with respect to C, SS, Fe and P fluxes;
e. the trapped P for all the sites was positively correlated with Fe (r=0.782), Mg (r=0.649) and Ca (r=0.768). The low concentrations of Mg and Ca in the trapped material and the near neutral pH of the water however implied that these two elements can only play a minor role in P sorption characteristics.

8.3.3.4 Comparison of trapped and bottom sediment material

Table 8.6 gives a comparison of selected variables for both the bottom sediment and the trapped material. From the table, it is seen that:

a. at all locations (except T4 400 and T4 400 Lake), the N:P ratios were lower for the bottom sediment than for the trapped material, possibly suggesting (i) different levels of mineralization and (ii) that the trapped material is not simply resuspended. The molar N:P ratio for emergent macrophytes is generally 10-30:1 (Gaudet, 1977; Gaudet and Muthuri, 1981; Boyd, 1978) while that in sewage is more variable at 3-75:1 (Kramer *et al.*, 1972; Stumm and Morgan, 1970). The molar N:P ratio for papyrus in this study was 15:1 (Chapter 5). The bottom sediment therefore generally indicated low N content while the trapped material indicated high N content;

Table 8.6 Nutrient content, mass ratios and iron content for trap versus bottom sediments. Molar Fe:P is higher for trapped than for bottom sediment except at T1 150 (old water path).

Location	Bottom sediment					Trap material		Trap:bottom ratios		
	N [mg/g]	P [mg/g]	Fe [mg/g]	Molar C:N:P	Molar Fe:P	Molar C:N:P	Molar Fe:P	N	P	Fe
T1 150	8±1	1.2±0.1	12±1	320:15:1	5	493:23:1	5	2.1	1.7	1.1
T1 259	7±1	1.2±0.1	16±1	350:15:1	7	-		-	-	-
T2 100	7±1	1.4±0.2	9±1	362:18:1	3	539:25:1	6	2.1	1.0	1.7
T2 225	10±1	2.3±0.3	7±1	134:5:1	2	496:23:1	6	2.2	0.8	2.7
T2 400	5±1	1.8±0.2	8±1	265:13:1	2	481:18:1	5	2.8	1.2	2.5
T4 100	4±1	2.8±0.3	4±0	106:3:1	1	-		-	-	-
T4 200	10±3	1.7±0.4	4±0	239:15:1	1	618:30:1	4	1.8	0.9	3.0
T4 300	10±1	1.3±0.2	4±0	350:15:1	2	-		-	-	-
T4 400	10±1	1.1±0.1	4±0	542:23:1	2	423:23:1	5	2.4	2.2	5.0
T4400 Lake	6±1	0.9±0.9	4±1	475:18:1	3	426:25:1	5	3	2.3	5

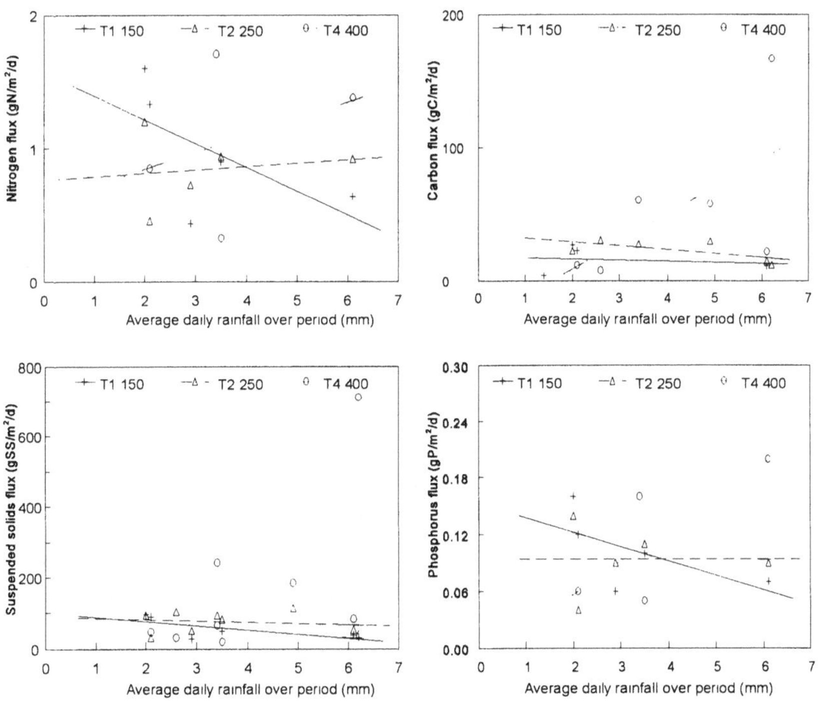

Fig. 8.12. Average daily vertical fluxes of material collected in the bottom traps as a function of the rainfall over the given periods for locations T1 150 (papyrus zone near the inlet), T2 250 (mixed vegetation zone in the middle of the swamp) and T4 400 (*Miscanthidium* zone at the swamp-lake interface). Curve fitting was done using SLIDE WRITE PLUS (for Windows '95) for indication of trends.

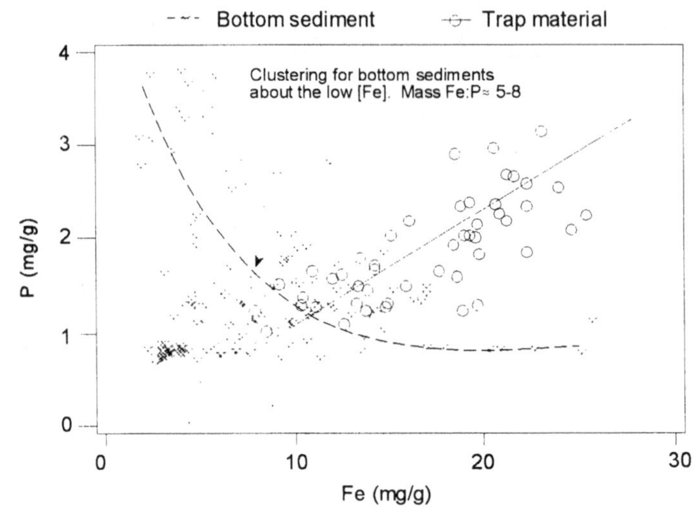

Fig. 8.13. Correlation between Total P and Fe for benthic and trapped sediments in the lower Nakivubo swamp. Bottom sediments: r=0.2, p=0.019, F=5.63, DF=134. For trapped sediment: r =0.77, p≤0.001, F=70.25, DF=49. The equation for the trap material is $P = 0.123\,Fe - 0.161$ (in mg/g).

b. the Fe content of the sediment was always lower than for the trapped material (trap:bottom ratio > 1), possibly showing that the swamp vegetation and mat were the major sources of iron or the concentration of Fe by the plants while the sediment Fe has been reduced in the anaerobic swamp sediments and flushed by the lake seiches. Iron was correlated with P in the trapped material but not so for the bottom sediment (r=0.77, mass P:Fe≈1/8; Fig. 8.13). This is possibly because the trap contains more freshly precipitated iron (e.g, $Fe(OH)_3$), with plenty of adsorption sites. On the other hand, it is speculated that the sediment may contain older Fe precipitates (e.g, $Fe_2O_3.nH_2O$ and $Fe_3(PO_4)_2$) and similar 'condensed' crystalline forms with less adsorptive capacity and hence lower P density per gram of sediment (Stumm and Morgan, 1970);
c. the nitrogen content (and the N:P ratio) was always higher in the trapped material, possibly suggesting lower decomposition states for trapped material;
d. the trap:bottom sediment nutrient contents were higher than 1 (except for P at some locations). The ratios indicate that the reduction in N:P from the trapped material to the bottom sediment was more due to N reduction than P reduction.

8.3.4 Fractional composition of sediments

Fractionation experiments in the laboratory were done to determine the main forms of P in the sediment with the view of identifying the contribution and spatial variation of the readily available P fraction to the total P pool.

Fig. 8.14 shows the percentage contribution of different P fractions. The main P form in all the tested sediments is the Al-Fe bound fraction (NaOH extractable P). The contribution of this fraction and that of organic P is higher near the swamp inlet and reduces towards the lake, as does the iron content. Beyond T1, there is more organic P in the sediments closer to the edges (T2 100 and T4 100) than in the central flow path (T2 400 and T4 400). The NH_4-Cl fraction is a very small fraction of the total-P (< 0.2%). A positive correlation between the ratio of organic P to non-organic P and of organic matter to non-organic matter (r=0.77, n=21), suggests that P dynamics are related to the carbon content of the sediment. The HCl-P fraction is less than 10% of the TP, a result of the low Ca and Mg content of the sediments.

8.3.5 Sorption kinetics and potential for Nakivubo swamp

Native Adsorbed Phosphorus, NAP

The phosphate desorbed from the sediment by distilled water (also called 'Native Absorbed Phosphate'), is a part of the reactive P liberated in the first step of the 'Hieltjes & Lijklema scheme' (Danen-Louwerse *et al.*, 1993). NAP (also called loosely-bound P) represents the P that is readily returned to the water phase. Fig. 8.15 shows the NAP from sediment samples between the 0-5 and 5-10 cm levels.

Although there was statistically no significant difference between the sediments in the upper and lower layers (p=0.816, n=30), the NAP seems higher in the upper 5 cm for all locations studied, except at T1 259 and T4 100. The locations at the edges of the swamp showed the

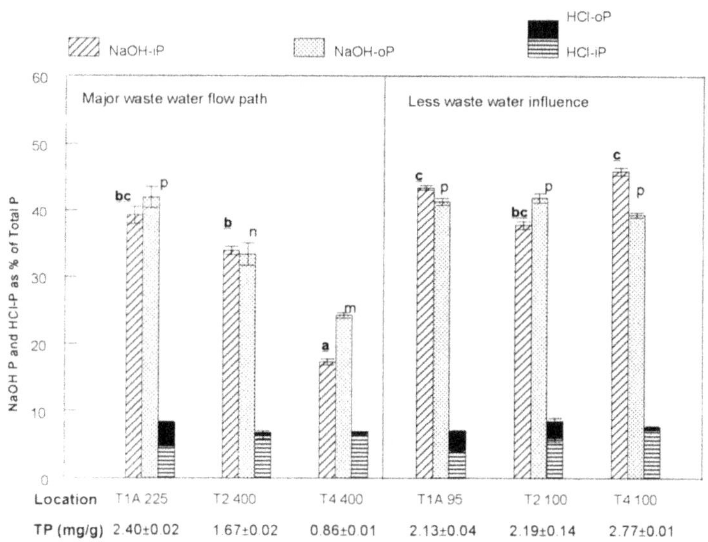

Fig. 8.14. Analysis of variance for the fractionated P as percentage of TP. iP and oP refer to inorganic and organic fractions, respectively. **a, b, bc** and **c** refer to sites whose total NaOH and NaOH-iP are not significantly different from each other ($p \leq 0.001$; F = 1111; DF=17 and $p < 0.001$; F = 253; DF=17, respectively). m, n and p show the sites whose NaOH-oP content are similar ($p < 0.001$; F = 27; DF=17).

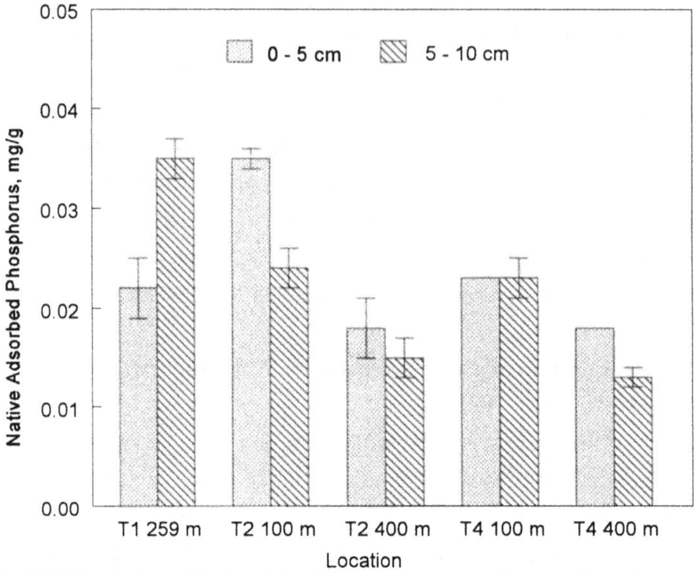

Fig. 8.15. Variation of Native Adsorbed P for swamp sediments. Samples were taken from the 0-5 cm and 5-10 cm core slices (n = 3). There is no significant difference in the NAP content of the two layers (p=0.816; F = 0.06).

highest amounts of loosely bound P in the upper sediment layer (0-5 cm). The locations at the sides of the swamp (100 m, and papyrus dominated) showed higher amounts of loosely bound P than those along the major flow path (400m, *Miscanthidium* dominated). There was a general decrease in the amounts of loosely bound P towards the lake, more in parallel to the P content in the water phase.

Adsorption isotherms

i) Approach to equilibrium

Figure 8.16 shows that equilibrium was reached after 8 days for all initial concentrations. Adsorption took place in two phases. In the first phase P was rapidly adsorbed; about 70% of the total P adsorption onto the 0.5 g sediment

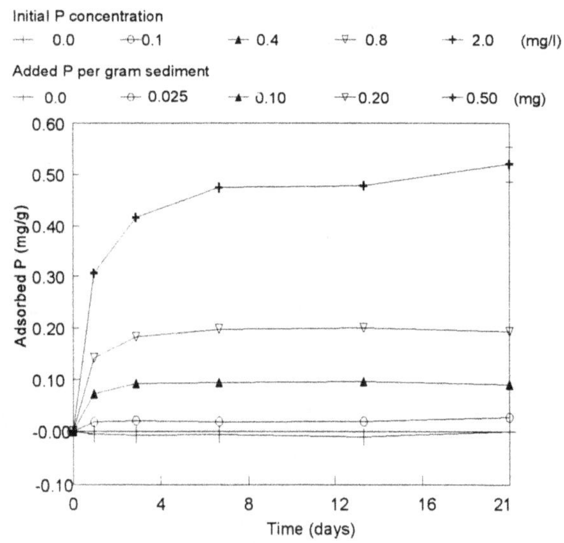

Fig. 8.16. Approach to equilibrium for adsorption. Sediment from the 0-5 cm core slice of T4 400 samples. The legend refers to initial P concentrations. By day 8, nearly all the added P is adsorbed for all concentrations.

occured within 24 hours. This is followed by a slow adsorption to equilibrium. The experiment also showed that nearly complete adsorption occurred for all added P and that higher amounts of P can be adsorbed onto the sediments. So less sediment (0.25 g) and higher concentrations of up to 10 mgP/l (5 mgP/g sediment) were used further for the Langmuir isotherms.

ii) Langmuir isotherms

The parameters C^{max} (maximum adsorption capacity) and K_d (the adsorption coefficient) in the Langmuir equation:

$$C_{ads} = \frac{K_d C^{max} C_{eqm}}{1 + K_d C_{eqm}} \qquad (8.3)$$

were iteratively estimated using nonlinear regression statistics (SPSS). C_{eqm} is the equilibrium concentration in the overlying water. C_{ads} is the solid phase concentration of P adsorbed onto the sediment. Fig. 8.17 shows the derived isotherms.

The maximum adsorbed concentration was highest at T2 400 m and T2 100 m (Table 8.7). The value of the coefficient of correlation (r^2) is high for all cases showing that P adsorption onto lower Nakivubo swamp sediments can be satisfactorily described by the Langmuir adsorption model. Adsorption capacities for sediments along T2 are not significantly different ($p=0.05$) from each other. They were however different from T4 and inlet sediments that, in

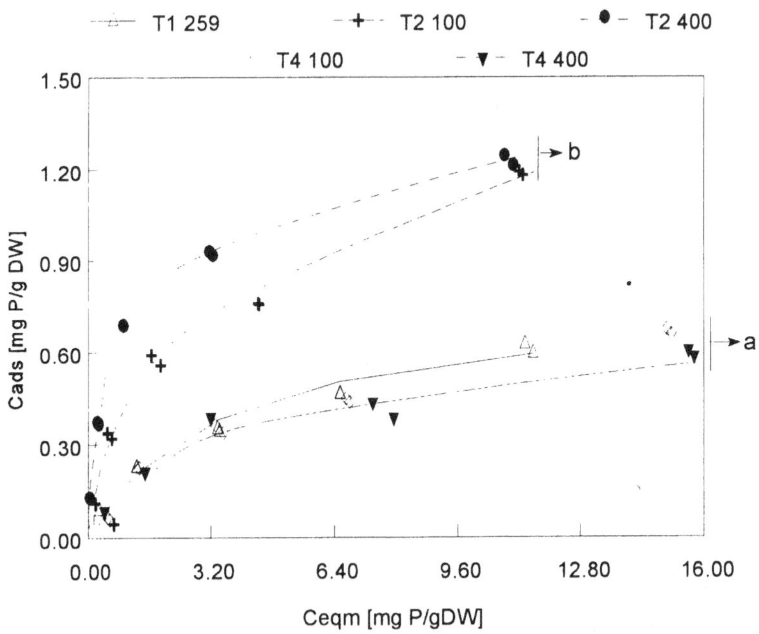

Fig. 8.17. Adsorption isotherms for sediments from different locations in the lower Nakivubo swamp. a and b each combine those locations where adsorption capacities were comparable. a and b are significantly different ($p<0.01$, DF=61).

turn, were similar (p =0.07, n=61). There was no direct correlation between maximum sorption potential and either the P, C or Fe content of the sediment. The maximum sorption potentials for T2 sites could be linked to, *i.a.*, the smaller particle size and the more moderate water flow velocities.

For all locations however, the sediments of Nakivubo swamp clearly still have the capacity to

Table 8.7 Parameter estimates for Langmuir-type isotherms (top 0-5 cm sediment layer; pools a and b in Fig. 8.16).

Sediment source	K_d [-]	C^{max}[mgP/gDW]	r^2
T1 259, T4 100 and T4 400	0.25 ± 0.04	0.76 ± 0.50	0.93 (n = 30)
T2 100 and T2 400	0.49 ± 0.13	1.38 ± 0.13	0.88 (n = 30)

adsorb additional P (under laboratory conditions; the initial concentrations were higher along the edges and inner swamp:e.g, T4 100, T2 100, T2 400 - Table 8.6). The sorption potential ranges from 0.6 to 1.6 mgP/g (say 1 mgP/g) sediment. If the active sediment layer is assumed to be 1 cm thick, then for the sediment density of 1830 kg/m^3, the active mass per m^2 is 1830 kg/m^3*1cm*1m^2 or 18.3 kg. For the study area of 1.15 km^2, this corresponds to a P load of 21.045 x 10^6 kg. Using an average daily load from the Nakivubo channel of 2 g/m^3 (or 206 kgP/d - Chapter 4), then the sediment of Nakivubo swamp would have a sorption capacity for about 100 days' load. During this time, more material would be deposited to the swamp bottom, further increasing the sorption capacity.This is the situation if all the incoming phosphate went to the sediment. However, about 23% is taken up by plants while about 50% leaves with the effluent (Chapter 9).

8.3.6 Phosphorus release from bottom sediments

i) Anoxic P release using column experiments

Experimental conditions were presented in Table 8.2.

Experiment I

The effects of differences in overlying water quality and vegetation at T4 400 were tested using intact sediment cores, filled with either swamp or lake water (Fig 8.18). Initial P sorption (possibly due to oxygenation of the sediment during set-up) for all columns was observed.

The sorption capacity for swamp water P was not significantly different from that of lake water P for both sites ($p \leq 0.08$, F=3.47, DF=17). However, that for lake water P was higher in the papyrus zone (p=0.025, F=6.15, DF=17). Whatever the water type, sediments from the papyrus vegetated zones had a higher sorption capacity than the *Miscanthidium* base sediments ($p \leq 0.01$, F=9.42, DF=35), closer to the lake.

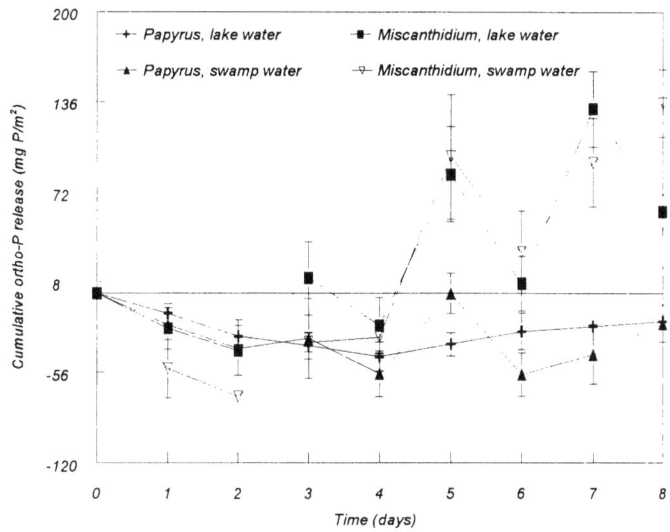

Fig. 8.18. Average cumulative ortho-P release from sediment at T4 400. Experiment with swamp and lake water for papyrus and *Miscanthidium*-based sediments (Experiment I), n=3.

The average (total) P adsorption for the papyrus-based sediments was 30 mg P/m^2 for both swamp water and lake water. Equilibrium was reached after about 3 days. This meant an average 3-day vertical flux of 10 mg P/m^2/d. For *Miscanthidium*-based sediments, there was an overall release of P into the water after the 4th day and by the 8th day, an average release of 100 mg P/m^2 was observed.

Experiment II

Experiment I was repeated for a comparison of central swamp sediments (at T2) and of the flux of P measured for sediment cores from T2 250 and T2 400. A net P accumulation by the sediments was observed (Fig. 8.19). Equilibrium was again reached after about 3 days for swamp water (TRP =1.1 mgP/l) and somewhat earlier for lake water (TRP= 0.2 mgP/l). The P transfer capacity for swamp water columns was different from that of lake water columns for either site ($p \leq 0.001$, $F=42$, $DF=39$). Similarly, the transfer capacity for lake water P is different for both sites ($p=0.028$, $F=5.67$, $DF=19$) but not so for swamp water P ($p=0.259$, $F=1.36$, $DF=19$).

The average 3-day sorptive flux for the four sets of cores was 17, 24, 81, and 103 mg P/m^2/d for the T2 250 lake water, T2 400 lake water, T2 250 swamp water and T2 400 swamp water cores respectively. The difference within sites can be due to a lower initial P for lake water columns. These values compare well with T4 values of Experiment I, namely higher sorptive capacity for P by the inner swamp sediments. Similar results were observed for the adsorption

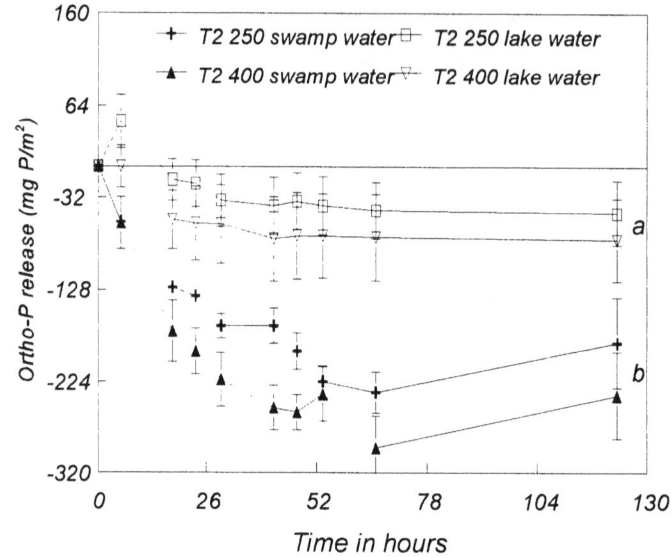

Fig. 8.19. Average cumulative ortho-P release from sediment at T2 250 and T2 400. Experiment with swamp and lake water (Experiment II), n=3.

experiment, where the P sorption capacity was higher for sediments from T2 400 than for T2 100 and T4 sediments.

General observations

Overall, the system reached equilibrium within about 3 days possibly implying 3 days retention time in the field for maximum adsorption. Unsterilized sediments from all sites were P sorptive although the system was not completely sealed from atmospheric gaseous exchange. Nevertheless, overall, the results of these experiments confirm the trends found in the adsorption experiments that the inner swamp sediments have a higher P sorptive flux. This might explain the high P concentrations in the sediments from T2. Under experimental conditions, the remaining adsorption potential ranged between 80 - 160 $mgP/m^2/d$ for inner swamp sediments versus 15 - 36 $mgP/m^2/d$ for the sediments near the lake.

ii) Anoxic P release with continuous flow

Results of the continuous flow anoxic P-release experiment are presented in Fig.8.20. An initial release of P from some sediments led to P increases in the water phase from initially 0.35 mgP/l to varying concentrations (possibly depending on pore water concentrations). This was followed by adsorption towards constant effluent concentration. The estimated adsorption fluxes for the 42 hours period are shown in the figure and are calculated from:

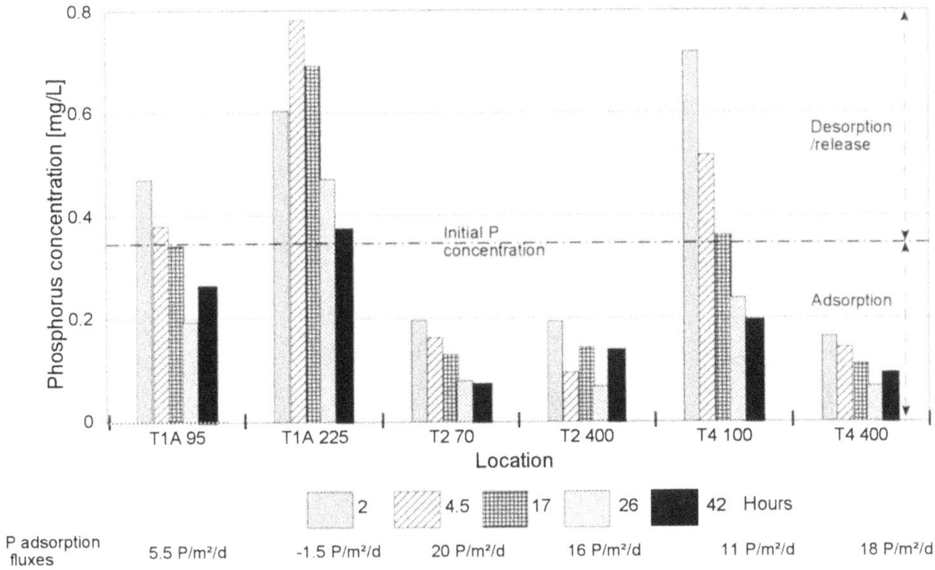

Fig. 8.20. Continuous flow anoxic P release experiment with continuous flow for sediments from different locations in the swamp. The initial P concentration in the feed water was 0.35 mg/L.

$$\frac{(initial\ concn.\ -\ 42\ hr\ concn.)}{sediment\ surface\ area\ *\ no.of\ days} * water\ volume \cdot \tag{8.4}$$

These observations for an anoxic release system support further the conclusion from column release and the adsorption experiment results that in general, the sediments of Nakivubo swamp still have ample unutilized adsorption potential. At 4.5 hours, P concentrations were highest for T1A 225, T4 100 and T1A 95, in that order. These are all papyrus regions. On the other hand, T2 70, T2 400 and T4 400 showed no initial P release but depicted a higher overall net P adsorption. It is probable that the more acidic conditions in and beneath the mat of the *Miscanthidium* zones is partly responsible for the higher P adsorption at these sites (Stumm and Morgan, 1970). There is a tendency for lower P sorption onto sediments closer to the inlet, possibly due to the continuous high P loads in the waste water.

8.3.7 Laboratory experiment on sediment resuspension

Figure 8.21 shows the plot for the suspended solids concentration against water flow velocity for sediments from T2 100 and T4 400, as measured in the micro-flume of Fig. 8.6. The plot shows that when water flows over the sediment, the initial SS reduction effect is due to flushing-out the residual SS concentration in the overlying water, replacing it with the lower SS concentration of tap water. When the flow velocity (non-linear) is increased however, a

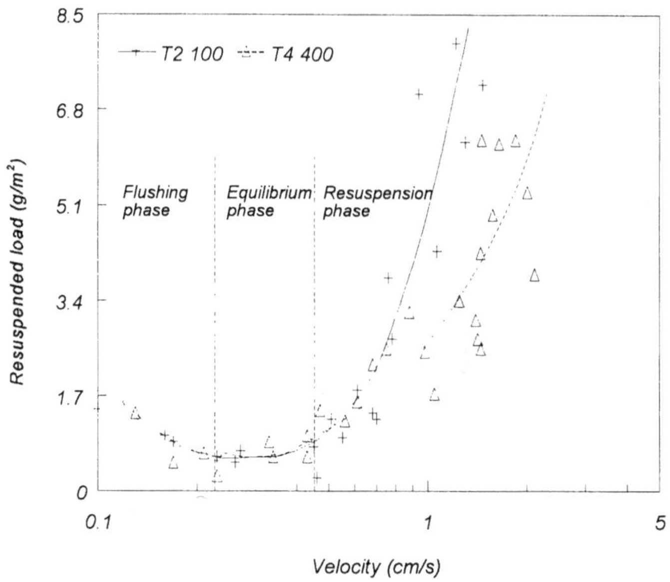

Fig. 8.21. Resuspension of settled material in the model 'micro-flume' for sediments collected from 100 m along transect 2 and 400 m along transect 4 (T2 100 and T4 400).

point is reached at which the settled material is resuspended and the suspended solids load in the effluent increases again, in proportion to the water flow velocity.

The velocity at which this minimum is reversed was termed the critical resuspension velocity of the sediment particles for the swamp. For the two sediments under study, there is little difference in the critical resuspension velocity (about 0.45 cm/s), although for velocities higher than 1 cm/s, the suspended solids concentration in the effluent is higher for T2 100 than for T4 400. This is because resuspension velocity varies with the particle size. In Section 8.3.2, it was shown that both soils are fine (>50% passing the 75μm sieve) and that there is no difference in their effective sizes, d_{10}. The particle size distribution curves (Fig. 8.8) showed that sediment from T2 100 had a higher fine fraction than that from T4 400 and therefore, it might have a lower resuspension velocity (although for an undisturbed sample the cohesivenss of the particles may increase with content of the fine (clay) particles).

The 0.45 cm/s could be a lower value for the critical resuspension velocity due to the disturbance caused while extracting the sediment, and is an indication of the minimum velocity at which particulate material in the swamp may be resuspended.

8.4 Discussion

Sediment particle size

The sediments of Nakivubo swamp are silty clays (over 90% less than 75 μm) underlying 5 -50 cm of muck (peat). The high surface area of the clays makes them effective adsorbents. In addition, adsorbing metals (Fe, Al and Ca) are more concentrated in such fine particles. Particle size also affects the content of bioavailable particulate P (Keulder, 1982; Stone and English, 1993), and indeed higher P content was found in the swamp interior sediments (T2; Table 8.6) that are also finer. The thicker peat/clay under papyrus compared with that under *Miscanthidium* suggests potentially higher adsorptive capacity because of a larger interactive surface area. The larger particle size on the eastern and western edges of the swamp may thus be partly responsible for the mostly lower P content of sediments in most of this region.

Sediment loads and accretion rates

Luzira flows did not appear to noticeably influence the solids load into the swamp. Low flows from the Luzira Prison enable the settling of its settleable matter on the north eastern edge of the swamp, resulting in the rapid development of rich loam soils on which *Enhydra fluctuans* thrives. This area is subsequently converted into agricultural land. Eroded material in the run off from the western (Bukasa) side of the swamp is also trapped by vegetation on the hillside and at the edge of the swamp.

The major external source of sediments into the water column of the swamp is the Nakivubo Channel with an average settleable load of about 700 tons/yr, highest during the rainy season due to erosion/flushing. Because of the low concentration and fineness of the particles, most of this load will not settle in the lower swamp. This SS load into the lower Nakivubo swamp is very low since most of the material is either filtered by the upper swamp or settles to the channel bottom during dry periods. The National Water and Sewerage Corporation (weekly data for January 1993 to May 1994) found an average suspended solids' concentration at the inlet (Fig. 8.1) of 23 mg/l (range = 6 - 31 mg/l with an outlier of 86 mg/l) compared with a concentration of 335 mg/l at a point upstream in the Nakivubo channel, after the Bugolobi Sewage Works outfall. The two points are separated by 2.1 km of channel and 1.1 km of swamp. The combined system reduced the average suspended solids concentration by 93%. During rainstorms however, the load into the lower swamp increases due to a reduced sedimentation time as well as increased flushing and scouring by higher flows in the channel and upper swamp. Even assuming that 100% of the allochthonous load settled in the lower swamp, the settleable matter flux (1.71 g $SS/m^2/d$) for the lower swamp is low. In comparison, autochthonous loads are about 30 times higher (gross sedimentation, = 24,700 tonnes/yr, paragraph 8.3.2) and mostly settle within the swamp, except in the *Miscanthidium* zones (where it more easily resuspended because of the higher flows) and the areas near the lake.

The large variability in SS flux of the autochthonous load at T4 400 and not anywhere else in the swamp suggested that the effect of seiches and rainstorms on internal SS flux was small, except at locations nearest to the swamp-lake interface. This is probably due to comparatively high flows during rainstorms that flush out the allochthonous material. The brown plume of run-off water that appears in the bay after storms due to transport of the eroded laterite soils from the swamp's catchment area bore this out. Larger loads at T4 400 compared with T4 200 were attributed to the presence of the water hyacinth in the bay whose decaying matter was transported into the swamp by seiches, and was occasionally found in the traps.

Contrary to earlier expectations and findings in some similar systems, there is virtually no peat development in the Nakivubo swamp. In the Kawaga swamp 20 km south west of Nakivubo on the shores of Lake Victoria, Gaudet (1979) estimated a 13-year deposition of 1.5 m (or 0.12 m/yr) from the depth of old rhizomes drowned during the storms of the early 1960s. Such a layer was not encountered in any of the augured cores in this study. In fact, a similar order of magnitude for the papyrus vegetation of Nakivubo swamp would mean that the swamp would be fully silted in about 15 years. In the Nakivubo swamp where the firm sediment is within 0.6 m of the peat surface, even an extravagant assumption of zero deposition in 1962 would yield a maximum net accretion rate of $0.6\ m\ /\ 32\ yr\ =\ about\ 0.02\ m/yr$. When the ratio of the sediment particle density to the bulk density[1] (*i.e.*, '1830:60 or approximately 30) is used to estimate the effect of compaction on sediment accretion rates, the results are about 0.01 m/yr and 0.009 m/yr for papyrus and *Miscanthidium* zones, respectively. From the low peat thickness in much of the swamp (< 1m for a swamp that is over 50 years old) and the P content of sediments in the old path of the Nakivubo channel (T1 150; paragraph 8.3.3.2 ad.ii.b) the accretion rates might indeed be ≤ 0.01 m/yr. This rate is within the range of reported common net sediment accretion rates for different marshes of 0.005 - 0.01 m/yr (Nixon, 1980; Fennessy *et al.*, 1994; Mitsch and Gosselink, 1993), and indicate that burial may not be a significant pathway for the loss of P in the Nakivubo swamp.

The low sediment accretion rates in the Nakivubo swamp are because:

(i) heavy storms induce occasional high flows, washing out much of the organic material as seen at the lake-edge (brown colour of the water due to the laterite soils in the catchment area);

(ii) the traps measured gross (and not net) sedimentation rates as the material trapped could not be resuspended or transported whereas in the field, resuspension and occasional flushing (i, above) could occur; and

(iii) the estimates did not take into account the effect of material decomposition that, due to favourable pH, temperature in tropical Africa and altitudes of less than 2000 m.a.s.l.

[1]The bulk density of the sediment refers to the dried mass of the sediment particles per 'bulk' volume of a sediment block, including pores, as compared to the particle density which excludes the volume occupied by pores.

(Howard-Williams, 1979; Thompson and Hamilton, 1983; Allan, 1995), is substantial in the swamp and causes large volume reduction. Indeed, in another Lake Victoria papyrus-*Miscanthidium* dominated wetland about 40 km southwest of Nakivubo swamp, Lind and Visser (1962) found no peat buildup and attributed this to rapid decomposition under tropical conditions and to export of digested material into the lake with the flow-through.

Although there was no significant peat layer beneath the *Miscanthidium* water column due to the faster flows in the major waste water flow path enhancing transport of the finer deposited material, the mat is a large reservoir of nutrients. The combined effect of low decomposition rates [Using the ratio of fulvic acid to humic acid as an indicator for the decomposition constant Visser (1964) found a 34% and a 97% reduction in plant litter weight after 20 months for *Miscanthidium* and papyrus respectively] and a tightly interwoven mat structure (Chapter 5) that allows very little fall-through lead to the build-up of a thick mat. Using a bulk density of 69.7 kg/m^3 and an average mat thickness of 1.4 m, it can be shown that for the average mat P content of 0.22% (from trapped material data, also Table 8.6), then the mat contains about 0.21 kg P/m^2 and similarly, 1.66 kg N/m^2. Corresponding contents for the about 0.4 m thick papyrus mat are 0.05 kg P/m^2 and 0.35 kg N/m^2.

Nutrient and mineral content of benthic and trapped sediments

The nutrient contents (C, N and to a lesser extent P) of the benthic sediment were independent of the main waste water flow stream. Unlike for biomass productivity (Chapter 5), there was no discernible relationship between the waste water flow path and the sediment nutrient content.

The P content of the Nakivubo swamp firm (benthic) sediment (1.0 - 2.6 mg/g) is higher than the average 0.61 mg/g estimated for Ugandan swamps (Gaudet, 1976) possibly because of the long term exposure to higher P concentration in the waste water and the sediment characteristics leading to higher adsorption. In a comparison of C:N:P ratios for different systems by different researchers (Table 8.8), the sediments of the Nakivubo swamp had a higher ratio (in addition to the higher absolute P) than other systems. The molar C:N:P ratios which are all greater than the Redfield ratio (106:16:1) in the trapped material but lower than that in the bottom sediment (low N component) suggest that in the Nakivubo swamp,

a. C and N are mineralized at a faster rate than P [the C:N and N:P ratios also reduced with increasing depth, an effect that could likewise be attributed to the recycling of N in the system compared to loss of carbon dioxide and methane, and the more rapid mineralization of N compared to P (Andersen and Jensen, 1992)];
b. in the event of complete elimination of external nutrient sources, N (and not P) is potentially the limiting nutrient for the firm sediments; and

c. the trapped material was not simply a product of resuspension, but originated from the vegetation mats. The N:P ratio for both sediment types is quite comparable to that of wetland plants.

The N:P ratios were in the same range for all sites except T2 225 and T4 100 (Table 8.6) which had low N and also relatively high P. These sites also had low carbon or organic matter content. The dependence of P release on organic matter concentration and composition was also reported by Patrick (1990). It was noted that no P release from decomposing plant material occurred when the C:P ratio is greater than about 200 possibly due to P limitation for bacterial metabolism (it was indeed observed that at higher ratios, bacteria mobilize P from soil particles in solution). Below this threshold, mineralization occurs and the subsequent release of CO_2 further reduces the ratio, resulting into P release. Indeed at T2 225 and T4 100, the C:P ratios were greater than 200, possibly explaining the high P content of sediments from these areas.

Poor correlation between TP and either total Fe or C in the bottom sediments suggested that neither variable may be solely responsible for the P content, since the anaerobic conditions

Table 8.8 Comparison of nutrient content ratios in different compartments of wetland systems.

System	Molar C:N:P	Reference
Plants		
Papyrus (live and dead)	-:29:1	Gaudet (1977)
Duck weed	85:10:1	Alaerts et al. (1996)
Typha spp.	264:6:1	Boyd (1978)
Aquatic plants	112:19:1	Kramer *et al.* (1972)
Emergent plants	-:17:1	Boyd (1978)
Miscanthidium v.	-:27:1	This study
Papyrus	-:14:1	This study
Papyrus	-:16:1	Muthuri and Jones (1996)
Waste water (raw, medium strength)	52:11:1	Metcalf & Eddy (1979)
Waste water (secondary effluent, Nakivubo channel)	47:27:1	This study
Particulate matter		
Trapped material (papyrus)	-:57:1	Gaudet (1980)
Sediments	96:2:1	Kramer *et al.* (1972)
Sediments	68:9:1	Burns and Ross (1972)
Duck weed	119:6:1	Alaerts et al. (1996)
Sediments (Papyrus)	_:41:1	Muthuri and Jones (1996)
Sediments (Papyrus)	290:15:1	This study
Sediments (*Miscanthidium*)	265:13:1	This study
Trapped material (Papyrus)	565:24:1	This study
Trapped material (*Miscanthidium*)	480:18:1	This study
Other		
Redfield ratio (algae)	106:16:1	Kramer *et al.*.(1972)

in the swamp sediments would have released Fe and Fe-bound P. The longer exposure of bottom sediments is also believed to have led to the formation of more crystalline metallic P forms with less P (Stumm and Morgan, 1970 - role of allochthonous loading on P removal in the swamp) as was borne out by the lower Fe:P of the bottom sediments. This reaffirms the role of sediment particle size and the clay content in P adsorption characteristics (Figures 8.8 and 8.15).

Both adsorption and nutrient release experiments(Figures 8.17 to 8.19) showed that the bottom sediments of the Nakivubo swamp still have a high additional P sorption capacity. For all initial P concentrations, inner swamp sediments (along T2) generally had the highest sorption potential while those from nearer to the inlet and closer to the lake had somewhat less sorption capacity. Overall, sediments with a lower C and N content had higher P sorption capacities and higher P contents, but not necessarily higher Fe contents. The low N and C content is an indication of more inorganic material/clay in the sediment and hence more P adsorption. There appears to be a strong influence of particle size, and the clay (or Al) content of the sediments.

There was a poor correlation between most “metals” (Ca, Mg, K, Na, Fe) and P ($r^2 = 0.48$, $p=0.86$) possibly due to their low contents of the former in the bottom sediments. In comparison, multiple linear regression showed a better correlation between P and the metals for the trapped material ($r^2=0.92$, $p=0.216$). This suggests that the P content of the trapped material may be a function of the contents of the plant and mat litter. The oxidised the vegetation mats could be offering more sites for adsorption.

The absence of a significant spatial difference for either N or C and the N:P ratios in the trapped material indicated that these elements were less dependent on the overlying water, whose quality had a clear spatial trend controlled by waste water flow and the overlying vegetation. The clear negative gradient of total Fe towards the lake (-0.012 mg Fe/g dry weight/m) may be attributed to allochthonous intake from the red laterite soils of the catchment area, and the continued flushing of reduced sediment Fe (and probably P) by seiches closer to the lake.

Role of overlying water

There was a poor correlation between the TRP in the water and the total P content of the sediment ($r=0.082$). The lower pH in the *Miscanthidium* mat could be responsible for the higher sorption potential of sediments from these zones, and the lower content of organic peat here may indicate that the sample contains more adsorbing clay. There is very little variation in water temperature, and it would not be an explanation for any noticed difference in sediment nutrient sorption.

Phosphorus forms

Fractionation experiments are done to try and characterize P compounds by their mineral composition, stoichiometric and structural properties, as well as their ecological significance and behaviour (Hieltjes and Lijklema, 1980), notably the biovailability and release potentials of P species. Some researchers relate the NaOH-P and NH_4Cl-P to bioavailable P (*i.a.*, Psenner and Pucsko 1988; Stone and English, 1993). Klapwijk *et al* (1982), on the other hand, suggested a higher correlation between algal-available P and total P than either the NaOH fraction or the NH_4Cl + NaOH fractions. The rapid release and cycling of non-refractory P forms in aquatic systems (Wetzel, 1975) may however suggest that the total non-refractory P pool might be a better indicator of bioavailable P.

Our experiments revealed that a very small percentage (< 1%) of TP was labile indicating that a minute fraction of TP is immediately available P and that interstitial P is a minor portion of the sediment P pool (Fig. 8.20). The bulk of sediment P (> 50%) was in the NaOH-P fraction adsorbed to metal (Al, Fe) and exchangeable against OH^- and P compounds soluble in bases (Psenner *et al.*, 1988). The organic to inorganic fractions of this component were in the ratio of nearly 1:1. These quantities decreased from the swamp inlet towards the lake, as did sediment Fe content, suggesting the high influence of Fe on P sorption in this system. Due to both the low Ca (and Mg) content of the sediments and the weakly acidic pH of the overlying water, less than 10% of the TP was bound to carbonates and apatite-P.

The large percentage (> 80%) of TP that was non-refractory suggested that a large portion of the TP pool can be bio-available (Wetzel, 1975) even after allochthonous loads are stopped. Thus a 10 cm layer (the layer that is involved in P dynamics: Wetzel, 1975; Kamp-Nielsen, 1989, and Lijklema, 1990) of the swamp sediments could yield the equivalent of 4.5 years'[2] supply of allochthonous P load. Though about 50% of the TP could be available, depth profiles revealed that much of it is not accessible but buried with the sediment, making burial an important P sink (in the event of no flushing). This could again be due to the constant supply of high loads of P from waste water and recycling by plants, availing it in excess of biotic requirements.

The narrow and low Fe:P molar ratio range (1-6; Table 8.6) for both trapped and bottom sediments suggests the predominance of non-occluded Fe-associated P. It eliminates the possibility of an adsorption mechanism in which ligand exchange of hydroxyls for ortho-P occurs (Faulkner and Richardson, 1989). If sorption to amorphous Fe was the main

---

2

Sediment mass (10 cm) $0.1 * 1.15 * 10^6 * 1830 = 2.105 * 10^8$ *kg*
Equivalent non refractory P mass $2 * 10^{-3} * 80\% * 2.105 * 10^8 = 336720$ *kgP*
Equivalent P supply period $= 336720/(2 * 10^{-3} * 103000 * 365) = 4.5$ *years*

mechanism, more sorbed P would be expected with higher times of exposure implying higher P content for the benthic sediment. This low range of Fe:P is consistent with discrete Fe phosphates like vivianite $Fe_3(PO_4)_2.8H_2O$ (molar ratio =1.5), with only a limited capacity for additional P sorption (Cooke *et al.*, 1992).

Co-precipitation and/or adsorption onto $CaCO_3$ precipitates is not a very important process in Nakivubo swamp because (i) Ca concentration is low, (ii) the pH is not high enough for apatite formation, (iii) fractionation experiments showed that Ca and Mg P forms are only about 10 % of the TP in the sediment.

Turbulent mixing and bottom dynamics

Resuspension increases the particulate P content in the water phase and the available surface for P adsorption and may hence exhaust the available P in the water (Håkanson and Jansson, 1983). The importance of resuspension and erosion and the resulting increased output of dissolved P from the pore water is limited in a wastewater receiving wetland due to the higher P content in the overlying water. Instead, the tendency in the Nakivubo swamp was for P to move from the water column where concentrations were higher (*eg*., at T2 400, TRP = 0.69 mgP/L) to the sediment layer (TRP interstitial water = 0.24 mgP/l; Chapter 4). Additionally, the changing of the particulate environment (Lijklema, 1980b; especially with respect to redox conditions in stratified systems, where the particles are recycled back into aerobic conditions that would oxidize the iron, thus co-precipitating P) may also have little impact in the Nakivubo swamp where conditions are (mostly) anaerobic (Chapter 4). Regions of the swamp subjected to turbulent mixing and frequent flushing of sediment (like T4) showed a higher nutrient content in the trapped material but not so for the firm bottom sediment.

The resuspension experiment showed that a velocity of about 4.5 x 10^{-3} m/s was adequate to resuspend the sediment particles in the swamp. However, the particles had been subjected to earlier disturbance during transfer into the flume perhaps lowering this critical velocity. On the other hand, a calculation of the resuspension velocity for sediment particles (in laminar flow conditions) (Appendix 8.1) showed that resuspension would not be expected till the water velocity exceeded 0.13 m/s. This value that is 30 times higher could be due to the fact that the actual critical particle size that may be much less than the 75 μm used in the calculations and the laminar flow assumptions made in the calculations. This, not withstanding the role of the cohesiveness of the sediments.

Water velocities of 0.07 m/s were encountered in the swamp both for the wastewater flow and during the seiches (Chapter 3). The ranges were about 0.03-0.25 m/s for the channelized flows at T1 and possibly lower for T4. These, together with the occasional floods mean that there is a strong likelihood of loose particles to be resuspended and transported (especially during occasional floods), possibly showing why there is less material accumulation in the major waste water flow path. This also points to the need to control flows in the swamp if it is to act as an effective trap of particulate matter and sediments.

8.5 Conclusions

The major materials source for the Nakivubo swamp sediments is the autochthonous load which was at least 30 times the allochthonous load. Internal productivity and cycling especially by macrophytes is the major source of both particulate and nutrient load in the Nakivubo swamp. Figure 8.22 summarizes the SS pathways in the Nakivubo swamp and quantifies the average fluxes.

Nutrient content (N, P and C) of the sediment was independent of the characteristics of the major waste water flow path. The longitudinal change of sediment P from inlet to the lake did not reflect the pronounced decrease in the external load of the overlying water from the influent to the lake. Assuming a uniform distribution over the swamp area, the total P-load into the swamp was 0.18 g/m^2/d. Rapid growth of papyrus coupled with a more loose floating raft and less water discharges through papyrus zones enable higher quantities of decomposing material to fall through and settle, forming a thick (up to 60 cm) peat layer. This layer is a large store of nutrients whereas the relatively thinner mat is a smaller reservoir (Chapter 9).

The tightly-woven *Miscanthidium* mat on the other hand, with its low bulk density (average

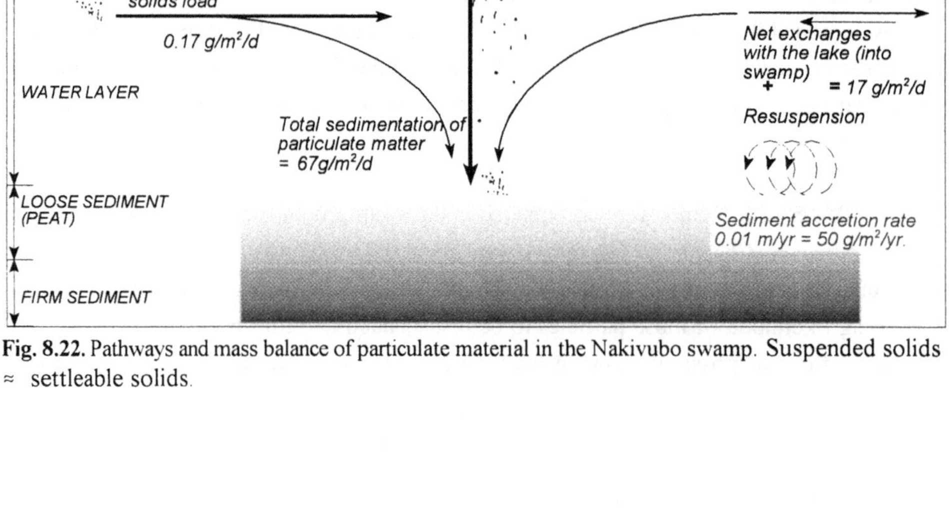

Fig. 8.22. Pathways and mass balance of particulate material in the Nakivubo swamp. Suspended solids ≈ settleable solids.

about 70 kg/m^3) that keeps it afloat, has the following effects on the mass and nutrient content in the system:

a. There is limited mixing between the underlying water and its nutrients with the interstitial water in the mat and to the live roots nearer to the surface;
b. very little material falls off the mat into the water layer leading to mat-build up and very low peat accumulation on the firm sediment; and
c. the thick mat is a large reservoir of nutrients.

The firm sediments are a large reservoir for P (about 80% is bio-available) and they have further potential for further adsorption and storage of P. However, access to them is limited due to mostly organic peat layers that overlay them.

The major P removal mechanisms from the water phase are adsorption onto non-occluded Fe /clay alumino-silicates and/or vivianite. The fineness of the sediments (more than 90% $\leq$ 75μm) means that they have a large surface area for adsorption. Co-precipitation and/or adsorption onto $CaCO_3$ precipitates is not a significant process in the swamp. Sediment resuspension and transport especially near the lake and in the major wastewater flow path influence both the suspended load and nutrient distribution. Internal cycling of nutrients seemed significant although and only about 30% of the incoming P remained in the system. There appeared to be no contact between the underlying firm sediment and the deep sediment, due to a negligible water exchange. Figure 8.23 summarizes the TP pathways in the Nakivubo

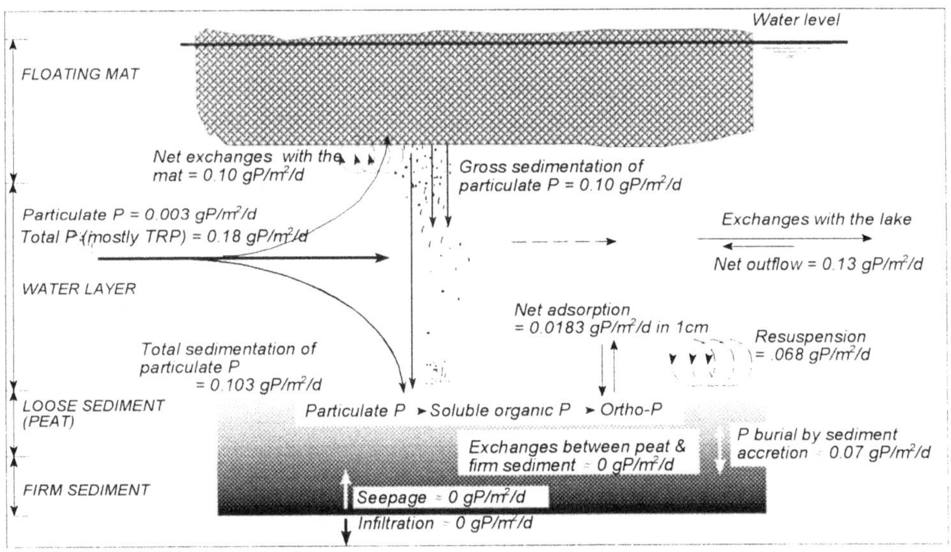

Fig. 8.23. Pathways and mass balance of particulate (and Nakivubo channel) P in the Nakivubo swamp.

swamp and quantifies the average fluxes.

8.6 References

Alaerts, G.J., Mahbubar, R., Kelderman, P. (1996). Performance analysis of a full-scale duckweed-covered sewage lagoon. *Wat. Res.*,30 (4): 843-852.

Allan, J.D. (1995). Stream Ecology: Structure and functioning of running waters. Chapman & Hall, pp. 388.

Andersen, F.O. and Jensen, HS. (1992). Regeneration of inorganic phosphorus and nitrogen from decomposition of seston in a freshwater sediment. *Hydrobiologia* 228 (1): 71-81.

APHA/AWWA. (1992). *Standard methods for examination of water and waste water.* Greenberg, E., *et al.*(eds.), Washington. D C.,18th edition.

ASTM. (1986) Annual book of ASTM standards: Soil and rock; building stones. Section 4, volume 04.08. D422:116-126.

Auer, M.T., Penn, M.R., van Orman, E.L. and Effler, S.W. (1993). Measurement and verification of rates of sediment phosphorus release for a hypertrophic urban lake. *Hydrobiologia* 253: 219-231.

Boyd, C.E., (1978). Chemical Composition of Wetland Plants. In: Good, R.E., Whigham, D.F., and Simpson, R.L., (eds.). *Freshwater Wetlands: Ecological Processes and Management Potential.* Academic Press, New York, pp. 155-167.

Burns, N.M., and Ross, C., (1972).Oxygen-nutrient relationshlps within the central basin of Lake Erie. Project HYPO: Canada Centre for Inland Waters Paper no 6 ; and Environmental Protection Agency Technical Report .

Cooke, G.J. (1992). Phosphorus removal processes in a wetland after a decade of receiving sewage effluent. *J. Environ. Qual.* 21 (4): 733-739.

Cooke, G.J., Stub, L., and Mora, N., (1992). Fractionation of phosphorous in the sediment of a wetland after a decade of receiving sewage effluent. *J. Env. Qual.* 21 (4): 726-732.

Danen-Louwerse, H., Lijklema, L. and Coenraats, M., (1993). Iron content of sediment and phosphorus adsorption properties. *Hydrobiologia* 253: 311-317.

Faulkner, S.P. and Richardson, C.J. (1989). Physical and chemical characteristics of wetland soils. In: Hammer, D.A. (ed.) *Constructed Wetlands for Wastewater Treatment: Municipal, Industrial and Agricultural.* Lewis Publishers, Michigan, pp. 41- 47.

Fennessy, M.S., Brueske, C.C. and Mitsch W.J. (1994). Sediment deposition patterns in restored freshwater wetlands using sediment traps. *Ecological Engineering,* 3 (4): 409-428.

Furumai, H. and Ohgaki, S. (1982). Fractional composition of phosphorus in sediments related to release. *Wat. Sci. Tech.*, 14: 215-226.

Gaudet, J.J. (1977). Uptake, accumulation and loss of nutrients in tropical swamps. *Journal of Ecology,* 50: 415-422.

Gaudet, JJ. (1979). Seasonal changes in nutrients in a tropical swamp: North swamp, Lake Naivasha, Kenya. *Journal of Ecology,* 67: 953-981.

Gaudet, J.J. (1980). Nutrient dynamics of papyrus swamps. In: Gopal, B. (ed.) *Proc. of first Int. Wetlands Confr.*, National Institute of Ecology and International Scientific publications, New Delhi-India, pp.305-319.

Gaudet, J.J. and Muthuri, F.M. (1981). Nutrient Relationships in Shallow Water in an African Lake, Lake Naivasha. *Oecologia*, 49 (1): 109-118.

Håkanson, L. and Jansson, M. (1983). *Principles of Lake Sedimentology*. Springer-verlag, New York.

Hieltjes, H.M.A., and Lijklema, L., (1980). Fractionation of inorganic phosphates in calcareous sediments. *J. Environ. Qual.* 9:(3) 405-7.

Howard-Williams, C. (1979). Interactions between swamp and lake. In: *Lake Chilwa studies of changes in a tropical ecosystem* (Kalk M.A.J *et al.* eds.). *Monographieae biological* 35: 233-245.

Kadlec, RH., and Knight, R.L., (1996). *Treatment Wetlands*. Lewis publishers.

Kamp-Nielsen, L. (1989). Sediment-water exchange models. In: Jϕrgensen S.E. and Groemic M.J.(eds.). *Mathematical submodels in water quality systems*. Elsevier, pp.371-401.

Kelderman, P. (1985). *Nutrient dynamics in the sediment of Lake Grevelingen (SW Netherlands)*. Ph.D. Thesis, Groningen University. Delftse Universitaire Pers., Delft.

Klapwijk, S.P., Kroon, J.M.W. and Meijer, M.L. (1982). Available Phosphorus in lake sediments in The Netherlands. *Hydrobiologia,* 92: 491-500.

Keulder, P.C. (1982). Particle size distribution and chemical parameters of the sediments of a shallow turbid impoundment. In: Sly, GP., (ed.). *Developments in Hydrology 9: Sediment freshwater interactions*. DW Junk Publishers, pp. 341-354.

Klute, P. (1986). *Methods of soil analysis, Part 1. Physical and mineralogical methods*. 2nd. edn. American Society of Agronomy.

Kramer, JR., Herbes, S.E., Allen, HE., (1972). Phosphorus: Analysis of water, biomass, and sediment. In: *Nutrients in natural waters*, John Wiley and Sons, Inc., New York, pp. 51-100.

Ku, C.W., DiGiano, F.A., and Feng, T.H. (1979). Factors affecting phosphorus adsorption equilibria in lake sediments. *Water Research*. 12: 1069-1074.

Lijklema, L. (1980a). Interaction of ortho-phosphate with iron(III) aluminium(III) hydroxides. *Environ. Science and Technology,* 14:537-541.

Lijklema, L. (1980 b). Eutrophication: The role of sediments. *Hydrological Bulletin,* 14: 98-105.

Lijklema, L. (1986). Phosphorus accumulation in sediments and internal loading. *Hydrobiological Bulletin*. 20 (1/2): 213-224.

Lijklema, L., Gelencser, P., Szilàgyi, F., and Somlyódy, L. (1986). Sediment and its interaction with water. In: Somlyódy and van Straten (Eds.), *Modelling and managing shallow lake eutrophication*. Springer-verlag, 157-203.

Lijklema, L. (1990). Mass balances and cycling of nitrogen and phosphorus in flooded soils. In: van Oorschot MMP. and Verhoeven JTA., (Eds.), *The Utrecht Plant Ecology News Report no.11*. Post Academic Course on Wetlands for Purification of Water. p.44-51.

Lijklema, L. (1993). Considerations in modelling the sediment-water phosphorus exchange. In: *Hydrobiologia 253*: 219-231. Proceedings of the 3rd. International workshop on phosphorus in sediments (Boers *et al.*, eds.). Kluwer Academic Publishers.

Lind, E.M., and Visser, S.A. (1962). A study of a swamp at the north end of Lake Victoria. *Journal of Ecology*. 50: 599-613.

Maidment, R.D (1992). *Handbook of Applied Hydrology*. McGraw Inc.

Metcalf and Eddy, Inc. (1979). *Wastewater Engineering, Treatment, Disposal and Reuse*. 2nd. Edn. Revised by Tchobanoglous. McGraw-Hill Book Company .

Mitsch, W.J. and Gosselink, G.J. (1993). *Wetlands*. van Nostrand Reinhold, 2nd. edn.

Muthuri, M.F. and Jones, B.M., (1996). Nutrient distribution in a papyrus swamp: Lake Naivasha, Kenya. *Aquatic Botany*. 56: 35-50.

Novozamsky, I., Houba, J.G., van Eck, R. & van Vark, W. (1983). A novel digestion technique for multi-element analysis. *Commun. in soil sci. plant anal.,* 143: 239-249.

Patrick Jr., W.H. (1990). Microbial reactions of nitrogen and phosphorus in wetlands. *In:* van Oorschot, M. M. P. and Verhoeven, JTA. (eds.). *The Utrecht Plant Ecology News Report,* Utrecht, pp.52-63.

Peck, D.E., Corwin, D.L., and Farmer, W.J. (1980). Adsorption-Desorption of Diuron by freshwater sediments. *J. Environ. Qual.,* 9 (1): 101-106.

Psenner, R., and Pucsko, R. (1988). Phosphorus fractionation: advantages and disadvantages of study of sediment phosphorus origins and interactions. *Schweizerbart'sche verlagsbuchhandlung*, D-7000. Stuttgart, pp. 43-59.

Psenner, R., Boström, B., Dinka, M., Pettersson, K., Pucsko, R., and Sager, M. (1988). Fractionation of P in suspended matter and sediments. *Arch. Hydrobiol. Beih. Egenbn.Limnol.* 30: 98-103.

Reddy, KR., Overcash, M., Khaleel, R., and Westerman, P. (1980). Phosphorus adsorption-desorption characteristics of two soils utilized for disposal of animal wastes. *J. Environ.Qual.* 9:1.

Stirling, HP., and Wormald, P. (1977). Phosphate /sediment interaction in Tolo and long Harbours, Hong Kong and its role in estuarine phosphorus availability. *Estuarine and coastal Marine Science* 5: 631-642.

Stone , M. and English, M.C. (1993). Geochemical composition, phosphorus speciation and mass transport of fine grained sediment in two L. Erie tributaries. *Hydrobiologia* 253: 17-29.

Stumm, W. and Morgan, J.J. (1970). *Aquatic chemistry: An introduction emphasizing chemical equilibria in natural waters*. John Wiley and Sons, New York.

Stumm, W., and Morgan, J.J. (1996). *Aquatic chemistry: Chemical equilibria and rates in a natural waters*. 3rd. edition. John Wiley and Sons, New York.

Thomann, V.R., and Müller, A.J. (1987). *Principles of Water Quality Modelling and* Control. Harper and Row Publishers, Inc.

Thompson, K., and Hamilton, CA. (1983). Peatlands and swamps of the African continent. In: *Mires: Swamps, Bogs, Fens and Moores.* Gore, APJ. (Ed.) Regional ecosystems of the world 4B, Elsevier.

US Dept. of Interior (1969). *Earth manual.* A WR Technical Publication, 5th Ed. Denver.

Visser, S.A. (1964). A study in the decomposition of *Cyperus papyrus* in the swamps of Uganda, in natural peat deposits, as well as in the presence of various additives. *East African Agricultural and Forestry Journal*, 29 (3): 268-287.

Wetzel, G. R. (1975). *Limnology*. W.B. Saunders Company, London.

Chapter 9

Mass balance of the Nakivubo swamp

Nalubega, M. and Kansiime, F.

9.1 Introduction

Wetlands are reported to maintain efficient biological cycles that allow these ecosystems to function as nutrient traps (Patrick and Reddy, 1976; Howard-Williams, 1985; Armentano, 1980) and biological systems that metabolize exogenous organic matter, among other functions (Chapter 2). The capacity of a wetland to trap nutrients depends on hydrological and climatic factors, system geomorphology, chemical composition of inflowing water, soil chemistry, vegetation and the overall eutrophic status of the system.

The capacity of natural wetlands to sustainably retain nutrients is however under contention. Gehrels and Mulamoottil (1989) found 122% of the influent's bioavailable ortho-phosphorus in the swamp effluent, although TP was reduced by 50%. A review of regional wetland nutrient retention by Richardson (1990) led to his conclusion that the notion of wetlands being traps may be unfounded. He argued that this belief is usually made from the large nutrient storage in the system (notably the peat layer component that has accumulated over hundreds of years) and not the very low sediment/peat accretion rates (1 -10 mm/yr [Lijklema, 1990; Mitsch and Gosselink, 1993]).

To assess the potential of a wetland to trap and sink nutrients therefore, long-term detailed studies of the effect of wastewater loading on the flora, fauna, water quality and the geochemistry of these systems are needed. This is essential to protect the diminishing number and area coverage of the world's natural wetlands and to avoid the mistakes made when on the initiation of the self-purification idea, it was assumed that rivers and lakes could handle unlimited loads. It was only found later that even this self-purification capacity is limited and surface water can actually be overloaded (Sloey *et al.*, 1978). Mass balances, either for inlet-outlet analysis (black box approach) or including identification of internal processes and their kinetics, is the way to find out the retention of nutrients in these ecosystems.

In wetlands with a potential to act as treatment or disposal for wastewater, mass balances are carried out to: (i) assess overall removal degree and rate of pollutants from wastewater as it flows through the system; (ii) identify the pathways of different nutrients that would allow optimization of removal process; and (iii) facilitate forecasting of wastewater loading effects for managerial purposes.

Comprehensive mass balances require extensive measurements of hydrological (water balance) and nutrient contents in the systems, and are labour intensive. Moreover, some components of the balance may be difficult to determine and often estimates are made. Mass balances have been made for constructed macrophytic treatment systems in different parts of the world (Guida and Kugelman, 1989; Moshiri, 1993; Alaerts *et al.*, 1996). Estimates of nutrient balances in natural systems are usually more difficult owing to their scale and complexity especially with respect to hydrology and morphology. Nonetheless, numerous studies on mass balances of natural and artificial systems have been carried out (Boyt *et al.*, 1977; Gaudet and Muthuri, 1981; Knight *et al.*, 1987; Gehrels and Mulamoottil, 1989; Mitsch and Reeder, 1991; Hosomi *et al.*, 1994; Alaerts *et al*, 1996).

The net effect of a wetland on water quality is the result of cumulative fluxes into the storage and transport compartments and into the receiving water and atmosphere (Klopatek, 1978). These compartments and the processes responsible for nutrient transport into and out of them in a floating wetland are illustrated in Fig.1. They are the macrophyte/periphyton compartment, the mat floating over a water layer and into which the periphyton is emergent, the water, the peat compartment which is made up of unconsolidated but more decomposed than the mat material that falls through or out of the mat and the bottom sediment compartments (over which the peat settles). The processes involved in nutrient transformation and transport are outlined by several authors (Gaudet, 1980; Richardson, 1990; Koerselman and Verhoeven, 1992; Sasser, 1994; Reddy and D'Angelo, 1997) and are summarised in Fig.1.

The mass balance terms commonly evaluated for the different compartments include:

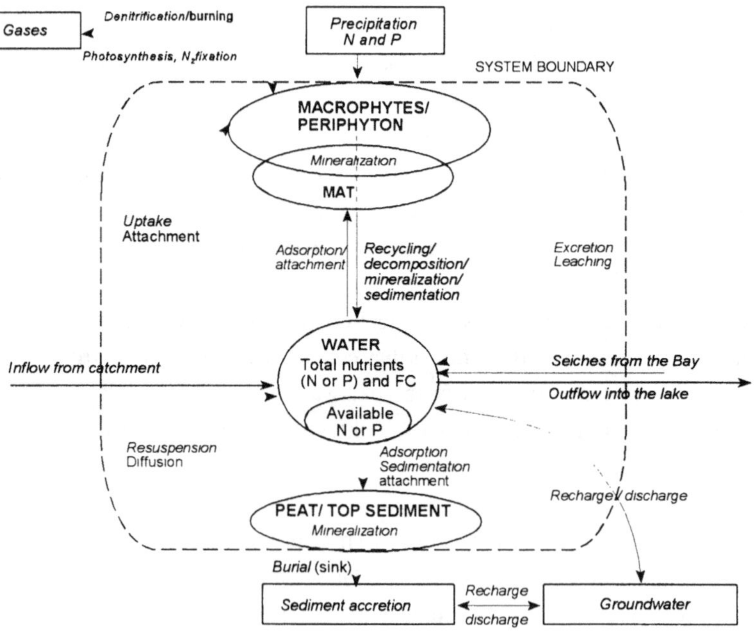

Fig. 9.1. Compartments and processes of nutrient transport and transformations in a floating wetland.

(i) biomass (macrophyte and periphyton) N and P pool, and their nutrient biomass uptake; exports of nutrients due to biomass harvest, retranslocation of nutrients from senescing plants (estimated from the difference in nutrient content between dead and living plant parts attached to plants), nutrient release from above ground litter upon death, and decomposition/ mineralization; (ii) hydrologic imports and exports of nutrients via surface waters (including wastewater discharges), precipitation, etc.; and fluxes into and from (iii) sediment/peat N and P pool with transformation processes including adsorption/desorption, sedimentation/ resuspension, decomposition/mineralization, ammonification, nitrification and denitrification.

A study of the processes involved in nutrient transformation, and (thus) factors affecting the mass balance in a wetland should include vertical and longitudinal profiles of nutrient allocations and fluxes.

The Nakivubo swamp in Kampala (Chapter 2) has received partially treated wastewater for over 30 years. The quantities and transformation of the wastewater BOD, nutrients and coliforms were not quantified. It was required that these quantities be assessed to protect the quality of the water in the Inner Murchison Bay (Kampala's water source), and to maintain the wetland ecosystem (Chapter 1).

The objectives of this part of the study therefore were to:

(i) evaluate nutrient accumulations in different compartments of the Nakivubo swamp (in relation to the loading in different vegetation zones, Chapter 2);

(ii) assess the relative importance of the different compartments as nutrient sinks/traps in the Nakivubo swamp, and;

(iii) evaluate the overall nutrient and faecal coliform removal efficiencies in the swamp.

9.2 Assessment of the wetland system components

This study is an integration of the work presented in Chapters 3 through 8. Detailed methods for data collection were discussed therein.

Plant contribution to below-ground mass

The measurement of below-ground biomass in floating wetlands (*i.e*, below the water level) is difficult and estimates are based on either simulation in hydroponic systems or from extrusions of small cores from a very heterogeneous system. The errors involved in extrapolation of the results to the whole swamp may be substantial. However, it is the best that could be achieved under difficult experimental circumstances and it does give an understanding of the variations in the system. Results of this study based on weight increases of plants grown in buckets (Chapter 5) gave a below-ground biomass of 45% of the total dry weight (DW) for papyrus and 25-30% for *Miscanthidium* (based on the results of potted plants in Chapter 5). Similar estimates for papyrus in Kenya were 25% (Muthuri *et al.*, 1989 - for a hydroponic system) and 30 - 40% for emergent aquatic vegetation (Thompson, 1985). For a mixed-vegetation floating marsh in Florida, Sasser (1994) reported 50% of the total productivity to be below ground. Klopatek (1978) reported a below-ground biomass of 33% for *Typha latifolia*. Howard-Williams and Gaudet (1985) gave an above ground biomass for papyrus swamps in East Africa of 52%. The values used in this chapter are those measured for our systems (55:45 for papyrus and 70:30 for *Miscanthidium* for the above:below ground portions respectively), and are well comparable with those measured by other researchers.

The nutrient content of the above ground biomass for papyrus was estimated from the contents of the individual plant parts (umbel and culm) multiplied by a factor taking into account the ratio of the weight of the umbel to the culm, which was found to be 0.45 (Fig. 9.2).

$$(C)_{above} = AG * \{(NC)_{umbel} * 0.45 + (NC)_{culm} * 0.55\}$$

where $(C)_{above}$ = nutrient content in above-ground biomass (g/m^2); AG = above ground biomass (g/m^2); (NC) = nutrient content of plant part (g/g DW). The same approach was used for the below-ground biomass for the *Miscanthidium* vegetated zones.

Background data

The total study area was found to be 1,150,000 m^2 of which 80% is papyrus vegetated and 20% is mainly *Miscanthidium* vegetated (Chapter 2). Because of the channelization of water

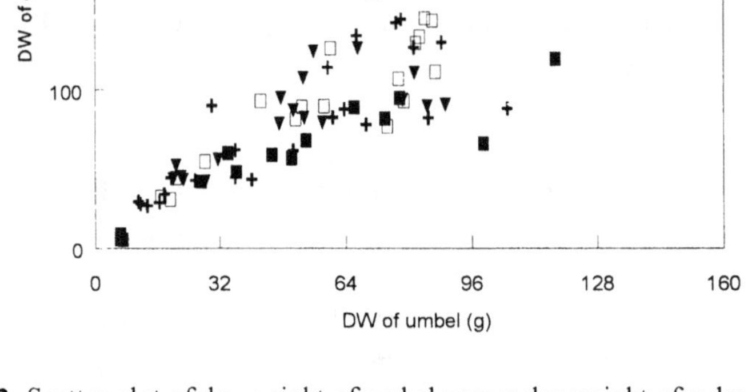

Fig. 9.2. Scatter plot of dry weight of umbel versus dry weight of culm for papyrus from different locations in the Nakivubo swamp. T1 to T4 are the study transects. $r_{(all)} = 0.82$, $p < 0.001$, n=81.

flows in the system due to morphological and vegetation differences, two flow zones (corresponding to different treatment efficiencies) were identified. These are the major flow paths whose cross-sectional area along T4 was 600 m x 1.5 m, and the minor flow path at both edges with an area of 250 m x 0.9 m. This cross-sectional area ratio of 4:1 for the major and minor flow paths was used in the determination of material flows through the two areas.

Values for the average field results of above-ground biomass growth rates from Chapter 5 were used {4,770 g/m^2/year or 13.1 g/m^2/d for papyrus, and 1,850 g/m^2/year or 5.07 g/m^2/d for *Miscanthidium;* equivalent biomass including below-ground parts for the two plant types are respectively 8,673 g/m^2 and 2,643 g/m^2).

Table 9.1 is a summary of the nutrient contents in the solid components of the swamp over a vertical profile. This data is used in the determination of masses in subsequent components.

For leaching of the plant material into the mat/water complex, 24 hours laboratory leaching results were taken. Changes after longer periods were found to be negligible (Chapter 5, Figures 5.18 and 5.19 respectively) and could be related to secondary decomposition. These quantities were assumed to be released once in the life cycle of the plants.

Based on the measurements in Chapter 5, the life cycle of both major vegetation types was taken as 8 months to 1 year. The maximum biomass was reached within 5 - 7 months for cropped plots, and was reported to be within 6 months by Muthuri *et al.* (1989). It was therefore assumed that in an undisturbed system, regeneration (die-off and replacement by juveniles) would occur within 1 year.

Table 9.1 Nutrient allocation in the vertical compartments of Papyrus and *Miscanthidium* vegetated zones. Shown are the areal densities of the N and P (using average depths) and the total content of these nutrients in tonnes over the study area (1,150,000 m^2)

COMPARTMENT (thickness [m] for papyrus, *Miscanthidium*)	Papyrus (80% of study area)				*Miscanthidium* (20% study area)			
	N		P		N		P	
	g/m^2	tonnes	g/m^2	tonnes	g/m^2	tonnes	g/m^2	tonnes
Plants	113	104	18	16.8	16.9	3.9	4	0.9
Mat (0.4, 1.4)	350	322	53	49	1660	381	206	47.4
Peat (0.5, 0.05)	1210	1113	165	152	114	26.2	21	4.8
Bottom/firm sediment (top 10 cm)	1064	979	267	245	1369	315	222	51
TOTAL	2637	2518	503	485	3160	726	453	104
Plant nutrient content	1.3 %		0.21%		0.64%		0.15%	
Mat bulk density (kg/m^3)	52.1				69.7			

9.3 Results

9.3.1 Water balance

A water balance of the swamp was evaluated on a daily basis for 17 months (April 1995 to August 1996; Chapter 3) and a summary of the findings is presented in Table 9.2. This water balance is used to estimate the mass balance of nutrients and coliforms. It should be noted that this is a 17 months' average and that there are indeed seasonal variations in the flows due to the distinct rainy seasons. This also influences the retention time of the wastewater in the wetland and channelization of water is evident (Chapter 3). The seiche effect however seemed to be less affected by the seasons. As a result, the average exchanges with the lake are therefore not greatly influenced by the daily fluctuations in flow.

Table 9.2 Summary of the water balance for the Nakivubo Swamp. (Values are ± SE)

Inflows	**m³/d**
The Nakivubo Channel	103,575 ± 11,520
Luzira sewage outflows	700 ± 70
Sub-surface flow from Luzira	2,315 ± 316
Sub-surface flow from Bukasa	3,060 ± 418
Precipitation	4,945 ± 1840
Outflows	
Deep ground water seepage	≃ 0
Evapotranspiration	5,750 ± 115
Net exchanges with the lake	108,845[a] ± 14,280

[a] =balance

9.3.2 Nutrient allocation in solid components of the vertical compartments

The average allocations of N and P over vertical profiles for the papyrus and *Miscanthidium* vegetated zones of the Nakivubo swamp are shown in Figs. 9.3 and 9.4. The higher nutrient contents in the papyrus vegetation than in the *Miscanthidium* vegetation are a salient feature. The higher biomass density of papyrus corresponds to a net uptake of 285±25 kg N/d[1] for the papyrus zones and 10.1±0.9 kg N/d for the *Miscanthidium* vegetated zones. Corresponding values for P are 46.0±4.8 kg P/d for papyrus and 2.5±0.2 kg P/d for *Miscanthidium*.

The nutrient content of the mat litter and detritus is high for both vegetation zones. The total N content of the papyrus mat is (0.40 m x 1 m^2 x 16.7 mg/g x 52 kg/m^3) = 0.35 kg N/m^2 compared with 1.66 kg N/m^2 for *Miscanthidium*. Equivalent values for P are 0.053 kg P/m^2 and 0.206 kg P/m^2 for the papyrus and *Miscanthidium* vegetated zones respectively.

---

[1]= {(13.1x100/55)/1000}x(1.3%) x80% x 1,150,000

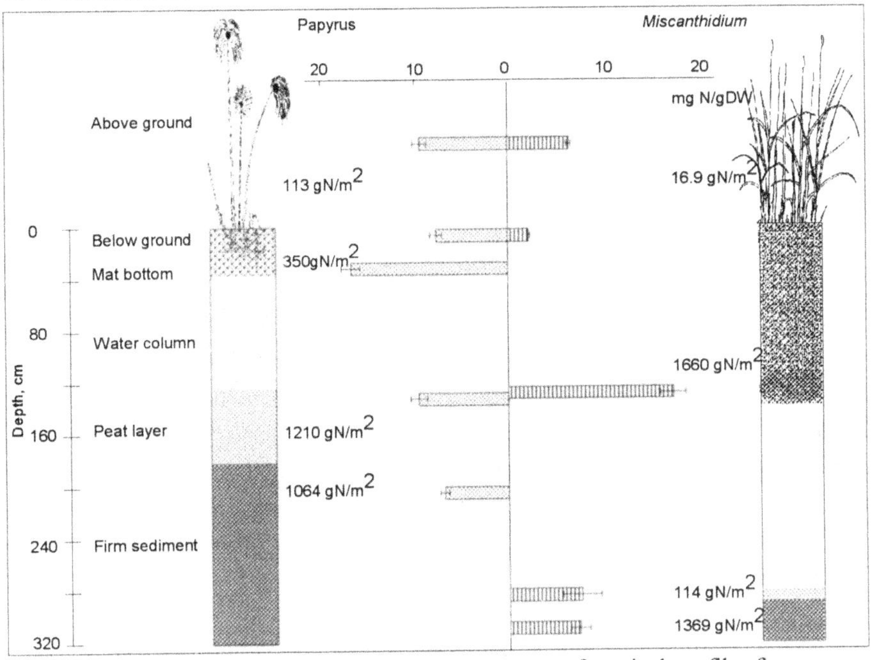

Fig. 9.3. Nitrogen allocations in the solid compartments of vertical profiles for papyrus and *Miscanthidium* in the swamp.

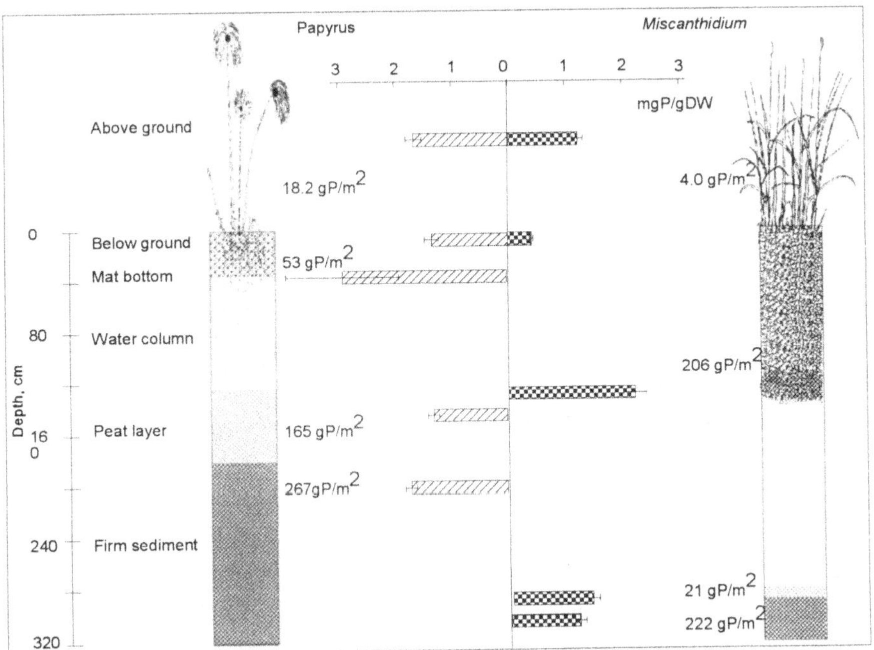

Fig. 9.4. Phosphorus allocations in the solid compartments of vertical profiles for papyrus and *Miscanthidium* in the swamp.

With respect to the peat compartment, papyrus vegetated zones have a much higher areal nutrient allocation than *Miscanthidium* vegetated zones. This is because of the very thin (≈5 cm) peat layers beneath *Miscanthidium* compared to the much thicker (average 50 cm) layers under papyrus. A layer of 10 cm thick bottom sediment can be in active nutrient exchange with the overlying water/peat (Wetzel, 1975). The equivalent nitrogen contents in the top 10 cm for both papyrus and *Miscanthidium* zones (1064 gN/m^2 and 1369 gN/m^2 respectively; density ≈1.850 kg/m^3) are of the same order of magnitude.

The net nutrient storage content is therefore about 2520 tonnes N and 500 tonnes P for papyrus and 730 tonnes N and 100 tonnes P for *Miscanthidium* vegetated areas (see Table 9.1). The high nutrient allocation in the *Miscanthidium* zones can be attributed to the high storage in the thick mat. This mat inhibits free mixing of swamp and mat water and therefore may limit nutrient fluxes and transformations within it (Chapters 4 and 5). Although the daily nutrient uptake is higher for papyrus than for *Miscanthidium* because of the difference in active biomass density, the present total amounts of nutrients, which is comparatively high in the *Miscanthidium* column is thus determined by long-term accumulation.

9.3.3 Nutrient allocations in the water phase of vertical profiles

Figures 9.5 and 9.6 show the nutrient content in the interstitial water of the different compartments. NH_4-N was the highest of total N (about 76%) in the influent. Fig. 9.5 shows a higher NH_4-N concentration in the water phase than either in the mat or peat for both vegetation types. This suggests that diffusive fluxes of nutrients would tend to occur towards the solid (mat and peat) components. The average NH_4-N content of compartments from *Miscanthidium* areas is higher than that for papyrus compartments possibly due to the channelization of the high nutrient wastewater as it flows through the swamp. The preferential flow path is towards the central *Miscanthidium* vegetated zones (Chapters 3 and 4).

The bulk of P in the water phase of the system is ortho-phosphorus (TRP; 74 - 90%). Phosphorus shows a somewhat similar picture to N with the exception that the mat surface water has high P concentrations (Fig 9.6). These can be explained by the leaching and excretion of plant material which is reported to occur continuously (Denny, 1987; Chale, 1989) and to lead to a more rapid release of P as opposed to N (Boyd, 1970; Wetzel, 1975; Prentki *et al.*, 1978; Verhoeven, 1986; Gopal, 1990).

Limited mixing of the water within the mat further prohibits the equilibration of P concentration over the mat depth. Phosphorus concentrations in the water are high for both vegetation types. Higher values are again seen in the *Miscanthidium* zone because of the preferential flow of wastewater in this area and possibly less interactions with sediment and plant material (and hence less removal). Lower concentrations in the interstitial water of both peat layers suggest possible diffusion of ortho-P towards the solid components.

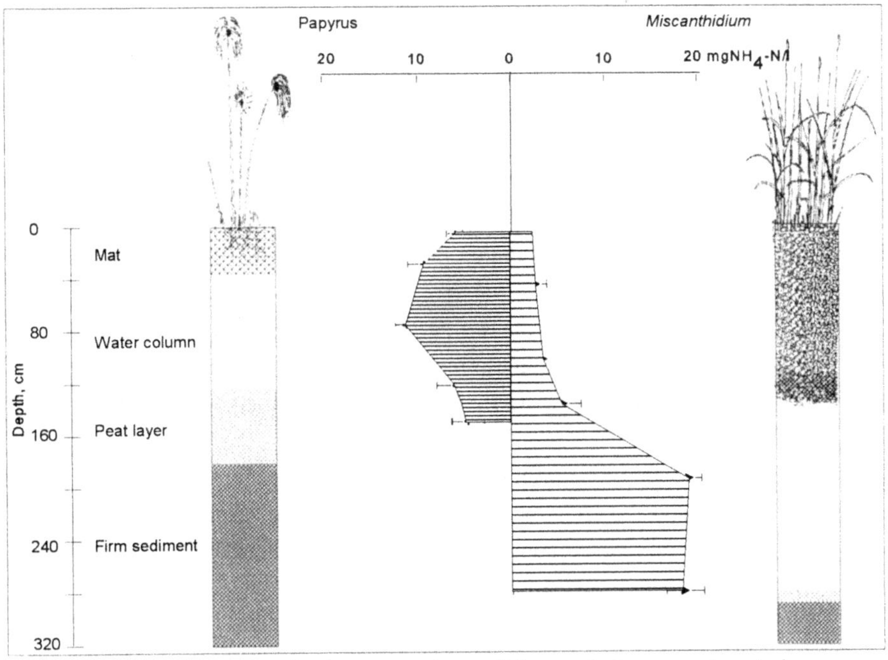

Fig. 9.5. Ammonia content in the water phase of the vertical compartments in papyrus and *Miscanthidium* vegetated parts of the Nakivubo swamp.

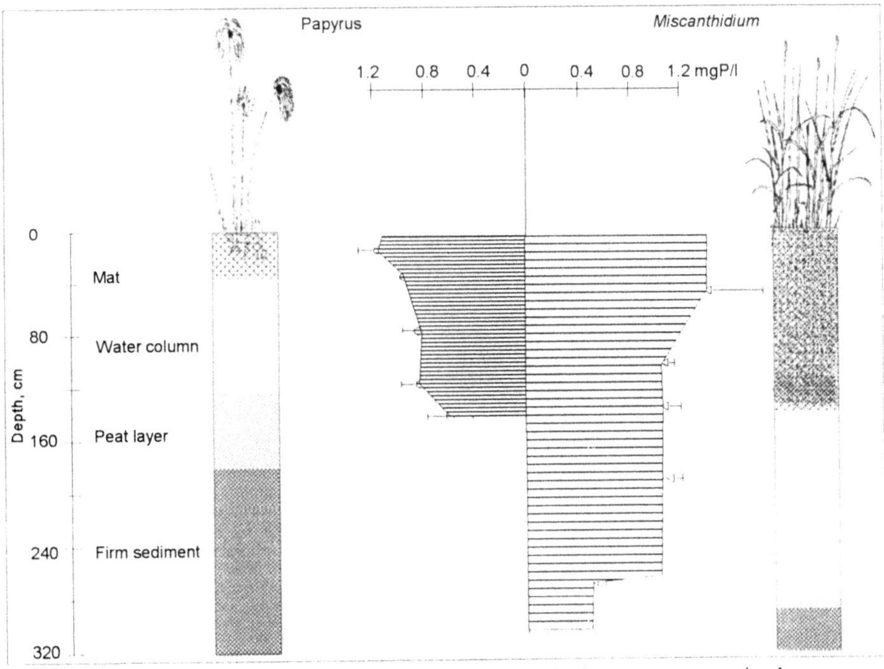

Fig. 9.6. Total P content in the water phase of the vertical compartments in the papyrus and *Miscanthidium* vegetated zones of the Nakivubo swamp.

9.3.4 Overall purification efficiencies

Figures 9.7, 9.8 and 9.9 show the pathways and balance of N, P and faecal coliforms as they traverse the swamp. Precipitation data used was the average of data for rainfall samples collected near the swamp during rainy periods. About 55% of NH4-N in the water was removed in the major flow path (inlet to T4-400) while 67% was removed in the minor path (inlet to T4 200). The major processes of removal are believed to be ammonification/ nitrification/denitrification, mineralization/decomposition, plant uptake and sediment accretion.

The pathways of P as the water moves through the swamp (Fig. 9.8) indicate that 33% and 67% TP is removed from the wastewater in the two paths, respectively. The dominant removal processes are sedimentation of organic P, adsorption, plant uptake, and accretion.

Faecal coliforms (Fig.9.9) are removed by up to 89.3% in the main wastewater flow path and 99.3% in the minor path. The dominant processes of removal are sedimentation and attachment to particulates, and natural die-off. The role of predatory protozoa, although not enumerated in this study, is envisaged to be significant.

A summary of the mass balance (Table 10.3) shows that the unaccounted for N and P amount to 40% and -3% of influent loads, respectively. The large value for N may be due to possible denitrification within the swamp (Chapter 7), the undetermined adsorptive term and possible

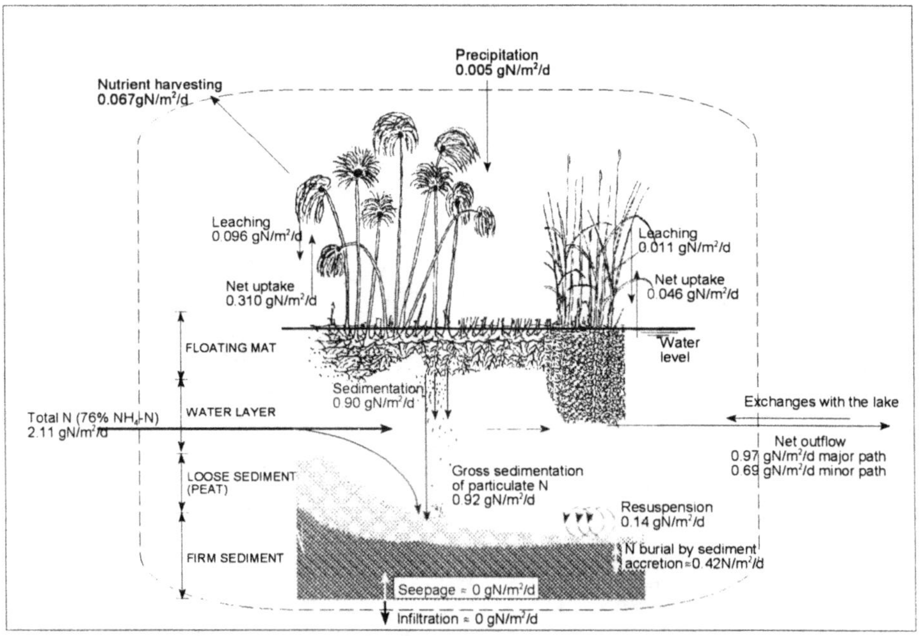

Fig. 9.7. Average N fluxes and balance for the Nakivubo swamp. Values are deduced from Chapters 4, 5 and 7.

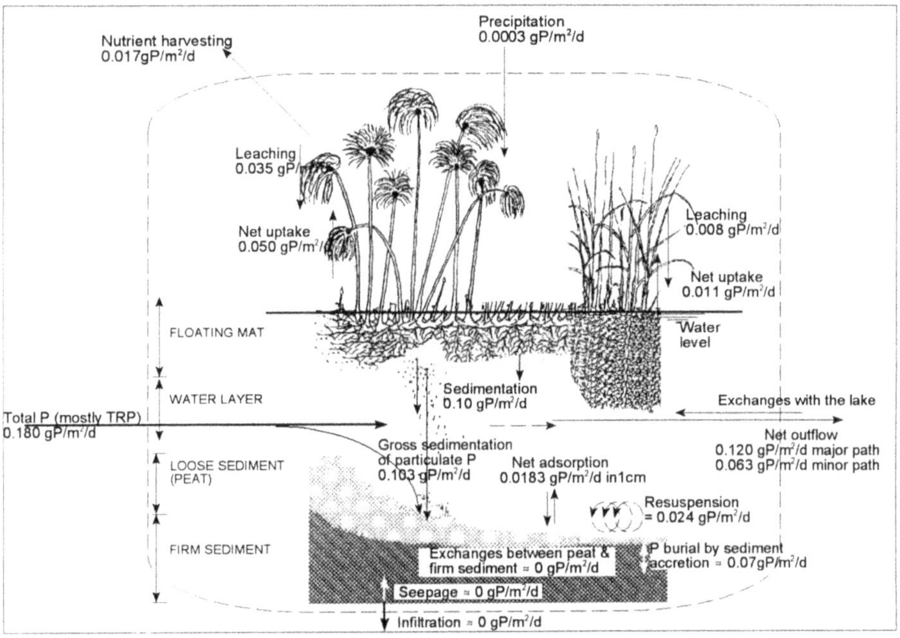

Fig.9.8. Average P fluxes and balance for the Nakivubo swamp. Values are deduced from Chapters 4, 5 and 8.

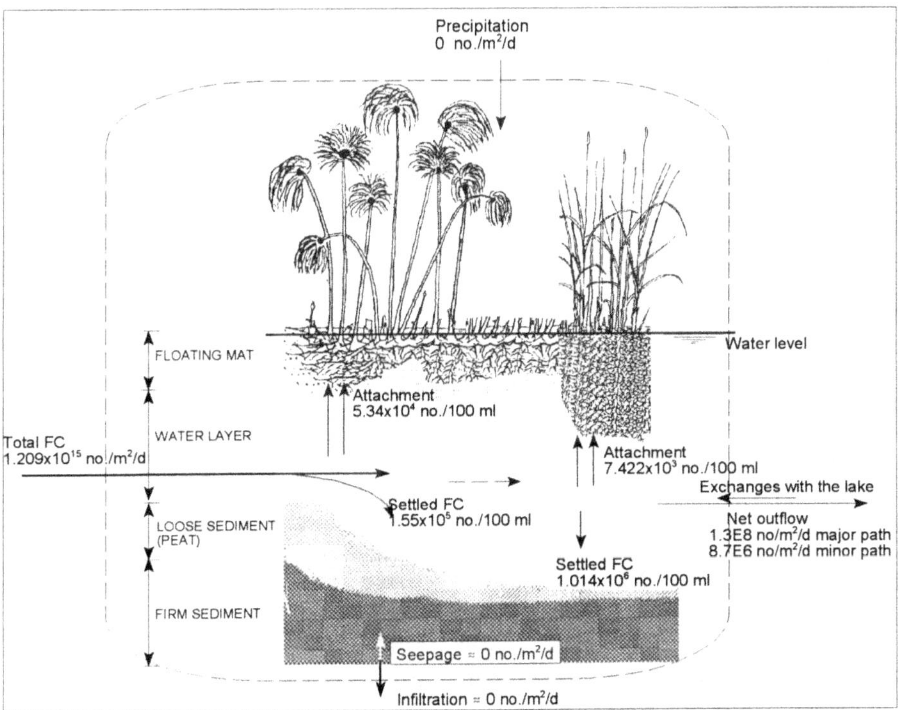

Fig.9.9. Fluxes and balance of faecal coliforms in the Nakivubo swamp. Values are deduced from Chapters 4, 5 and 6.

Table 9.3. Summary of mass balance for Nakivubo swamp **(A)** and a quantification of the internal processes **(B)** that partly account for the mass "remaining" in the swamp (row 5 of **(A)**). The values presented here are weighted averages that have taken into account the area covered by *Miscanthidium* and papyrus vegetation

	N $g/m^2/d$	P $g/m^2/d$	FC $no/m^2/d$
A. Mass balance			
Inflows (positive terms)			
Wastewater	2.11	0.180	1.21E9
Rainfall	0.01	0.0003	0
Outflows (negative terms)			
To the lake	0.91	0.109	1.06E8
Harvesting	0.07	0.017	---
Sediment accretion	0.21	0.035	---
Mass "remaining" in the swamp (X)	*0.93*	*0.019*	*1.10E9*
B. Internal processes			
Additive (to water; positive terms)			
Leaching	0.08	0.015	---
Resuspension	0.14	0.024	---
Reductive (from water; negative terms)			
Net sedimentation	0.02	0.003	(3.27E5 no/100ml)
Adsorption/attachment	---	0.018	
Plant uptake	0.26	0.042	(5.10E4 no/100ml)
Totals (Y)	*-0.06*	*-0.024*	---
Unaccounted for masses (X + Y)	**0.87**[1]	**-0.005**	---

[1]Amount mostly attributed to loss by denitrification

overestimation of input terms, like accretion. Denitrification in wetland systems can account for as much as 30% of the influent N. A mass balance carried out for Kenyan constructed wetlands with diverse vegetation types yielded unaccounted for N (and attributed to denitrification) equivalent to 1/3 of the influent amount (Nyakang'o and van Bruggen, 1998). In a green house papyrus wetland in Delft, Mujobeki (1994) gave denitrification equivalents of 20-50% of the influent for C:N ratios between 0.7 and 2. If the 0.87 $g/m^2/d$ N in Table 10.3 is attributed to denitrification, then it would account for about 40% of the influent N, a value that the above 2 studies seem to support.

The P value was less, possibly due to its conservative nature (no gaseous exchanges) which led to a better estimate of the terms. Rainfall contributed a minor amount to the overall nutrient load (0.2% for both N and P).

It should be noted however that floods had the effect of accelerating channelization of flow especially in the major flow path. Hence higher loads of pollutants would rapidly reach the lake. In addition, settled material in the lake would most likely be flushed into the lake as could be deduced from the very small layer of peat in the whole swamp (although this was mostly pronounced in the major flow path.

9.4 Modeling water quality in the Nakivubo swamp

A one dimensional, multi-sectional model was used to simulate the quality of water in the swamp with respect to nitrogen, phosphorus and conductivity. In addition to the initial and boundary conditions plus external variables used in the flow model in Chapter 3, the parameters used to calibrate this quality model were the dispersion coefficient and the decay rate constants (Appendix 9.1).

In the Nakivubo swamp, it was assumed that conductivity is a conservative variable and therefore any changes were a result of dilution. Based on this assumption, it was shown that the efficiency of removal of a pollutant (or its conversion) can be given by equation 9.1 (the derivation of this equation is presented in Appendix 9.2).

$$\mathcal{E}(\%) = \frac{X}{C_1 V_1} \cdot 100 = [\ 1 + \frac{C_2}{C_1}(F_d - 1) - \frac{C}{C_1}F_d] \cdot 100 \tag{9.1}$$

where $\mathcal{E}$ is the purification efficiency

X is the material that has been transformed in the system (sinks)

C and V are pollutant concentrations and volume of water at different locations as in the figure of Appendix 9.2

F_d is the dilution factor (Appendix 9.2) and is given by the expression: $F_d = (V_1 + V_2)/V_1 = (E_1 - E_2)/(E - E_2)$, where E is the electrical conductivity.

Hence the dilution factor F_d has been shown to be equivalent to the ratio of the conductivities of the influent (E_1) and the point of concern (E) from which lake concentrations have been corrected. Equation 9.1 can be used for assessing the pollutant removal from different sections in the swamp.

It was assumed that the decay rate of pollutants in the swamp follows the first order rate equation:

$$\frac{dc}{dt} = -kc \qquad \text{from which} \tag{9.2}$$

$$\ln\frac{c_t}{c_o} = -kt \qquad \textit{giving} \qquad k = -\frac{1}{t}\ln\frac{c_t}{c_o}$$

Where c_t is the concentration of the pollutant at time t (or a distance x away), c_o is the initial concentration and k is an overall decay rate constant. The value of c_t/c_0 is the expression of the fraction of the pollutant remaining in the system and is equivalent to (1- efficiency, $\mathscr{E}$ %).

Nutrient concentration data from Chapter 4 was used to estimate the value of this rate constant for N and P, over the different sections in the swamp. Travel times shown in Table 3.5.4 of Appendix 3.5.5 were applied. Table 9.4 shows the derived values for k. These values were used as first estimates for the decay rate constant in the model. Hence the k value takes into account the correction for the dilution in equation 9.1.

The conductivity was first calibrated using the dispersion coefficient as an adjustable parameter.Under the existing flow conditions, values for d >2 m^2/d in the central flow path created instabilities in sections 40 and 41. Close to the lake (sections 44 and 45), relatively high (>1 m^2/d) dispersion coefficient values led to a rapid decrease in conductivity contrary to what is seen in the field. This is because high dispersion in this area of high flows (as a result of large volumes of water entering the swamp with the seiches) leads to high dilutions. Close to the inlet however, very low dispersion factors led to increases in pollutant concentrations.

After calibration of the model for EC, the calculated decay rate constants in Table 9.4 were used to simulate the concentrations of ammonia nitrogen and the total reactive phosphorus at different points in the swamp. It was however found that these values of the decay rate led to decreases in the nutrient concentrations to levels below observed values. Hence the decay rates were adjusted to values that were about 20% of the calculated values (k_p = 0.01/d and k_n = 0.02/d) This change in the k values did not affect the variability of the simulated concentrations but only led to more stable nutrient concentrations. Figure 9.10 shows the simulated output for conductivity, ammonia nitrogen and reactive phosphorus for the major flow path.

Table 9.4 Calculated decay rate constants for N and P in the Nakivubo swamp from equations 9.1 and 9.2, and nutrient and conductivity values in Chapter 4 and in Appendix 9.2. .The average travel time was calculated as shown in Appendix 3.5.5

Section	$\mathscr{E}_n$ (%)	$\mathscr{E}_p$ (%)	Average time (d)	k_n (d^{-1})	k_p (d^{-1})
Bukasa route	48.6	54.0	17.5	0.04	0.04
40 to 42	26.2	13.6	1.85	0.16	0.07
43 to 45	9.9	11.6	1.35	0.08	0.09
40 to 45	37.7	25.2	3.20	0.15	0.09

$\mathscr{E}$ is the average efficiency of pollutant removal in the section.

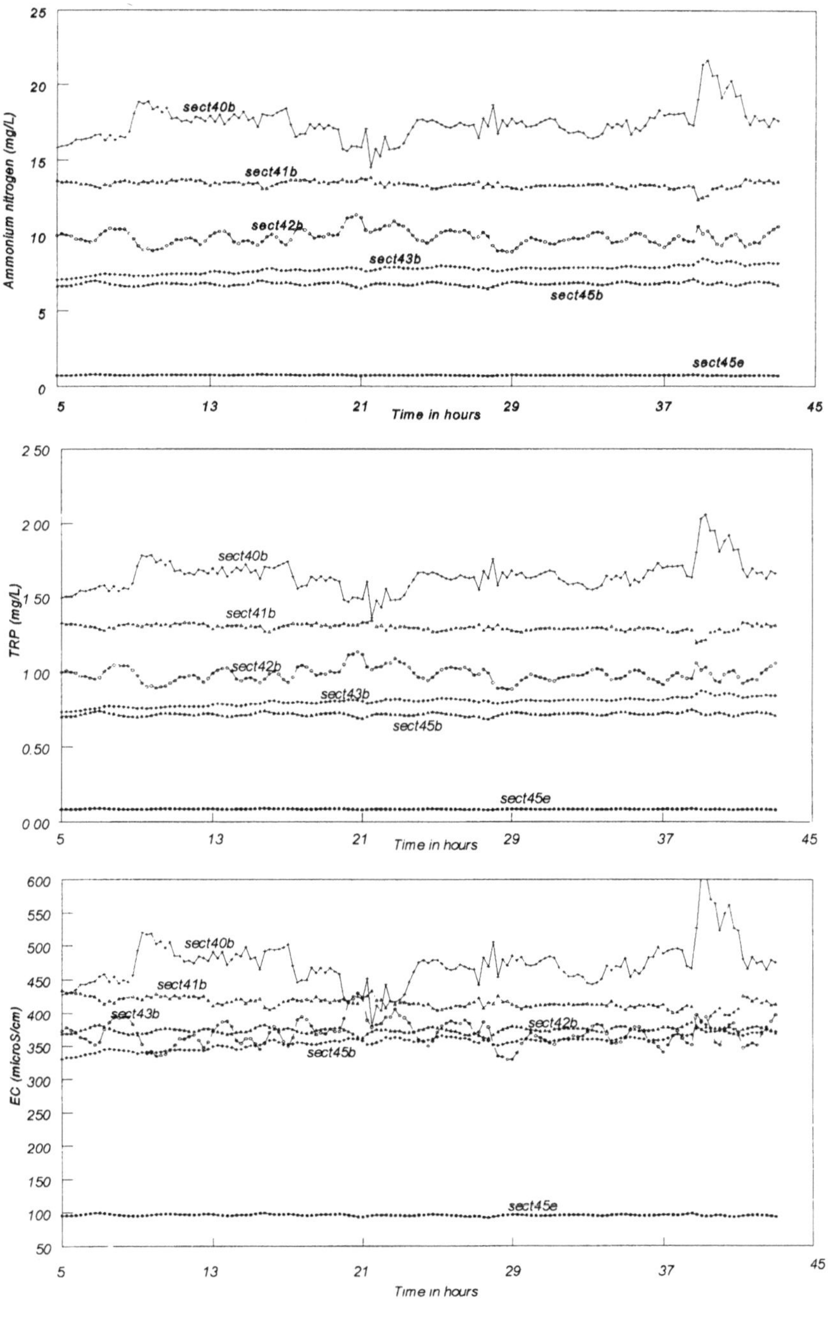

Fig. 9.10 Simulated variation of conductivity, nitrogen and phosphorus concentrations at different sections in the swamp.

The justification for the use of conductivity as a tracer of the wastewater flow patterns was the fact that conductivity and the nutrient concentrations were found to be correlated (Chapter 4). Figure 9.11 depicts this relationship at a selected node 45. Water level variation in the swamp was shown to be directly correlated to the conductivity (and nutrient concentrations) in Chapter 4. The model also shows this relationship as seen at different sections along the major flow path (Fig. 9.12).

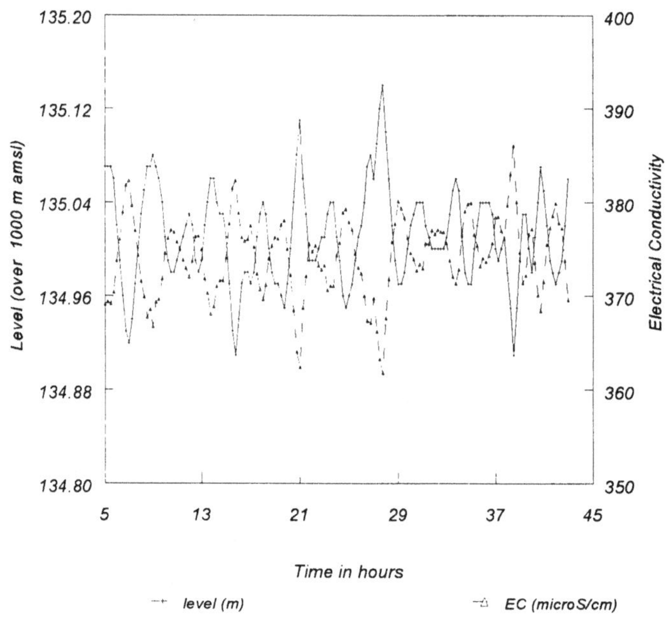

Fig. 9.11 Relationship between water levels and conductivity at node 45. Compare with Fig.3.16.

The model attempts to simulate the behaviour and distribution of pollutants within the swamp. The initial values of the simulation were important since the variation of the simulated concentrations with time was quite low, and hence after the selection of an appropriate decay factor (which led to non-falling concentrations), average measured values were used. Simulated concentrations varied along these initial values.

A comparison of the measured and simulated data showed that the model was a fair representation of the concentrations at the inlet sections but was not so good for the other sections of the swamp (figures 9.13 and 9.14). Figure 9.13 shows that the simulated conductivity is more varied than the measured one but varies to within $\pm$ 50 μS/cm at the inlet and by a slightly larger range for the swamp/lake interface. For TRP, again the measured and simulated data were in better agreement at the inlet compared to the swamp/lake interface.

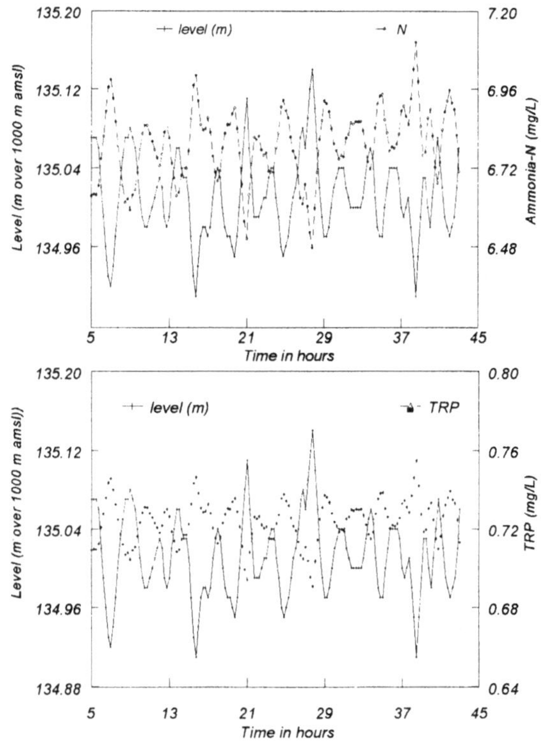

Fig. 9.12 Relationship between water levels and pollutant concentrations at node 45.

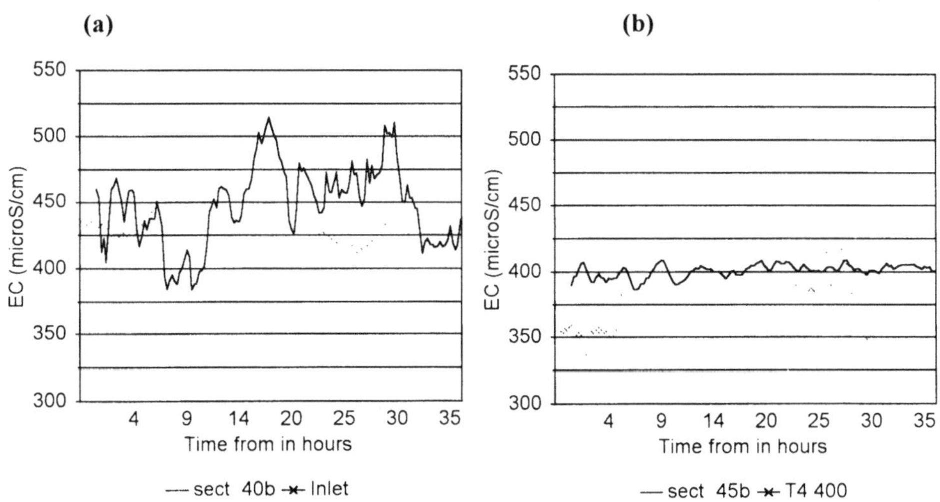

Fig. 9.13 Comparison of simulated and measured electrical conductivity (EC) at node 40 (inlet; **(a)**) and at node 45 (T4 400; **(b)**).

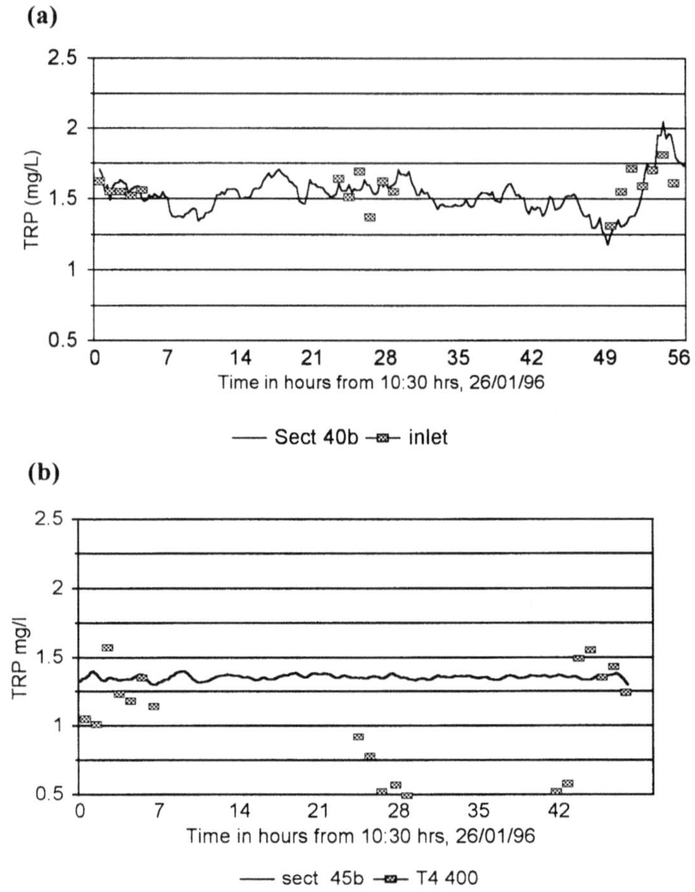

Fig. 9.14 Comparison of the simulated and measured TRP at node 40 (inlet; **(a)**) and at node 45 (T4 400; **(b)**).

9.5 Discussion

Water balance and flows

Most of the wastewater in the lower Nakivubo swamp flows through a central (major) path (average width is about 400 m, increasing from about 200 m near T1 to about 600 m along T4, closer to the lake). The major source of both wastewater and nutrients is the Nakivubo channel, contributing more than 90% of the allochthonous nutrient load into the swamp. High channel flows and the free floating vegetation in the centre of the swamp are responsible for the channelization of water. The effect of seiches from the lake was suggested to be one of modification of retention times and limited as opposed to full-scale dilution (Chapter 3).

Oxygen brought into the system by these seiches contributed to the nitrification capacity of the swamp (Chapter 7).

The role of vegetation

Nutrient uptake and excretion/recycling by wetland vegetation play a big role in the overall nutrient apportioning and balance. In Lake Chilwa, Howard-Williams (1975) attributed 90% of the net biomass production to macrophytes (*Phragmites*) compared with 6% by periphyton and 3% by algae. For the Nakivubo swamp the net average biomass production averaged 35.6 $g/m^2/d$ for papyrus and 10.1 $g/m^2/d$ for *Miscanthidium* The values for papyrus were close to the estimate by Muthuri *et al.*, (1989) of 41 $g/m^2/d$ for fertile sites. The maximum rates in the Nakivubo swamp (42.1 $g/m^2/d$) suggest that the swamp is indeed fertile possibly due to the nutrient load in the wastewater.

The only effective means of nutrient removal from swamps via macrophytes is harvesting. This is however limited as only the above ground biomass is harvestable. Sloey *et al.* (1978) estimated that P removal through harvesting in fresh water wetlands accounted for <0.014 g $P/m^2/d$. The harvest period is important since depending on the age of the plants, only 5% to 20% of the total nutrients may be stored in harvestable parts (Wetzel, 1975). Estimating that about 50% of the study area is annually harvested (Chapter 10), the quantities of nutrients removed through harvesting are therefore only about 3% (14.1 tonnes N/yr.) and 9% (3.6 tonnes P/yr.) of the total N and P in the influent, respectively.

Most of the nutrients taken up by plants would be recycled back into the system after the growth cycle if no harvesting takes place. Nutrients are reportedly excreted from plant parts into the water during plant growth (Denny, 1987). During senescence, nutrients are translocated into below-ground tissues (Thompson, 1985) and upon death nutrients are rapidly leached into the water (Wetzel, 1975; Prentki *et al.*, 1978; Nelson *et al.*, 1982; Kadlec, 1987; Chale, 1989; Gopal, 1990). On average 0.079 g $N/m^2/d$ and 0.03 g $P/m^2/d$ or 28% of N and 60% of P uptake by plants was recycled through leaching. Values reported by Gaudet and Muthuri, (1981) for leaching and decomposition of papyrus tissues in swamps of 0.041 g $P/m^2/d$ and 0.087 g $N/m^2/d$ are comparable.

Death of macrophytes leads to accretion and decomposition/mineralization of plant litter. Depending on the conditions of the swamp water column (depth, suspended solids content, water velocity, etc), the thickness/density of the mat and the decomposition rates, the decaying material may lead to accretion of mat material, increase the suspended solids load in the swamp water, or to peat build up. No significant peat build up occurs in the Nakivubo swamp, although highly decomposed material of up to 50 - 60 cm thick has resulted from the sedimentation of highly decomposed material from the mat bottom to the surface of the sediments, especially in papyrus zones.

The presence of large quantities of material in the mat and peat compartments suggests that these are large nutrient stores. The *Miscanthidium* mat, for example, accounted for about 76% and 71% of the total N and P allocations while the peat layer in papyrus is responsible for 46% N and 33% P. The thick mat of *Miscanthidium* limits free mixing of the water in the mat and minimises the potential of these macrophytes to get nutrients from the wastewater beneath. This is partly responsible for the high gradient in nutrient (especially N) concentrations in the water column of *Miscanthidium* vegetated zones compared to the mat. Lower N compared to P in the interstitial mat water suggests N limitation.

The role of decomposition and mineralization on nutrient transformations is again depicted by the nutrient contents of the mat compartments for both vegetation types. Mineralization is reported to be the most important source of N and P (71% and 83%) especially for systems not fed with wastewater (Koerselman and Verhoeven, 1992). High P content in the mat suggests an accumulation of this nutrient possibly because of leaching, elution, lower mixing with swamp waters and mineralization. Lower N content could be due to possible nitrification/denitrification within the rhizosphere. Denitrification in freshwater wetlands may account for as much as 0.35 g N/m^2/d (Sloey et al., 1978) which is a large portion of N removal from wetland systems. The role of this in the Nakivubo swamp is limited due to the high organic load and the lack of dissolved oxygen.

Mineralization of N in organic matter does not occur unless C:N < 20 by weight (Prentki *et al.*, 1978; Sasser, 1994; Thomann and Bayley, 1997) due to poor nutritive capacity for microbial activities. The C:N ratios of sediments and peat material for different portions of the swamp (Chapter 8) suggest that for both trapped material (from the mat) and the bottom sediments, the critical value is not exceeded, and hence mineralization would occur in the swamp.

The hydraulics of waste water flows and the chemistry of the swamp waters may impact the nutrient content of the mat and peat water through physico-chemical processes like mineralization, adsorption, diffusion and resuspension/sedimentation. These processes were thought to have a significant effect on the nutrient storage in the swamp. Net sedimentation of particulate P and N to the bottom and subsequent accretion accounted for 38% P and 20% N reduction. About 50% of the bottom sediment P content is organic (Chapter 8). Under anaerobic conditions, this P form is relatively resistant to enzyme hydrolysis (Reddy and D'Angelo, 1997) and forms an important sink in the Nakivubo swamp.

The model and overall nutrient and coliform removal efficiencies from the wastewater

It was shown that the decrease in EC within the swamp is proportional to the dilution and that EC values can be used for the assessment of the decay levels and decay rates for nutrients and pollutants. The model however only fairly represented the travel of nutrients in the swamp.

The calculated values for the decay rates however had to be adjusted in the model because the pollutant concentrations were continuously falling with time to levels much lower than those observed in the field. This could be due to the use of averaged values in the calculation of the decay rates and the average retention times of the water in the swamp. Overall however, the simulated values were within about 20% of the measured values for sections close to the inlet and within a somehow higher range in the swamp interior.

Relatively low nutrient and FC retention capacities (e.g. compared to facultative and maturation ponds, see Table 11.2) were recorded in the Nakivubo swamp. Such efficiencies for N removal (55% and 67% in the major and minor paths, respectively) and for P removal (33 and 65%, respectively for the two paths) in the water column, are not uncommon for such systems. In a constructed tidal marsh in New Jersey , TP removals were found to be 30-73% although NH_4 removal was higher at 69-98% (Guida and Kugelman, 1989). Reddy and De Busk (1987) reported 16-75% N and 12-73% TP removal in aquatic based systems. In a managed and harvested duckweed based system in Bangladesh, Alaerts *et al.* (1975) reported N and P removals of 74-77%.

There was better removal of all variables in the minor flow path dominated by papyrus, compared to the major flow path (*Miscanthidium* dominated). The factors for this difference were, *inter alia*, the lower flows and hence higher retention times in the former zone, higher suspended material load (peat) that promoted attachment and adsorption, and better nutrient uptake capacity of papyrus.

Of the nearly 890 tonnes N/yr and 75 tonnes P/yr that enter the swamp system from external sources, 384 tonnes N/yr and 45 tonnes P/yr leave the system into the lake. This is likely to cause eutrophication of the Murchison Bay water. The relatively low treatment efficiency for N and P could be attributed to overloading of the system with respect to both wastewater discharges (which promote channelization and lower retention times) and nutrients (beyond the plant uptake and sediment adsorption potentials). Retention time is considered to be a primary causative factor influencing P removal rates (Knight et al., 1987).

The balance has a possible error of $\pm 10\%$ due to the estimates made in the determination of the water balance, identification of the major flow paths, the sediment accretion rates and the averaging of fluxes from selected sites over the whole swamp.

9.6 Conclusions

The water balance and flows, the suspended matter content in the water column and the vegetation type are among the major variables that affect the wastewater purification performance in the Nakivubo swamp. Nutrient and FC removal efficiencies from the wastewater are variable and relatively low (average 57% for N, 40% for P and 91% for FC. Large quantities of pollutants therefore enter the Murchison Bay via the swamp.

About 37% N and 72% P are removed through the processes of plant uptake and sedimentation/accretion. However, nutrient recycling via leaching, decomposition and mineralization and desorption are responsible for sustaining high nutrient concentrations in the water phase. Such nutrients are released from the plants or the solid compartments of the swamp (mat, peat and sediment). The major storage compartment in the swamp is the mat (for *Miscanthidium* vegetated zones) and peat (for the papyrus vegetated zones).

The major macrophyte for wastewater purification is papyrus because of its high nutrient uptake rates, more loose and thinner mat that allows nutrients to travel to the plants from the water column, and higher SS load in the water column that promote adsorption. It is also a marketable plant that which is sustainably harvested (Chapter 10). This increases the nutrient removal potential from the swamp. However, its potential would only be realized if the flows into the swamp were controlled (to avoid overloading) to increase the retention time and the active area and if the water depth of the water column could be regulated to maximize the contact between the wastewater and the mat/plant complex. To this effect, there would be need to redistribute the water flows to minimize short-circuiting. The effect of this modification for a system that is also subjected to regular seiches is, however, doubtful.

9.7 References

Alaerts, G.J., Mahbubar, R., Kelderman, P. (1996). Performance analysis of a full-scale duckweed-covered sewage lagoon. *Wat. Res.*,30 (4): 843-852.

Armentano, V.T. (1990). Soils and ecology: Tropical wetlands. In: Williams, M. (ed.). *Wetlands: a threatened landscape.* Basil Blackwell, pp. 115-144.

Boyd, C.E., (1970). Losses of mineral Nutrients during decomposition of *Typha latifolia. Archiv fur Hydrobiologie.*, 66 (4): 511-517.

Boyt, F.l., Bayley, S.E. & Zoltek, J. Jr. (1977). Removal of nutrients from treated municipal waste water by wetland vegetation. *Journal WPCF.,* 5: 789-799.

Chale F.M.M. (1989). Nutrient release from swamp water, *C.papyrus* L. Organs and swamp sediments. *Hydrobiol. Bull.*, 232, pp.195-199.

Denny, P. (1987). Mineral cycling by wetland plants - a review. *Arch Hydrobiol. Beih.*Vol. 27, pp. 1-25.

Gaudet, J.J. (1975). Mineral Concentrations in Papyrus in various African Swamps. *J. Ecol.*, 63 (2): 483-491.

Gaudet J.J., (1980). Nutrient dynamics of papyrus swamps. In: Gopal, B. (ed.) *Proc. of first Int. Wetlands Confr.,*. National Institute of Ecology and International Scientific publications, New Delhi-India. pp.305-319.

Gehrels, J. and Mulamoottil, G. (1989). The transformation and export of phosphorus from wetlands. *Hydrological Processes*, 3:365-370.

Gopal, B., (1990). Biology and ecology. In: Patten, B.C.(Ed.). *Wetlands and shallow continental water bodies, Vol. I. Natural and Human relations*. SPB Academic Publishing bv, The Hague.

Guida, V.G., Kugelman, I.J. (1989). Experiments in Wastewater Polishing in Constructed Tidal Marshes: Does it Work? Are the Results Predictable? In: Hammer, D.A. (ed.) *Constructed Wetlands for Wastewater Treatment: Municipal, Industrial and Agricultural.* Lewis Publishers. Michigan. pp. 727-734.

Hosomi, M., Murakami, A., Sudo, R., *et al.* (1994). A four- year mass balance for a natural wetland system receiving domestic wastewater. *Water Sci. Technol.* 30 (8): 235-244.

Howard-Williams, C. (1985). Cycling and retention of nitrogen and phosphorus in wetlands: a theoretical and applied perspective. *Freshwater Biology* 15: 391-431.

Howard-Williams, C. and Gaudet, J.J. (1985). The structure and functioning of African swamps. In: Denny, P. (Ed.). The ecology and management of African wetland vegetation. Junk Publishers. Dordrecht.p 153-175.

Kadlec R.H., (1987). Nutrient dynamics in wetlands. In: Reddy K.R. & Smith W.H (eds.). *Aquatic plants for waste water treatment and resource recovery*. Magnolia, pp.393-419.

Klopatek, J.M. (1978). *Nutrient Dynamics of freshwater riverine marshes and the role of emergent macrophytes.* National Technical Information Service, Springfield , VA.

Knight, R.L., McKim, T.W., Kohl, H.R. (1987). Performance of a Natural Wetland Treatment System for Wastewater Management Water Pollution Control Federation. *JWPF.* 59 (8): 746-754.

Koerselman, W. and Verhoeven J.T.A., (1992). Nutrient dynamics in mires of various trophic status: nutrient inputs and outputs and the internal nutrient cycle. In: Verhoeven (ed.). *Fens and bogs in the Netherlands: Vegetation, history, nutrient dynamics and conservation. Geobotany 18.* Kluwer Academic publishers. Dordrecht.

Lijklema, L. (1990). Mass balances and cycling of nitrogen and phosphorus in flooded soils. In: van Oorschot M.M.P and Verhoeven J.T.A., *The Utrecht Plant Ecology News Report* no.11. Post Academic Course on wetlands for Purification of Water. pp. 44-51.

Mitsch, J.W. and Gosselink, G.J. (1993). *Wetlands*. 2nd. Ed. van Nostrand Reinhold, New York.

Mitsch, W.J., Reeder, B.C. (1991). Modelling nutrient retention of a freshwater coastal wetland: estimating the roles of primary productivity, sedimentation, resuspension and hydrology. *Ecological Modelling,* 54 (3/4): 151-187.

Moshiri, A.G. (1993). *Constructed wetlands for water quality improvement.* Lewis Publishers.

Mujobero, P. (1994). *Nitrogen dynamics in a Cyperus papyrus constructed wetland.* M.Sc. Thesis EE 150. IHE Delft, The Netherlands.

Muthuri, F.M., Jones, M.B., and Imbamba, S. K. (1989). Primary productivity of papyrus *Cyperus papyrus* in a tropical swamp, Lake Naivasha, Kenya. *Biomass,* 18: 1-14.

Muthuri, F.M. and Jones, M.B. (1997). Nutrient distribution in a papyrus swamp: Lake Naivasha, Kenya. *Aquat. Bot.,* 56: 35-50.

Nelson, L.M. (1982). Streamflow and Sediment Transport in the Quillayute River Basin, Washington USGS Open-File Report pp. 82-627.

Nyakang'o, J.B. and van Bruggen, J.J.A. (1998). Combination of a well functioning constructed wetland with a pleasing landscape design in Nairobi, Kenya. In: *Proceedings of the 6th International Conference on Wetlands Systems for Water Pollution Control.* Sao Paulo, Brazil.

Prentki, R.T., Gustafson, T.D. and Adams, M.S. (1978). Nutrient Movements in Lakeshore Marshes. In: Good, R.E., Whigham, D.F., and Simpson, R.L., (eds.). *Freshwater Wetlands: Ecological Processes and Management Potential.* Academic Press, New York. pp. 169-194.

Reddy, K. R., and DeBusk, T. A. (1987). State-of-the-Art utilisation of aquatic plants in water pollution control. *Wat. Sci. Tech* , 19 (10): 61-79.

Reddy, K.R. and D'Angelo, E.M. (1997). Biogeochemical indicators to evaluate pollutant removal efficiency in constructed wetlands. *Wat.Sci.Tech* 35 (5): 1-10.

Richardson, C.J. (1990). Wetland processes, biogeochemical cycles: regional. In: Patten, B.C. (ed.). *Wetlands and shallow continental water bodies,' Vol. I. Natural and Human relations.* SPB Academic Publishing bv. The Hague. pp. 259-280.

Richardson C.J. & Davis J.A. (1987). Natural and artificial wetland ecosystems: ecological opportunities and limitations. In: Reddy K.R. & Smith W.H (eds.). *Aquatic plants for waste water treatment and resource recovery*, pp .819-853.

Sasser Sasser, C.E., Gosselink, J.G. and Shaffer, G.P. (1991). Distribution of Nitrogen and Phosphorus in a Louisiana Freshwater Floating Marsh. *Aquat. Bot.,* 41 (4): 317-331.

Sloey, W.E., Spangler, F.L., Fetter, C.W.Jr. (1978). Management of Freshwater Wetlands for Nutrient Assimilation. In: Good, R.E., Whigham, D.F., and Simpson, R.L., (eds.). *Freshwater Wetlands: Ecological Processes and Management Potential.* Academic Press, New York, pp. 321-340.

Thompson, K., 1985. Emergent plants of permanent and seasonally-flooded wetlands. In: Denny, P. (ed.). *The ecology and management of African wetland vegetation.* Dr. W. Junk Publishers, Dordrecht, pp. 43-108.

Thormann N.M., and Bayley, E.S. (1997). Decomposition along a moderate-rich fen-marsh peatland gradient in Boreal Alberta, Canada. *Wetlands.* 17(1):123-137.

Verhoeven, J.T.A. (1986). Nutrient Dynamics in Minerotrophic Peat Mires. *Aquat. Bot.* 25 (2): 117-137.

Wetzel, G. R., 1975. *Limnology.* W.B. Saunders Company, London.

Chapter 10

Socio-economics of the Nakivubo swamp

Kansiime, F. and Nalubega, M.

10.1 Introduction

The Nakivubo swamp connects Kampala City to Lake Victoria at the Inner Murchison Bay. The swamp receives wastewater carried by the Nakivubo channel from the City which includes treated or untreated, municipal sewage, and rainfall run-off. The swamp is made up of the upper and lower swamps which are separated by the Kampala - Port Bell railway line (Fig 10.1). Initially, the swamp stretched from Bugolobi Hill up to Inner Murchison Bay, covering a total distance of about 4.8 km. With increased subsistence farming in the swamp, the total length has reduced further to about 3.0 km (Taylor, 1991).

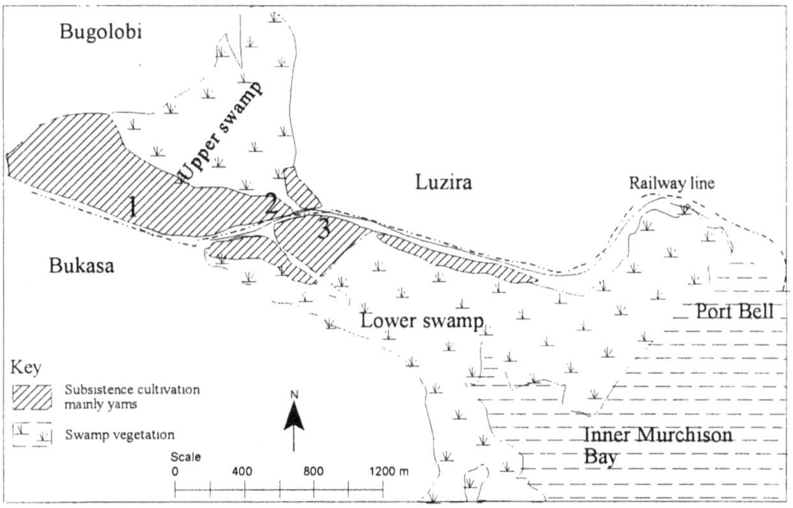

Fig. 10.1. Map of the upper and lower Nakivubo swamp showing the extent of subsistence farming. 1, 2 and 3 are study plots.

In addition to wastewater treatment and agriculture, the swamp has other anthropogenic uses. These include biomass harvesting, sand mining and brick making. The swamp is also a habitat for both fauna and flora, a source of fish and is used for cattle grazing. The harvested papyrus biomass has several social uses like thatching of houses, fencing and craft making (Table mats, baskets, carpets etc.).

Wetland plant communities may have a diversity of vertical structure (canopy, and ground cover) and horizonal diversity through the plant zones and intermixtures of open water and emergent plants. These types of structural diversity are generally correlated with increased faunal diversity and abundance (Hammer and Bastian, 1989). Animal diversity in natural wetlands includes micro-invertebrates, macro-invertebrates, fish, reptiles, amphibians, birds and mammals.

A diversity of plant structures and food types can also result in a variety of niches available for supporting animal species. Plants have different adaptations for growth and reproduction with some species producing fruits and seeds that attract animals as part of symbiotic mechanism for survival (Knight, 1997). Large stands of aquatic macrophytes provide covers for many wetland animals.

Wetlands receiving wastewater pose possible health risks to people working in them or living around them. The toxic metals and organics in treatment wetlands are of great concern. A portion of these toxic chemicals that are retained by treatment wetlands may be incorporated into biological tissues and eventually reach toxic levels through biomagnification. This may have disastrous effects at higher trophic levels (Knight, 1997). Other possible hazards include pathogens, biting insects, dangerous reptiles, and the potential for drowning.

10.2 Problem statement

Although the Nakivubo swamp has been used for tertiary wastewater treatment for more than 30 years now, this swamp has received little attention with respect to its management and conservation. Nakivubo swamp, like a number of wetlands in Uganda, is under serious pressure from human activity. The swamp is being reclaimed for agriculture. This will reduce the area covered by macrophytes and hence the effective treatment area for wastewater water discharged into the swamp. Since crops are grown in areas where wastewater flows, uptake of nutrients from the wastewater by the crops could lead to nutrient removal through harvesting. In this way, crops provide tertiary treatment of wastewater to some degree.

The faecal coliforms in Nakivubo channel range between 10^5-10^7 per 100 ml and this is far above the limit of WHO guidelines (1000 faecal coliforms / 100 ml) recommendations for the reuse of wastewater in agriculture. This puts people working in the swamp at great health risks since they come in contact with wastewater which containing vectors of diseases.

Since the Nakivubo swamp in addition to wastewater treatment has other socio-economic functions like providing habitats for fauna and flora, and it acts as a source of building material and fish, there is need to use it sustainably. Also, animals of commercial value and national importance reside in the swamp. These include, among others, the Crowned (Crested) Crane (Uganda's National bird) and Sitatunga (*Tragelaphus spekei* Sclater- also commonly known as an aquatic antelope). It is hoped that when people appreciate the multiple functions of the swamp, there will be concerted effort to preserve it. This study was carried out to highlight the major functions and uses of the Nakivubo swamp.

10.3 Objectives of the study

This study was aimed at establishing the anthropogenic activities in the swamp and more specifically to:

i. gather information on the general uses and functions of the swamp.
ii. establish the effect of agriculture on the status of the swamp and on nutrient removal from the swamp via harvesting of the crops,
iii. assess the health risks of people working in the swamp.

10.4 Approach to the study

Information was collected mainly through interviews, questionnaires and observations. The role of nutrient removal through farming in the swamp concentrated on the cocoyam (*Colocasia esculenta* (L.) schott var *esculenta*), as it is the major crop grown in the swamp. Furthermore, this plant grows in areas where the wastewater carried by Nakivubo channel flows. Concentrations of major nutrients (N and P) in the water where the plants were growing were measured. In addition, N and P in plant parts were measured and its standing biomass estimated and used to calculate the overall nutrient removal from the swamp through harvesting of this crop. Faecal coliforms were also determined (as described in chapter 6) in the water and those detached from the surface of yam tubers.

10.4.1 Nutrient removal by cocoyam

To estimate nutrient uptake and eventual nutrient removal, three plots (3 x 4 m^2 each) were secured from farmers and demarcated. One plot (plot 2) was in the Nakivubo channel flows (at the swamp inlet), whereas the other two (plots 1 and 3) were far from the channel but occasionally flooded during rainy seasons (Fig. 10.1). Weekly measurements of ammonia and orthophosphate were made on water samples collected from the plots over a period of 2 months. After this period randomly selected plants from each plot were harvested (by uprooting), separated into individual plant organs (viz. leaves, stalk, tubers and roots) and N and P were determined. The tuber is the only part of the plant that is taken out of the swamp and used as food, and it is also the plant part that was used to estimate nutrient removal from the swamp. Measurement of nutrients was done as described in Chapters 4 and 5. Average number of plants per plot was used to estimate the overall biomass in the swamp.

10.4.2 Questionnaires/interviews

Questionnaires were used to gather information about the anthropogenic uses of the swamp. Questions were posed to the respondents through an interpreter and the responses noted. People interviewed were from different areas around the swamp. In total, 56 people (54 % female) were interviewed. The information collected included: (i) the activities of people around the swamp, and whether they have any activity in the swamp or use swamp resources; such categories include people who go to the swamp to dig, harvest biomass, mine sand or make bricks, or who reside near the swamp, (ii) the health aspects of people living around the swamp or working in the swamp, including, occurrence of diseases like diarrhoea and the possible relationship of these diseases with the use of swamp water or by working in the swamp and malaria and other relevant swamp associated diseases; and (iii) the types of crops grown in the swamp and their socio-economic value.

10.4.3 Biodiversity

The biodiversity study of the swamp was restricted to birds and mammals. Observations were made with the assistance of an expert, Julius Arinaitwe of Makerere University Institute of Environment and Natural Resources. The floristic features of the swamp were covered in Chapter 2.

10.4.4 General observations

For activities in the swamp, which were either unpredictable or unsystematic, observations were made and information was collected there and then. Such activities included fishing, hunting, and harvesting of *Miscanthidium violaceum.*

10.5 Results

10.5.1 Status of the Nakivubo swamp

The 1958 map of Kampala showed that the Nakivubo swamp extended and partly surrounded Bugolobi Hill. In 1991 the remaining swamp area was estimated at 2.8 km^2 (Taylor, 1991) and by then was remaining between Kitintale and Bugolobi whereas the portion between 5th Street (Industrial area) and the Railway embankment had been converted into agricultural land through subsistence farming. Through drainage the swamp has been reduced from the original 4.8 km (in 1958) to less than 3 km length, with an area of 1.9 km^2 (this study). The farming used to be localized on the upper part of the swamp, but it is currently proceeding towards Inner Murchison Bay across the railway line (Fig 10.1).

10.5.2 Farming in the swamp

Farming in the swamp provides people with food especially those who do not own land outside the swamp. Most of the people cultivating yams are rural-urban migrants. The main crop cultivated is the cocoyam and accounts for 70 - 80% of the crops grown in the swamp. Other crops include cassava, potatoes and sugar canes. The only restriction to extensive swamp reclamation is the water in the swamp which becomes deeper towards the Inner Murchison Bay.

Farming is also existent on the Bukasa side of the swamp, but more so on the edges. As a result the rate of encroachment is not as fast as that on the upper part of the swamp towards Bugolobi. Crops on the Bukasa side of the swamp are dominated by sugar cane and cassava. Another difference regarding farming on the Bukasa edges compared to that within the swamp, is that people living on the Bukasa side own the land bordering the swamp. Here people, as seasonal opportunists, encroach on the swamp to grow crops that are not drought resistant, like potatoes, to back up their food supply during the dry season. However, there are indications of the farming on the Bukasa side beginning to extend into the swamp especially for growing yams. The Luzira side of the swamp is dominated by yams and farming is done by prisoners.

From the interviews with local people, it transpired that yams are marketable and have better yields compared to other crops like cassava and potatoes. Yams are also easy to cook, can be eaten at all meals including breakfast, making it versatile especially for low income earners who cannot afford expensive staple foods like matooke (bananas). This could be one of the reasons why most people in the swamp are growing yams. Additionally, other factors like water tolerance, elimination of weeds through shading and lower vulnerability to attack by pests and diseases, make the yam an ideal crop for cultivation in the swamp.

Yams also appear to have a higher commercial value compared to other crops like cassava. The yams may sell for up to UG Shs 500 (0.5 U$) per tuber compared to potatoes and cassava that may sell for only 0.1 U$. The only problem with yam cultivation is the intensive labour needed for the initial clearing of the swamp to remove the papyrus plants and rhizomes and the making mounds of soil where the yams are planted.

10.5.3 Nutrient removal through farming

The concentration of ammonium-nitrogen and phosphorus in water collected from the yam plots in the Nakivubo swamp and the biomass yield of the plots are presented in Table 10.1.

Table 10.1. Nutrient concentrations in water in the study plots, the Nakivubo channel and the aerial biomass. [1]

Plot no/ Location	NH_4-N (mg/l)	TRP (mg/l)	Above-ground biomass (gDw /m^2)	Below-ground biomass, tubers (gDw /m^2)
1	1.6 ± 0.05	0.66 ± 0.05	257	717
2	4.5 ± 1.1	0.08 ± 0.001	1075	5514
3	nd	nd	1011	2645
Nakivubo channel	13.6 ± 2.9	1.08 ± 0.08	-	-

[1] Values are ± standard error of the mean (n= 5). nd = not determined.

Plot 2, which was close to the Nakivubo channel had the highest concentration of nutrients in the interstitial water (in the soil where the yams are growing) and the biomass yield of yams was also high. This plot gets nutrient from wastewater percolating through the soil from the channel and overflows during rain seasons or from nutrients adsorbed to soil as the wastewater percolates through. The other two plots which were not close to the channel, had lower concentrations of nutrients in the interstitial water and also the biomass yield was low.

The N and P concentrations in plant parts are shown in figures 10.2 and 10.3 respectively.

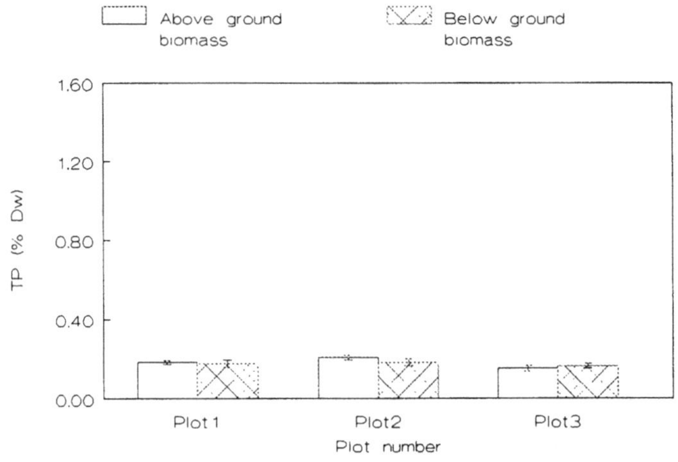

Fig. 10.2. Total phosphorus in above ground (leaves and stalk) and below ground (tuber and roots) of yams in given plots. n= 3.

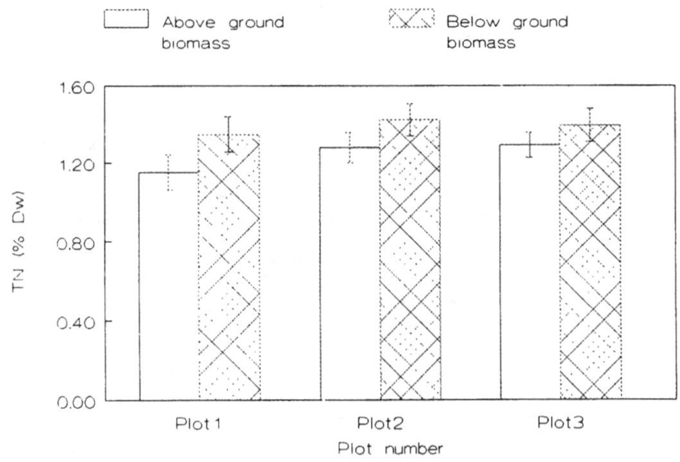

Fig. 10.3. Total nitrogen in above ground (leaves and stalk) and below ground (tuber and roots) of yams in given plots. n=3.

The concentration of nitrogen and phosphorus was more or less the same for both above and below ground biomass. However, the level of nitrogen was higher in the below ground biomass though the difference was not significant. Despite the fact that plot 2 which was close to the Nakivubo channel and had the highest biomass (Table 10.1), the nutrient concentrations in the plant parts were not different from those observed in other plots.

The tuber part of the yam which is eaten as food, is the only part of the plant that is taken out of the swamp. Usually the stalks are replanted after removing the tuber. They regenerate and produce new tubers and leaves. The average biomass density of tubers was 2958 g DW/m^2 and the total area under yam cultivation (and which in one way or another is under the influence of wastewater), was estimated at 5.65 $*10^5$ m^2. This amounts to a yam yield of 1.67* 10^9 g from the cultivated area per year. Considering the average nutrient content of 0.17 % P and 1.3% N (as dry weight), this results in the removal of 2.84*10^6 g P/yr and 2.17*10^7 g N/yr respectively. Compared to the load entering the swamp through the Nakivubo channel (Chapter 4), nutrient uptake by crops may account for 5.33 % P and 3.54 % N removal from the load into the swamp.

10.5.4 Health aspects

The numbers of faecal coliforms in the interstitial water and those detached from the tubers of the yams are presented in Fig 10.4. The numbers of coliforms detached were correlated with those measured in the interstitial water.

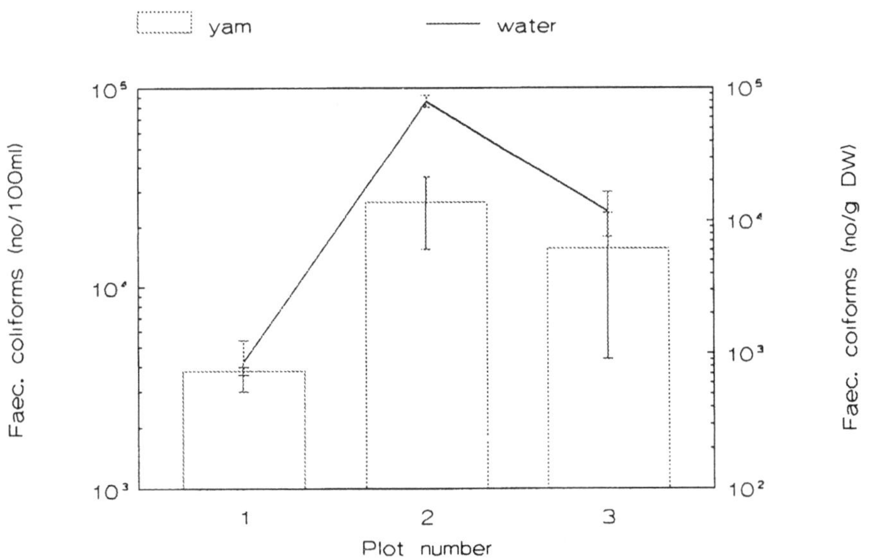

Fig. 10.4. Numbers of faecal coliforms in the interstitial water in the soil in which the yams are growing and coliforms detached from yam tubers.

It was not possible to establish the occupational health hazards of farming in the Nakivubo swamp which receives raw and secondary treated effluent. The information obtained from people working in the swamp was not conclusive. There were only 2 cases of diarrhoea reported in Bukasa Village during our survey and these people had not worked in the swamp. Of the people interviewed, and particularly those working in the swamp, none of them acknowledged having caught any of the diseases related to wastewater discharge into the swamp.

It appears that people do not get infected by eating crops grown in the swamp. This is probably because the crops are washed and cooked before eating. People are aware that Nakivubo channel water is polluted and so do not use it for domestic purposes. Most people collect water from protected springs and wells whereas others have tap connections in the neighbourhood. Most, but not all people, boil their water for drinking.

The health problem related to the swamp as acknowledged by almost all people interviewed, is malaria. It is spread by mosquitoes which use the swamp as a breeding ground. Most people, even those who do not go to the swamp, were of the opinion that the swamp should be cleared or sprayed to get rid of mosquitoes. However, clearing the swamp would produce more open water surfaces which result in more breeding grounds. The swamp also has biting flies which are prevalent early in the morning and late in the evening.

10.5.5 Sand mining and brick making in the Nakivubo swamp

Sand mining and brick making are prevalent on the edges of the swamp closer to the railway. The clay type of soil found in this area is known to be good for making bricks. The sand and bricks are sold to house constructors and the sand miners earn a living from it. Sand mining and brick making leave behind big deep holes that are dangerous to human and animal movement. The holes are also potential breeding grounds for mosquitoes as they contain stagnant water.

Another environmental problem is the firing of the bricks. This requires much fuel wood and can leave the catchment area without trees. Despite all this, it is unrealistic to recommend cessation of local brick making less until suitable substitute arrangements or alternative building material can be found. Mass cuttings of trees should be prohibited and instead afforestation should be encouraged, particularly the planting of as many trees as possible in the catchment area. Another alternative is to harvest papyrus in a sustainable way and make briquettes.

10.5.6 Papyrus harvesting

Papyrus harvesting is dominant activity in the swamp. Papyrus plants are harvested by villagers and sold for craft making, thatching houses and making fences. The practice of cutting and clearing swamp vegetation for domestic use is widespread and has been on the increase since imported building materials are expensive. Iron sheets for roofing houses, for example, has often been replaced by papyrus thatching especially for the low income earners.

The harvesting is done in rotation after every 6 - 8 months. However, this period is shorter than the 12 months or more suggested by Muthuri *et al.* (1989) and Ndyabarema, (1991). The villagers especially women harvest papyrus for fuel. On average, a person can harvest 12 bundles of papyrus each containing approx. 160 culms. From each bundle of papyrus, up to 14 carpets can be made each selling at 3000 UG Shs. This results in an income of 144, 000 UG Shs per month for at least six months. However, harvesting is only done during dry seasons as the harvested papyrus is left in the swamp to dry.

In addition to the income obtained from selling papyrus products, harvesting removes nutrients from the swamp. The area of the swamp that is usually harvested was estimated at $5.75*10^5$ m^2. With the average culm biomass of 2860 g DW /m^2, this results in a total harvest of $1.63 * 10^9$ g of papyrus culms per year. Using the nutrient content of the culm of 0.60% N and 0.13% P and considering the yearly nutrient loading into the Nakivubo swamp (Chapter 4), harvesting of papyrus culms would remove only 3 % N and 5% P per year. These nutrients are usually incorporated into plant biomass from the wastewater, through plant uptake. Up to 7% of N and 15.7% P of the annual load entering the Nakivubo swamp can be maximally removed from the swamp through papyrus harvesting (Chapter 5).

10.5.7 *Miscanthidium* harvesting

Miscanthidium was seldom observed to be taken out of the swamp. It has fewer anthropogenic uses than papyrus. In the initial part of the study (early 1993), *Miscanthidium* was occasionally harvested to make temporary housing structures for soldiers. It is more convenient than papyrus especially if it is to be used immediately to repair leakages in houses during the rainy seasons.

Miscanthidium is seasonally burnt by villagers in an attempt to flush out Sitatunga and other games that migrate to the grassy patches from nearby islands on Lake Victoria. Sitatunga and other animal like to graze on regenerating burnt *Miscanthidium* because of the delicacy of the young shoots.

In other areas of Uganda like Masaka, *Miscanthidium* is additionally used as mulch for banana and coffee plantations. In such areas where agriculture is extensive, *Miscanthidium* is sold to those who can't do the harvesting themselves.

10.5.8 Fishing

Nakivubo swamp shores (near the lake where DO > 0), support large numbers of fish, which is one the main protein source for people living near the swamp. Fishes are commonly seen at the swamp-lake interface where they come to breed. This zone is well inhabited with fish, especially Nile tilapia. The catfish and lungfish are found inside the Nakivubo swamp. Some of these fishes grow to a considerable size and are fished by the local people. The importance of Lake Victoria wetlands and the ecology of fish (Nile tilapia) has been documented (Balirwa, 1998). These wetlands were reported to be important for sheltering and feeding for the Nile tilapia especially the juvenile ones.

10.5.9 Hunting

Hunting is commonly practised by people living around the swamp. Some of the hunters come from other villages like Kauku to join the Nakivubo team. Sitatungas are commonly hunted as they come to the swamp to feed on *Miscanthidium* and crops around the swamp. The hunting is led by village elders, who are experienced and own hunting ropes. The kill is shared among hunters. Through hunting the villagers get an extra source of protein and earn some income by selling the animal skins. Most teenagers on the other hand see hunting as a mere sport especially if they own dogs.

10.5.10 Biodiversity

This biodiversity study of the Nakivubo swamp was restricted to birds and mammals. The mammals and birds of the Nakivubo swamp are largely influenced by the type of habitats and these include:

(i) the open water which is referred to as the Inner Murchison Bay and stretches from Gaba waterworks up to Port Bell. In the last seven or so years, there has been an increasing colonisation of this habitat by the water hyacinth *Eichhornia crassipes;*

(ii) the open shore which is restricted to areas near Gaba waterworks. These are shore areas that are bare or are covered with scanty vegetation;

(iii) the papyrus swamp which forms uniformly dense cover that is unbroken; the swamp is a habitat for highly specialised animal species most of which are endemic to this habitat.

The birds of the Nakivubo swamps

Table 10.2 summarises the distribution of waterbirds in the different habitats of the Nakivubo swamp.

Table 10.2. The migration status and conservation importance of waterbirds at Nakivubo swamp

Habitat	Palaearctic migrants	Others	Total	Species of global conservation concern	Species of regional conservation concern
Water hyacinth	0	4	4	0	0
Open water	3	10	13	0	1
Open shore	6	8	14	0	0
Nakivubo swamp	7	16	23	2	9
Cultivated swamp	2	3	5	0	0

Table 10.3. Birds of the Nakivubo swamp and its surroundings

Location of associated birds and their English names		
Nakivubo papyrus swamp	Inner Murchison bay	Open shores of Nakivubo swamp
African Marsh Harrier	Egyptian Goose	African Pied Wagtail
African Finfoot	Fish Eagle	Black-winged Stilt
Angola Swallow	Garganey	Cattle Egret
Basra Reed Warbler	Greater Cormorant	Goliath Heron
Black Crane	Knob-billed Duck	Hadada
Black-crowned Night Hero	Lesser Black-backed Gull	Kittliz's Plover
Blue-headed Coucal	Little Grebe	Kittliz's Plover
Common Moorhen	Long-tailed Cormorant	Little Egret
Crowned (Crested) Crane	Osprey	Little Stint
Great Reed Warbler	Pink-backed Pelican	Ringed Plover
Great White Egret	Purple Gallinule	Ruff
Gull-billed Tern	White Pelican	Sacred Ibis
Malachite Kingfisher	Yellow-billed Duck	Saddle-billed Stork
Papyrus Gonolek		Spotted Redshank
Pied Kingfisher		Yellow Wagtail
Purple Heron		Yellow-billed Duck
Rufous-bellied Heron		
Sand Martin		
Sedge Warbler		
Shoebill		
White-backed Night Heron		
White-winged Black Tern		
Yellow-backed Weaver		

Little is known about the birds in the Nakivubo swamp area; especially the waterbirds. The analysis in (Tables 10.2 & 10.3) above is a result of observations made in the swamp and by using the knowledge on bird-habitat associations.

The Nakivubo swamp is a host to migrant birds including Palaearctic migrants that breed during the northern summer in the Palaearctic region. Examples are the Sedge Warbler and Wood Sandpiper. Other migrant birds breed from elsewhere in the Afrotropical region but spend the non-breeding season at swamps like the Nakivubo swamp. Examples are Knob-billed Duck and Woolly-necked Stork. Other species are rather sedentary and breed from the swamp. The Nakivubo swamp hosts up to 16 species of waterbirds with 90% of their global population within East Africa (and whose populations are declining). Furthermore, it supports two species of waterbirds which are globally recognised as threatened viz. the Shoebill and Papyrus Gonolek (Collar *et al.*, 1994). Therefore, the importance of Nakivubo swamp as a habitat for birds is not only a national concern but is a global issue. The Nakivubo swamp also supports papyrus endemic birds. These birds live only in papyrus swamps. An example found in Nakivubo swamp is the Papyrus Gonolek.

In addition, other bird species are likely to use the Nakivubo swamp. These include swifts and other species that come to the open water to drink, grassland birds that drink from swamp edges, and others that feed from and take shelter in the swamps such as the Speckled Mousebird. The common birds of in the Nakivubo swamp and its surroundings are shown in Table 10.3.

Mammals of the Nakivubo swamp

Little is known about the mammals of Nakivubo swamp. Some of the species expected to be residing in the area are summarised in Table 10.4 below. However, most of the animals are shy and rarely seen.

Table 10.4. Some common species of mammals habiting in Nakivubo swamp

Common name	Habitat
Vervet monkey	Edges of swamps (often seen near crop plantations)
African Clawless otter Spotted-necked Otter White-tailed mongoose Marsh mongoose Banded mongoose	Open water and surrounding swamp
Striped ground squirrel Common and Lesser cane rats Bats and other small mammals Sitatunga	Swamps and surrounding dry land

10.5.11 Potential for tourism

Many species of birds and mammals found in the Nakivubo swamp are naturally shy and are more often heard than seen (e.g., most Warblers, small mammals). However, some species are spectacular and quite rare (e.g., the Shoebill and Sitatunga). These species can attract eco-tourists who can brave the swampy conditions to see such special animals. The open water on the other hand supports species that are more conspicuous such as Ducks and Jacana. This habitat is also frequented by Otters, which are declining on a global scale.

However, several reasons make tourism in Nakivubo swamp at best a difficult venture, including:

(i) the location near Luzira Prison would not appeal to visitors and it may not compliment the security of the prison,

(ii) the swamp is a repository for domestic and industrial sewerage which is a potential health hazard,
(iii) the Inner Murchison Bay is still the source of water for residents of Kampala; therefore, tourist activities in the bay are likely to cause further pollution and degradation with serious consequences.

10.6 Discussion

Drainage and clearing of the Nakivubo swamp wetland should be halted until better management strategies are put in place. Whereas this wetland could be used sustainably for agriculture, the current practice only leads to draining the swamp. The Nakivubo swamp receives wastewater from most of the developed part of Kampala City. As a result, it receives high loads of pollutants from both the industrial and domestic discharges. This is compounded by encroachment on the swamp by urbanisation and agriculture. The result of these factors is a reduction in the potential of the swamp to purify incoming wastes (Taylor, 1991).

There is a National Policy for the Conservation and Management of Wetland Resources in Uganda (Ministry of Natural Resources, 1995) which aims at curtailing the rampant loss of wetland resources and ensuring that benefits from wetlands are sustainable and equitably distributed to all people of Uganda. Despite this policy, wetlands continue to be drained. Worse still, some wetlands are currently being converted into industries and factories; for instance Nakawa and Ntinda wetlands. Although there is already serious degradation of the integrity of Nakivubo swamp, it is not too late to draw up and implement mitigation measures for the continued survival of the swamp.

The aim must be to protect the wetlands from over exploitation and provide a welcome heritage for generations to come. However, any conservation and management issues must be viewed in the context of their place in the economies of people living around the swamp. A moratorium on wetland development would be quite unacceptable; a sympathetic, coordinated development, in keeping with the ideals and aspirations of the local people, is most appropriate. One of the most effective ways of conservation and management is to demonstrate benefits which accrue to the various users of the swamp.

If the area of the swamp covered by aquatic macrophytes is reclaimed, then the effective area responsible for treating wastewater is reduced resulting in lower treatment efficiency. The possible consequence is that all pollutants discharged into the swamp would eventually reach the Inner Murchison Bay with negligible treatment. However, it should be noted that the treatment efficiency of the swamp also depends on the amount of wastewater and the concentration of pollutants (load), which should not exceed the carrying capacity of the swamp. So, to protect the swamp and use it sustainably, efforts should not only concentrate on halting reclamation but also reducing the loads of effluents / pollutants being discharged into the swamp.

Municipal wastewater is a valuable resource which should be used wherever this is possible with adequate health guidelines. The advantage of such use is the consequent reduction in environmental pollution as well as increasing agricultural production. In Nakivubo swamp, the crops grown are washed and cooked before eating, so if strict hygiene is achieved there is limited threat to public health. However, farmers must be advised to protect themselves when working in the swamp to avoid health risks associated with wastewater that is discharged into the same swamp. Faecal coliforms numbers in the interstitial water in which the yams were growing varied between $4.2*10^3$- $8.6*10^4$ no / 100ml. This is well above the 1000 faecal coliforms / 100ml recommended for wastewater reuse for agriculture (WHO, 1989). Faecal coliforms attached to yam tubers were also high varying between $7.42 *10^2$ - $1.37*10^4$ no/gDw. Despite the fact that yams are cooked before eating, the high numbers of coliforms in the interstitial water indicate that farmers may become infected through contact with the water and through handling of yams.

Most human activities in the swamp are not supportive of biodiversity. Cultivated areas supports the least number of birds (and probably other forms of biodiversity) as compared to other habitats and all those are the common ones under no immediate threat as far as can be predicted. Natural swamps on the other hand not only support many species, but include those whose immediate continued survival requires conservation effort.

People working in the swamp should be educated on the importance of using protective clothing while working in the swamp and be advised to adhere to high levels of hygiene. However, this may be difficult to achieve in the Nakivubo swamp where peoples income is low and to whom sparing money to buy boots is not a priority. Also public health is given less priority except when someone falls sick. Furthermore, the deep water in the swamp and its floating nature, would make putting on boots hazardous, as it could lead to drowning especially when the boots are filled with swamp water.

10.7 Conclusions

The Nakivubo swamp has multiple functions, ranging from agriculture, provision of raw materials, food and as habitats for fauna and flora. An integrated management and conservation plan of the Nakivubo swamp should be developed. Such plan should consider all stakeholders in the Nakivubo swamp, understand and appreciate the different functions and uses of the swamp. This plan should consider all functions holistically. A particular use of the swamp should not be done at the expense of other uses. For instance, growing crops at the margins of wetlands will not affect other benefits like wastewater treatment and biomass harvesting. Furthermore, discharging appropriate wastewater loads and may be the redistribution of wastewater at the inlet points into the swamp will not affect the biodiversity.

10.8 References

Collar, N. J. Crosby, M. J. and Stattersfield, A .J. (1994) *Birds to Watch 2: The world list of threatened birds*. BirdLife, Cambridge.

Balirwa, J. S. (1998). *Lake Victoria wetlands and the ecology of the Nile Tilapia, Oreochromis niloticus Linneé*. PhD Thesis, Wageningen Agricultural University, The Netherlands and International Institute of Infrastructural, Hydraulic and Environmental Engineering, Delft, The Netherlands.

Hammer, D. A. and Bastian, R. K. (1989). Wetland ecosystems: natural water purifiers? *In:* Hammer, D. A.(ed.). *Constructed wetlands for wastewater treatment*. Lewis Publishers, Inc. Michigan, pp 5 - 19.

Knight, R. L. (1997). Wildlife habitat and public use benefits of treatment wetlands. *Wat. Sci. Tec. 35*: 35 - 43.

Muthuri. F. M., Jones, M. B. and Imbamba, S. K. (1989). Primary productivity of papyrus (*Cyperus papyrus*) in a tropical swamp; Lake Naivasha , Kenya. *Biomass*, *18*: 1- 14.

Ministry of Natural Resources, (1995). National policy for conservation and management of wetland resources. Kampala, Uganda.

Ndyabarema, R. (1991). *An investigation into sustainable use of papyrus (Cyperus papyrus) in Nakyetema swamp, Mpigi district, Central- Uganda*. M.Sc. Thesis, Makerere University, Kampala - Uganda.

Taylor, A. R. D. (1991). *Status of Nakivubo and Luzira swamps in Kampala*. Report prepared for the Ministry of Environment Protection, Natural Wetlands Conservation and Management Programme, Kampala- Uganda.

WHO, (1989). Health guidelines for the use of wastewater in agriculture and aquaculture. Technical report series 978 WHO, Geneva.

Chapter 11

Summary and Conclusions

Nalubega, M. and Kansiime, F.

11.1 Recapitulation

Background

Increased urbanisation and industrialisation from the riparian towns of Lake Victoria in Uganda have led to higher pollutant loads. However, wastewater treatment still ranks low on the list of priorities for national/institutional budgets. Natural wetlands are thus seen as suitable alternative low cost and low maintenance systems for wastewater disposal, in the mostly unsubstantiated expectation that this also means treatment. The Nakivubo swamp found in the capital city (Kampala) has received partially treated sewage from central Kampala for more than 30 years, and stormwater for much longer. The effluent from this swamp flows into the Inner Murchison Bay, 4 km downstream of which Kampala's water supply is abstracted. Influent/effluent studies, based on pollutant concentrations (not loads) on the ability of this mostly floating swamp to polish nutrients and faecal coliforms (FC) from the wastewater have reported efficient performance (Kizito, 1986; Taylor, 1991; and Gauff, 1988). However, they quantified neither the contribution of dilution effects of seiches from the lake nor that of true removal processes within the swamp. To evaluate the present wastewater treatment performance of the Nakivubo swamp and to design an optimum management strategy for enhanced performance, the processes in the swamp interior that are responsible for the retention of nutrients and pathogens were studied.

Study Objectives and approach

The objectives of this study were therefore to:

(i) evaluate the potential of the Nakivubo swamp to polish wastewater of nutrients and pathogens in a sustainable way;

(ii) describe and quantify the pathways, transformation and budgets of nitrogen, phosphorus and pathogens; and

(iii) comment upon the sustainable use of this wetland for wastewater treatment while maintaining its ecological functions.

To achieve these objectives, quantifying the hydrological characteristics of the swamp and nutrient loads into the wetland was necessary. Water balance terms of channel discharges, rainfall, subsurface flows, evapotranspiration and seiches were measured or calculated from existing hydrometeorological data to form a water balance. Nutrients (N and P) and pathogens transformations in the swamp were studied from four transects cut across the swamp. Vertical and longitudinal profiles of nutrients and pathogens were constructed. Laboratory simulations were done to estimate nutrient fluxes into the plant and sediment compartments and the removal mechanisms of FC from the water column. In this study differences in the morphological, hydraulic, physico-chemical, floristic and overall wastewater treatment performance between areas covered by the two major vegetation types *Cyperus papyrus* L. and *Miscanthidium violaceum* Robyns (about 80% and 20% of the study area, respectively) were elucidated.

11.2 Water flows, nutrient and coliform pathways and budgets

The Nakivubo channel is the major source of wastewater, nutrients and coliforms into the Nakivubo swamp contributing about 90% of point discharges into the swamp. Owing to the combined (stormwater and sewage) nature of the flows, discharges are dependent on rainfall. The dry weather flows (DWF) average 50,000 to 60,000 m^3/d while the rainy weather flows may be as high as 500,000 m^3/d (though over short periods). The 17 months' average is about 103,575 m^3/d. Water flow is highly channelised and hydraulic retention times in the swamp during the rainy periods may be as low as 18 hours and over 5 days during the dry period. Seepage is negligible. Water quality variations within the swamp indicated that wastewater is not evenly transported to all parts of the swamp as it passes through. There was a low concentration of nutrients and low values of physical variables at the edge of the swamp attributed to less wastewater flowing into this area due to the morphology. The bulk of the water funnels through the middle part, whereas the edges of the swamp receive low wastewater flow or none at all. Two major flow paths were identified within the swamp namely: (i) the central path going through the *Miscanthidium* island and (ii) the edge paths where the water traverses papyrus vegetation. About 80% of the flow was estimated to pass through the central path.

Seiches from the lake with periods of 2¼ hours and 4 hours, amplitude up to 10 cm and exhibiting a plug flow character, further complicated flows in the swamp. The wastewater in the swamp is thus pushed backwards during the landward phase and forwards into the lake during the lakeward phase (some mixing occurs since the plug flow is not ideal). This flow pattern affects the wastewater residence time distribution in the swamp. In the wet season however, higher flows resulted into rapid flushing out of the system.

Of the two major vegetation types in the Nakivubo swamp, papyrus occupies 80% of the total swamp area. It is a rooted emergent crop near the edges of the swamp where the water level is more affected by the seasons (rainfall). It floats towards the centre and closer to the lake. The loose rhizomatous floating raft of papyrus allows for fairly free fall-through of plant debris and decomposing matter onto the sediment via the water column resulting in high suspended solids content in the underlying water. This possibly slows down the water flow somewhat in some areas. Due to the lower flows closer to the edges, a thick (up to 60 cm) layer of peaty material is formed. The loose mat facilitates vertical mixing between the interstitial mat water and the water beneath the mat during the rise and fall of water/mat levels. This leads to a less steep gradient of nutrients over the vertical profile and facilitates nutrient uptake from the water column by papyrus vegetation.

In comparison, *Miscanthidium* vegetated zones are characterised by a very thick (0.9 to 1.6 m) mat with highly interlaced roots, but low bulk density (60 - 300 kg/m^3, surface to bottom). The thick mat facilitates the retention of falling plant debris on to its surface, where low rate decomposition and further mat accretion take place. The combination of material retention onto the mat surface and high water flows beneath results into a clearer water column and a very thin peat layer (maximum 10 cm) of poorly decomposed plant material. Further, the mat structure prevents free vertical and lateral mixing of the mat water with the water column beneath. This leads to reduced interactions of the plants with wastewater in these zones, and therefore less nutrient abstraction by plants from the wastewater in these zones.

The nutrient load into the swamp was 770 $gN/m^2/yr$ and 66 $gP/m^2/yr$. Different nutrient uptake rates and above-ground plant tissue contents (N=1.3%, P=0.21% for papyrus and N=0.64% and P=0.15% for the *Miscanthidium* vegetated zones) as well as the above structural differences in flows and retention times are partly responsible for the disparate purification efficiencies between the vegetation zones. In the papyrus vegetated zones, the average removals (based on loads) were 67% N, 67% TP and 99.3% FC while in the *Miscanthidium* vegetated zones, they were lower at 55% N, 33% TP and 89.3% FC. The lower discharges (about 20%) and velocities that went through the papyrus vegetated zones enabled higher retention times for these areas. The major mechanisms of nutrient removal in papyrus vegetated zones were identified to be plant uptake for N and P, and attachment onto particulates followed by sedimentation, for FC and P.

Predation and natural die-off of FC may be high especially in the root zones where micro-aerobic zones exist (mostly in papyrus zones). In the *Miscanthidium* vegetated zones, less removal occurred due to:

a. higher flow rates and hence lower retention times (Chapter 3);
b. less vertical mixing of water and therefore less chance of attachment to the roots and particulates, due to the compact mat;
c. low SS content of the water column implying less adsorption sites for pathogens and P; and
d. lower nutrient requirements by *Miscanthidium* and therefore less nutrient demand from the wastewater;
e. lower N:P ratio (4.3 compared to 6.2 of papyrus). This could limit growth; and
f. a slower rate of organic matter decomposition from *Miscanthidium* and eventual build up of the mat.

The thick mat of *Miscanthidium* limits the number of live roots that can reach the water column to get nutrients from there. Since the bulk (80% near the lake) of the wastewater goes through this zone, then it means that the overall (swamp-wide) nutrient and pathogen removal efficiency from the wastewater is low (56% N, 40% TP and 91% FC; heavy flushing during storms probably reduces these percentages).

Very low levels of oxygen were observed in the Nakivubo swamp (and if any, very infrequently) due to the high oxygen demand exerted by decomposing organic matter in the swamp. Mostly, either hypoxic or anoxic conditions existed in most compartments of the swamp limiting nitrification despite the fact that most physical and chemical variables were in the range that would favour the survival of nitrifying bacteria. In the *Miscanthidium* mat, the low pH also possibly limited the viability and the activity of the nitrifiers in this zone.

The sharp decline in the concentration of pollutants from the swamp interface to the open waters of the Inner Murchison Bay can be explained by mixing and dilution in the lake. Combined effects of solar radiation, temperature, pH, biocides and the grazing protozoa may also be responsible for the lower FC numbers. Values of physical variables and coliforms recorded for offshore sites (2500 m from the swamp edge), were in the range reported earlier (WHO, 1970; Kizito, 1986; Gauff, 1988; Taylor, 1991).

11.3 Management of the Nakivubo swamp

11.3.1 Introduction

The wastewater discharges as well as the effective area of the swamp in its present state indicate that the system is grossly overloaded and that on its own, with the present level of human encroachment, it is unable to prevent the loading of the lake with substantial amounts of nutrients and to thoroughly protect the quality of Kampala's raw water source.

To optimally use the Nakivubo swamp as a wastewater treatment plant that can guarantee a satisfactory raw water source for Kampala while ensuring the protection of the lake's ecosystem and biodiversity of the swamp, a management plan must be set up. This plan should take into account the different users/uses of the swamp including the need for enhanced pollutant removal from the wastewater, biodiversity protection as well as resource recovery. It must be noted however that some of these interests conflict with each other and hence the uses have to be prioritised. Presently, wastewater treatment is recognised by the National Wetlands Programme ,and perhaps by the National Water and Sewerage Corporation, as the most important utility function of the Nakivubo Swamp.

11.3.2 Possible Management Strategies for the Nakivubo swamp

Scenarios 1 to 7 on the possible future alternatives for the management of the Nakivubo swamp as a wastewater recipient are presented. The order in which they appear does not reflect the preferences of the authors. **Scenario 1** is a situation where nothing is done to alter the status quo. In **Scenario 2**, reduction in the wastewater load by the improvement of the sewage collection and treatment systems is presented. **Scenario 3** considers the possibility of substituting the wetland with treatment lagoons. **Scenario 4** is a case in which the effective wastewater treatment area of the swamp is improved by minimizing channelization of the flow. An identified present threat to the swamp's integrity is subsistence farming. The cultivation of coco-yam with the accompanying draining of the swamp is on the increase. **Scenario 5** is a case for the facilitation and encouragement of the replacement of the wetland vegetation with the coco-yam.

In **Scenario 6**, the capacity of the wetland to take up nutrients while providing an alternative source of income to the farmers is presented. This would imply an extension of the influent distribution channels discussed in Scenario 1. Also, part of the upper swamp could be planted with similar trees and the water distributed over larger areas to improve nutrient retention. Figure 11.1 shows the proposed wastewater redistribution and agroforestry coverage for the lower swamp. Still, care should be taken to guard against flooding of the city. Some typical examples of trees that can be planted in the Nakivubo swamp are presented in Table 11.1. **Scenario 7** considers the possibility of altering the wastewater recipient from Kampala's water supply source to the Lake Kyoga basin which is partly found in northern Kampala.

At the end of this presentation, a discussion is given on the possible 'best' alternative (s).

Scenario 1 - No action is taken

Target/strategy

Wastewater treatment is too expensive and at present not a national priority. Nature will take care of its own.

Action(s)/requirements

None.

Implications/Outcome

Continued and perhaps increasing pollutant loads are discharged into the swamp and finally into the bay. This may lead to further proliferation of aquatic weeds including the water hyacinth and deterioration in water quality especially, with pathogens.

Increased encroachment onto the swamp area by subsistence farmers, and hence further reduction in effective treatment area. There may also be increased incidences of diseases due to increased direct contact with wastewater.

(Remark: This is an undesirable situation that should be avoided).

Scenario 2 - Improvement of wastewater collection/treatment in Kampala

Target/strategy

Wastewater nutrient loads reduced to levels that can be assimilated by the swamp (1.1 $gN/m^2/d$ and 0.07 $gP/m^2/d$ or average concentrations of 12 mgN/L and 0.8 mg P/L at the swamp inlet). This is about 50% of the present load (Chapter 9).

Action(s)/requirements

Concerted efforts by NWSC and the Uganda government to improve both the sewage collection within the city as well as its treatment. Very high costs will be incurred to improve the present sewage collection coverage from about 10% to include all premises that presently discharge untreated wastewater into the Nakivubo channel and to increase treatment from about 16,000 m^3/d to over 50,000 m^3/d. Improvement of the sewerage system although underway is still no guarantee that the wastewater presently discharged directly into the channel will all be collected. It is therefore suggested that to minimise costs, a treatment system (possibly an engineered lagoon) be put upstream, including part of the Upper Nakivubo swamp to treat the wastewater further. This system should treat all the DWF, but allow for an "overflow" such that excess water runs directly into the swamps during the storms.

With increasing populations and expansion of the city of Kampala, there is need to prioritize wastewater management if the health of the people and the general environment are to be protected.

Implications/Outcome

It is possible, in principle, to make use of part of the area presently covered by the Upper swamp (Fig 2.1) and beyond to design and construct an additional wastewater treatment plant that will treat the bulk of the wastewater from central Kampala and suburbs while minimizing costs for expanding the sewerage system. Care should however be taken to protect the city from flooding by the provision of overflow weirs and such structures as may be necessary. The proposal is very costly.

Scenario 3 - Replacement of the wetland with lagoons/stabilization ponds

Target/strategy

Replace (part of) the wetland with engineered systems that are efficient and can be controlled.

Action(s)/requirements

A designed system with barriers to keep off lake water/seiches/rises in level (e.g, can be over 2 m during high rains), and to still allow storm water to flow rapidly into the lake. A lot of dredging required on the upper part and edges.

Implications/Outcome

A very expensive venture that will also completely alter the swamp's ecosystem and perhaps not cater for the needs of the local community.

Scenario 4 - Increase the effective treatment area of the swamp

Target/strategy

To increase the nutrient removal potential of the swamp vegetation from the wastewater (need to address the present channelization and short-circuiting of wastewater in the swamp).

Action(s)/requirements

It is proposed that the wastewater flow into the lower swamp be redistributed over much of the swamp area. This could be done for example, by construction of a number of distribution channels/baffles perpendicular to the longitudinal axis of the swamp, creating a serpentine-like flow. However, the more practicable approach is to dig an open channel along the perimeter of the swamp from which laterals are constructed to direct water into the swamp at given intervals. Again, the design should be such that DWF can be distributed over the swamp but storm water can be allowed to flow through the swamp as fast as possible into the lake, to minimize the flooding of streets in the city upstream.

Sustainable management/harvesting of papyrus (9-12 months for Eastern Africa; Jones, 1991; Ndyabarema, 1991) in the swamp would be required to remove the nutrients, minimize leaching/translocation effects and ensure continued high productivity.

Implications/Outcome

This situation will increase the potential of pollutant removal from the waste water due to the increased direct contact between the wastewater and the more productive papyrus vegetation. However, as shown in Chapter 5, about 58% NH_4-N and 54 % TRP would still leave the system into the lake untreated.

Scenario 5 - Replacement of wetland vegetation with the coco-yam

Target/strategy

To replace the present indigenous swamp species with the coco-yam.

Action(s)/requirements

Let and facilitate the subsistence farmers their continued draining of the swamp and replacement of the swamp vegetation with the coco-yam.

Implications/Outcome

N and P removal by the coco-yam in the lower Nakivubo swamp approximates 38 $gN/m^2/yr$ and 5 $gP/m^2/yr$ (Chapter 10). This implies that the coco-yam would potentially remove as much as 31% N and 23% P as would papyrus, and somewhat higher proportions for *Miscanthidium*. However, the way yams are grown requires that the wastewater is drained away from the plants, leaving them to use most of their nutrients from the surrounding soils. Moreover, the yam, being a tuber crop, is not the most suitable species for this type of wastewater whose FC exceeds 1000 no./100 mL (WHO, 1989). It is therefore believed that the continued replacement of the indigenous wetland vegetation by the coco-yam compromises the swamp's major function of wastewater treatment while subjecting man to occupational health hazards. On the other hand, such direct effects of human encroachment interfere with wetland biota. Full scale cultivation into the swamp centres should therefore be stopped.

Scenario 6 - Introduction of agroforestry into the lower Swamp

Target/strategy

To provide an alternative income-generating crop for coco-yam farmers, minimize short-circuiting of the wastewater in the lower swamp and perhaps improve nutrient removal from the wastewater.

Action(s)/requirements

Agroforestry to be practised along the periphery of the swamp, while the central path is left to indigenous vegetation. A good reconnaissance survey of the acceptability of the programme by the community and the market value of the different possible tree species must be investigated. This should be coupled with a study of the potential nutrient uptake and productivity rates of the different species. The project should be a joint venture between stakeholders like the National Environmental Management Authority (NEMA), Kampala City Council (KCC), the National Water and Sewerage Corporation (NWSC) and the National Wetlands Programme.

Implications/Outcome

Potentially higher pollutant removals and provision of an alternative source of income to the subsistence farmers. The project will also introduce wetland vegetation species that can serve useful values for man, eg., medicinal, firewood and raw material for crafts. A water quality monitoring programme springing from this project will be an added advantage.

Scenario 7 - Transfer of wastewater to another recipient (L. Kyoga?)

Target/strategy

To stop worrying about the potential contamination of the city's raw water source and to let the subsistence farmers go on with their activities.

Action(s)/requirements

Complete collection of the wastewater from Kampala followed treatment and an alternative recipient of the effluent from the treatment plant. The nearest alternative is the Lake Kyoga basin.

Implications/Outcome

This may be a very expensive option and although desirable, may not be attainable in the near future. Moreover, the wastewater management problem will be only exported to other communities, not solved.

The best scenario is one that would still meet the values of the Nakivubo swamp namely protecting the water quality in the bay, flood abatement and erosion control, provision of aesthetic value, protection of wild life, food/fibre production and science. No single scenario will be wholly satisfactory. Moreover, Scenario 7 (for example) apart from being very expensive will translate the problems of poor wastewater discharge to an even poorer community. Hence in the foreseeable future, wastewater discharge into the Nakivubo swamp is likely to continue although the upgrading and expansion of the wastewater treatment plant should be encouraged too.

Efforts are made to present an integration of the above scenarios that will lead to better management of the Nakivubo swamp. In view of the limited national GDP, scenarios that are foreseen to be cheap will be given precedence.

The primary function of the Nakivubo swamp at the national and town level is the treatment of wastewater from Kampala thereby protecting water quality in the bay and the lake in general. It is emphasized however, that the swamp has limited capacity for pollutant retention and should therefore be used as a tertiary treatment plant for nutrient (and coliform) polishing. Furthermore, the Nakivubo channel and swamp complex is the main conveyor of storm water from Kampala. The two functions obviously have conflicting hydraulic retention time requirements. On the local scene however, there is increasing encroachment onto the wetland area by subsistence farmers, indicating that this function may be the priority. The best management strategy will therefore be one that will cater for the interests of the different levels satisfactorily and at affordable costs, that is to say, the interests at the different levels should be complementary.

Implementation of a managed forest buffer system combined with the maximum use of the vegetated wetland area will be an improvement to the system's performance as a wastewater treatment 'reactor'. The buffer would be planted along the edges of the swamp (Fig. 11.1). The design should however take into account the need to rapidly transport the storm water away from the city centre. An overflow weir should therefore be installed at the inlet point (railway culvert) to allow the excess waters to bypass the diversions leading into the forested zones.

Forest buffers function similarly to wetlands by serving as filters, sinks, and transformers of suspended and dissolved nutrients (Richardson, 1989). The forest buffer in the upper swamp will help to strip nutrients, retain sediments and organic matter, thus reducing the load into the lower swamp and the lake.

Advantages of trees over other vegetation types include considerable conversion of nutrients into biomass, ability to withstand sediment deposition and inundation during periods of high water levels, their deep spreading roots that can withstand erosion, ability to stimulate biological and chemical processes and draw nutrient from deep within the soil. They also produce high carbon contents which are an energy source for bacteria during nitrification, in case oxygen is present as well. Their effectiveness will depend on the species type, structural attributes, age and understory (Todd, 1995). Studies of forest buffer performance on the coastal plain of Maryland showed reductions of up to 88% of nitrate (NO_3^-) and 76% of phosphorus (P) after agricultural runoff passed through a forest buffer (Peterjohn and Correll, 1984). For the Nakivubo swamp with high nutrient loadings and different climate, there will be need to carry out investigations into the nutrient uptake rates and productivity of different wetland 'trees' (see also Table 11.1).

Table 11.1 Some of the proposed trees that could be planted in the Nakivubo swamp

Species	Uses/advantages	Disadvantages
Eucalyptus	▪ firewood ▪ timber (low-value?) ▪ wide experience in Asia with its use for sewage farms	▪ not a native species ▪ not a true swamp species ▪ perhaps little returns on annual basis
Phoenix palm and raffia palm	▪ wood ▪ mat-making ▪ can support crafts industry ▪ indigenous swamp species	▪ less food value ▪ cannot grow on floating swamps
Ficus spp.	▪ wood ▪ indigenous swamp species	▪ no food value ▪ can grow on floating swamps

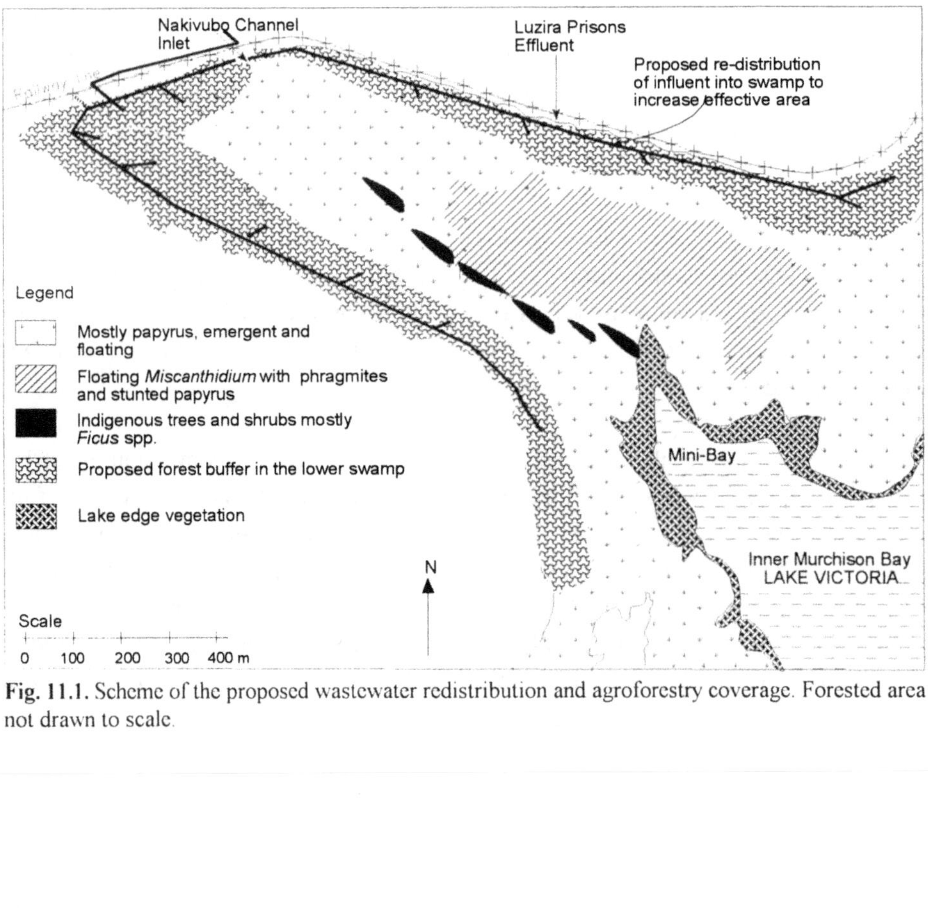

Fig. 11.1. Scheme of the proposed wastewater redistribution and agroforestry coverage. Forested area not drawn to scale.

The financial requirements of this system include costs for:

- an overflow weir;
- diversion and distribution system;
- a wetland tree species nursery;
- training of farmers/local community (for sensitization and proper management of the forest buffer); and
- setting up a water quality and forest buffer/wetland monitoring programme to optimize the systems performance.

11.4 Conclusions

As a wastewater treatment system, the Nakivubo swamp compares quite well with other systems, especially in view of the fact that it is a low cost system and that the loading rate is high (Table 11.2). In general, wastewater treatment is expensive. The cost of construction of artificial wetlands for wastewater treatment varies. Bingham (1994) while studying wetlands for stormwater treatment in urban landscapes argued that due to the great number of construction variables to be considered (e.g., excavation, planting, pretreatment or erosion control devices, land acquisition, etc.), determining an accurate cost is not realistic. However some reports have given the approximate costs of theses systems as a function of the area they cover while others have presented them as a function of the daily discharge they handle.

Reports indicate that the cost of a constructed wetland for removing sediment and nutrients vary considerably. Hammer (1993) reported costs varying from $\$1.4/m^2$ to $\$20/m^2$. Wieder *et al* (1993) gave a cost range for wetlands constructed by the Tennessee Valley Authority for the treatment of mine drainage of between $\$3.58/m^2$ and $\$32.03/m^2$. A large part of these costs went to materials and labour. Overall costs will also depend on the socio-economic status of the community. If a part of the lower swamp equivalent to the study area 1.15 km^2 is to be reverted into a constructed wetland for the present functions of sediment and nutrients removal, then the costs will be anywhere between \$1.6 million and \$23 million. This is quite costly considering that it does not include the costs of sewage conveyance. In fact from costs of 21 constructed wetland systems in the Czech Republic, Vymazal (1997) concluded that the costs (including pretreatment but excluding sewerage) were similar to those of conventional sewage treatment, although the operation and maintenance costs were 5 to 50 fold lower. Hence it is worthwhile to finance efforts that will improve the performance of this relative maintenance free and "no cost" wastewater treatment facility.

Table 11.2 Performance of different wastewater treatment systems

Contaminant	Primary Treatment	Actinated Sludge	Trickling filter	Carbon adsorption	Anaerobic ponds	Facultative ponds	Maturation ponds	Constructed Wetlands	Constructed Wetlands	Nakivubo (natural swamp)	Forest buffers	Duckweed lagoon
BOD	25-50	>50	>50	>50	30-60	50-70	20-50	90-100	73	-	-	90-95
TSS	>50	>50	>50	>50	50-70	Higher	20-30	85-100	69	-	>85	-
NH_4-N	25	>50		25-50				80-90	44	56	88, NO_3	90-99
TP	25	25-50		>50				90-100	55	40	76	77
FC		>50	25	>50	>90	>99	>99.9	-	-	91	-	-
References	← Metcalf and Eddy, 1979 →				← Veenstra *et al.* (1997) →			Hammer (1993), from various	Knight *et al.* (1993) averages for 69 sites, N America	This study	Peterjohn & Correll (1984)	Alaerts *et al.* (1997) Skillicorn *et al.* (1994)

The dominant indigenous vegetation of the swamp, namely papyrus, was shown to have a comparatively high standing crop density (Table 5.4) and high biomass production (Howard-Williams and Gaudet, 1985; in this study the highest was 5800 g/m^2 for papyrus). It therefore has a high potential for nutrient removal from aquatic systems compared to other emergent macrophytes like *Phragmites* and *Typha*. Hence actions necessary to enhance the capacity of natural wetlands with this type of vegetation to protect the aquatic environment from pollution should be sought and enhanced. Further drainage and clearing of the Nakivubo swamp wetland should be halted until better management strategies are put in place.

Within the Nakivubo swamp, nutrients were variably removed from the wastewater via plant absorption and adsorption to particulates. Some denitrification is also thought to occur, although this is limited by oxygen due to the perpetual hypoxic and anoxic conditions in the swamp. Faecal coliforms were reduced due to attachment to particulates and sedimentation, natural die-off and possibly due to predators. There was less purification in the central flow path where the vegetation consisted of a large tract (about 230,000 m^2) of floating *Miscanthidium* and associated vegetation with very poor interaction between the vegetation roots and the wastewater. It is hypothesized that the thick floating mat in *Miscanthidium* zones is a limiting factor for effective nutrient uptake from wastewater and that the nutrients taken up by these plants are from the decomposed and mineralized contents of the mat (rather than from the wastewater beneath). Moreover, lower resistance to flow led to faster flows beneath the mat leading to lower retention times and lower material sedimentation in these areas.

Papyrus' mat structure on the other hand allowed for better contact between the wastewater and the plant roots and enabled the removal of pollutants . Removal efficiencies were 67% NH_4-N, 67% TP and 99.3% FC compare to 55% NH_4-N, 33% TP and 89.3% FC in the *Miscanthidium* dominated major flow path.

If the area of the swamp covered by aquatic macrophytes is reclaimed, then the effective area responsible for treating wastewater is reduced resulting in a lower treatment efficiency. (Other values of the swamp like biodiversity, fish, waterbird habitat, *et cetera* will also be lost). The possible consequence is that all pollutants discharged into the swamp would eventually reach the Inner Murchison Bay with minimal treatment. However, it should be noted that the treatment efficiency of the swamp also depends on the amount of wastewater and the concentration of pollutants (load), which should not exceed the carrying capacity of the swamp. So, to protect the swamp and use it sustainably, efforts should not only concentrate on halting reclamation but also reducing the loads of effluents/pollutants being discharged into the swamp. Figure 11.2 is a schematized presentation of the integrated proposed activities that would help improve the watewater treatment capacity of the swamp while protecting the wetland ecosystem.

From the mass balances and the treatment efficiencies, for the swamp to be in equilibrium with respect to its capacity to trap nutrients, it would be necessary that the nutrient loading rates should not exceed 1.1 $gN/m^2/d$ and 0.07 $gP/m^2/d$ (Chapter 9). This would necessitate the reduction of the wastewater load onto the swamp by at least 50%. But this increase in wastewater treatment efficiency can be costly. Hence the creation of an additional buffer

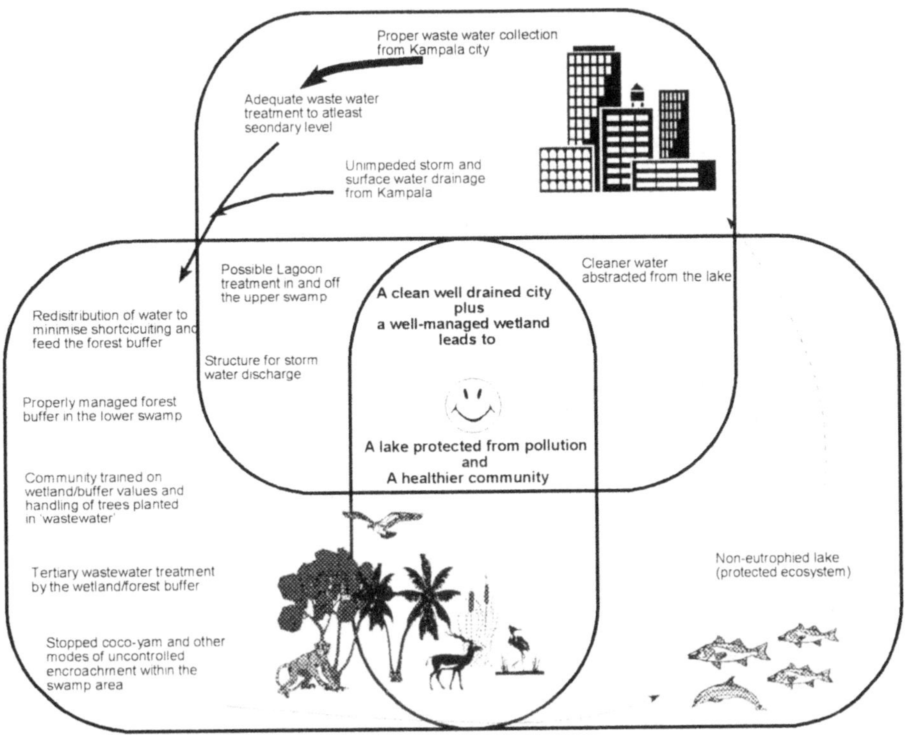

Fig. 11.2. Proposed integrated swamp management for better wastewater treatment and protection of the wetland ecosystem.

system in the form of a forest wetland with its resource and aesthetic values for the local communities, scientific values for researchers and managers and its enhancement of the water treatment capacity is should be considered. The costs for this system are most likely much lower than for alternative systems and yet the benefits will be many. The need to involve, sensitize and train the local communities before establishment of this option is emphasized.

Overall, although different processes effectively remove substantial portions of both nutrients and FC from the water, high nutrients loads reach the bay. Low concentrations in the bay can be attributed to dilution, precipitation/sedimentation, phytoplanktonic uptake and possibly natural die-off due to more sunlight (UV) in the case of coliforms. It is argued that exchange between the water in the bay and that in the main lake at the Gaba Narrows is partially responsible for the reported "purification capacity". The ratio of N:P in the Murchison Bay is more than that for fresh water lakes (Jorgensen and Vollenweider, 1988), suggesting that phosphorus may be limiting the productivity in the Inner Murchison Bay.

11.5 References

Alaerts, G. J., Mahbubar, M. D. R. and Kelderman, P. (1996). Performance analysis of a full- scale duckweed-covered sewage lagoon. *Wat. Res.*, *30*: 843 - 852.

Bingham, R.D. (1994) Wetlands for Stormwater Treatment. In: Kent, M.D., *Ed. Applied Wetlands Science and Technology*. CRC Press. pp 243-262.

Gauff, H. P. (1988). *Kampala water supply expansion project. Water quality studies.* Report prepared for the Ministry of Water and Mineral Development and National Water and Sewerage Corporation; Kampala - Uganda.

Hammer, D. A. (1993). Designing constructed wetlands systems to treat agricultural nonpoint source pollution. In: Eds. Olson, K.R. *Created and Natural Wetlands for Controlling Nonpoint Source Pollution*. CRC Press, Inc.

Higgins, M.J., Rock, C.A., Bouchard, R. and Wengrezynek, B. (1993). Controlling agricultural runoff by use of constructed wetlands. In: Moshiri, A.G Ed. *Constructed Wetlands for Water Quality Improvement*. Lewis Publishers, Inc. pp 359 - 368.

Howard-Williams, C. and Gaudet, J. J. (1985). The structure and functioning of African swamps. *In:* Denny, P. (ed.). *Ecology and management of African wetland vegetation,* Dr. Junk Publishers, Dordrecht. pp 153 - 175.

Jones, M. B. (1991). *Papyrus study*. Contract No. EARO/9363/45. Uganda Wetlands Conservation and Management Programme.

Jorgensen, S. E. and Vollenweider, R. A. (1988). *Guidelines of Lake management: Vol 1. Principles of lake management.* International Lake Environment Committee and United Nations Environment Programme.

Kizito, Y. S. (1986). *The evaluation of pollution levels in Nakivubo Channel, Kampala-Uganda.* M.Sc. Thesis, Makerere University, Uganda.

Knight, R. L., Ruble, R.W., Kadlec, R. H. and Reed, S. (1992). Wetlands for Wastewater Treatment Data Base. In: Moshiri, G. A, Ed. *Constructed Wetlands for Water Quality Improvement.* CRC Press, Inc. 35-58.

Metcalf and Eddy, Inc. (1979). *Wastewater Engineering, Treatment, Disposal and Reuse.* 2nd. Edn. Revised by Tchobanoglous. McGraw-Hill Book Company .

Ndyabarema, R. (1991). *An investigation into sustainable use of papyrus (Cyperus papyrus) in Nakyetema swamp, Mpigi district, Central- Uganda.* M.Sc. Thesis, Makerere University, Kampala - Uganda.

Peterjohn, W.T. and Correll, D.L. (1984). Nutrient dynamics in an agricultural watershed: Observations on the role of a riparian forest, *Ecology* 65(5):1466-1475.

Richardson, C.J. (1989). Freshwater wetlands: Transformers, filters, or sinks? *FOREM* 11(2), Duke Univ. School of Env. Studies, Durham, NC.

Skillicorn, P., Spira, W. and Journey, W. (1994). *A New Aquatic Farming System for Developing Countries.* The World Bank.

Taylor, A. R. D. (1991). *Status of Nakivubo and Luzira swamps in Kampala.* Report prepared for the Ministry of Environment Protection, Natural Wetlands conservation and management programme, Kampala- Uganda.

Todd, H.A. 1995. *The role and function of forest for non-point source management in the Chesapeake Bay basin.* Technical Notes - The NCSU WQ Group Newsletter. No. 70.

Veenstra, S., Alaerts, G.J. and Bijlsma, M. (1997). Technology Selection. In: Helmer , R. and Hespanhol, I. Eds. *Water Pollution Control: A guide to the use of water quality management Principles.* UNEP/WSSCC/WHO. E&FN Spon. 5-85.

Vymazal, J. (1997). Subsurface horizontal flow constructed wetlands for wastewater treatment: the Czech experience. In: Vymazal, J., Ed. *Nutrient recycling and retention in wetlands. Wetlands Ecology and Management* 4: 199-206.

Wetzel, R. G. (1993). "Constructed wetlands: scientific foundations are critical." In Constructed Wetlands for Water Quality Improvement. Ed., Gerald A. Moshiri. Boca Raton, FL. Lewis Publishers, Inc.

WHO/Resources Group London. (1970). *Master plans for Water Supply and Sewerage for greater Kampala and Jinja Areas (1970 - 2000).* Special studies Reports ``B'' Water quality;``C'' and ``D'' Lake investigations.

WHO, (1989). *Health guidelines for the use of wastewater in agriculture and aquaculture.* Technical report series 978 WHO, Geneva.

Wieder, K.R., Tchobanoglous, G. and Tuttle, W.R. (1993). Preliminary Considerations Regarding Constructed Wetlands for Wastewater Treatment. In: Hammer, A.D. Ed., *Constructed Wetlands for Wastewater Treatment: Municipal, industrial and Agricultural.* Lewis Publishers. 297 -305.

APPENDICES

APPENDIX 3.1 **Table A3.1 Sample spreadsheet for the calculation of potential open water evaporation and evapotranspiration**

Jan'95	T_{max}	T_{min}	RF	RH	SSS	U_z	T_{avg}	$P_w^{sat}(T_{avg})$	T_{dp}	$P_w^{sat}(T_{max})$	$P_w^{sat}(T_{min})$	$P_w^{sat}(T_{dp})$	R_a	R_s	R_b	R_n	Δ	φ	K.λ	λ.E_w	E_w	λ.ET	ET	ET/E_w
Date	°C	°C	mm	%	%	m/s	°C	kPa	°C	kPa	kPa	kPa	--------MJ/m²/d-------				kPa/°C			MJ/m²/d	mm/d	MJ/m²/d	(mm/d)	(-)
1	31	17.8	0	51	65	0.81	24.4	3.06	13.7	4.52	2.03	1.56	36.2	20.8	4.4	11.6	0.16	0.71	9.98	13.26	5.4	18.9	7.7	1.4
2	31	17.9	0	44	68	0.59	24.2	3.02	11.2	4.39	2.05	1.33	36.2	21.4	4.9	11.6	0.15	0.69	8.6	13.03	5.3	17.5	7.1	1.3
3	30	16.9	0	43	75	1.08	23.6	2.9	10.3	4.31	1.92	1.25	36.2	22.6	5.4	12	0.14	0.68	11.68	15.14	6.2	24.3	9.9	1.6
4	32	19	0	48	69	0.78	25.4	3.24	13.6	4.7	2.19	1.55	36.2	21.6	4.7	11.9	0.16	0.71	9.8	13.86	5.7	19.8	8.1	1.4
5	30	19.7	0	51	40	0.24	24.8	3.13	14.0	4.24	2.29	1.6	36.2	16.3	2.9	9.6	0.15	0.7	6.36	9.92	4	10.2	4.2	1
6	24	19.6	0	88	0	0.51	21.7	2.59	19.6	2.95	2.28	2.28	36.2	9.1	0.5	6.5	0.16	0.71	8.05	5.36	2.2	5.7	2.3	1.1
7	26	18.8	0	73	0	0.25	22.5	2.72	17.5	3.41	2.17	1.99	36.2	9.1	0.5	6.4	0.16	0.7	6.39	6.03	2.5	6.2	2.5	1
8	29	18.8	0	66	38	0.33	23.7	2.92	17.0	3.9	2.17	1.93	36.2	16	2.5	9.8	0.17	0.71	6.95	9.21	3.8	9.9	4	1.1
9	27	19.9	0.3	78	16	0.32	23.3	2.85	19.2	3.49	2.32	2.22	36.2	11.9	1.2	8	0.17	0.72	6.84	7.02	2.9	7.3	3	1
10	31	18.5	0.3	54	61	0.56	24.8	3.12	14.9	4.52	2.13	1.69	36.2	20.1	4	11.5	0.17	0.71	8.37	12.1	4.9	15.2	6.2	1.3
11	29	19.2	0	51	38	0.51	24.3	3.03	13.5	4.09	2.22	1.55	36.2	15.8	2.8	9.4	0.15	0.69	8.1	10.51	4.3	13.5	5.5	1.3
12	31	19.9	0	62	66	0.64	25.5	3.27	17.7	4.54	2.32	2.03	36.2	21	3.8	12.3	0.18	0.73	8.9	12.36	5	15.3	6.2	1.2
13	32	18.5	0	45	68	0.49	25.4	3.24	12.6	4.84	2.13	1.46	36.2	21.3	4.7	11.6	0.16	0.7	7.97	12.97	5.3	16.4	6.7	1.3
14	29	19.3	4.8	59	50	0.62	24	2.98	15.5	3.95	2.23	1.76	36.2	18.1	3.3	10.7	0.16	0.7	8.77	10.97	4.5	14	5.7	1.3
15	30	17.7	2	60	44	0.56	23.6	2.91	15.4	4.14	2.02	1.75	36.2	17	3	10.2	0.16	0.71	8.37	10.46	4.3	13	5.3	1.2
16	31	17.4	3.2	45	53	0.46	24.3	3.04	11.7	4.57	1.98	1.37	36.2	18.7	4	10.5	0.15	0.69	7.72	11.76	4.8	14.7	6	1.2
17	31	17.4	0	48	65	0.68	24.2	3.01	12.5	4.49	1.98	1.45	36.2	20.8	4.5	11.5	0.15	0.7	9.14	12.97	5.3	17.8	7.3	1.4
18	32	16.9	0	42	68	0.69	24.5	3.07	10.8	4.79	1.92	1.29	36.2	21.3	5	11.4	0.15	0.69	9.21	13.74	5.6	19.6	8	1.4
19	32	18.3	0	50	73	0.73	25	3.16	13.9	4.68	2.1	1.58	36.2	22.2	4.8	12.3	0.16	0.71	9.46	13.66	5.6	18.8	7.7	1.4
20	32	18.3	0	44	73	0.98	25.3	3.22	12.2	4.84	2.1	1.42	36.2	22.3	5.2	12	0.16	0.7	11.07	15.22	6.2	23.8	9.7	1.6
26	30	18	0.8	50	25	0.45	24	2.98	13.0	4.26	2.06	1.49	36.2	13.6	2.1	8.3	0.15	0.69	7.68	9.7	4	12.2	5	1.3
27	28	19	8	74	13	0.45	23.5	2.9	18.6	3.79	2.19	2.14	36.2	11.3	1.1	7.6	0.17	0.72	7.69	7.31	3	8.3	3.4	1.1
28	28	18	3	60	35	0.70	23	2.81	14.9	3.79	2.06	1.69	36.2	15.4	2.5	9.3	0.15	0.7	9.27	10	4.1	13.4	5.5	1.3
29	28	17	4.9	62	29	0.38	22.3	2.69	14.7	3.7	1.93	1.67	36.2	14.3	2.2	8.9	0.15	0.69	7.22	8.69	3.5	9.8	4	1.1
30	30	18	0	55	68	0.61	23.9	2.97	14.4	4.21	2.06	1.63	36.2	21.3	4.4	12	0.16	0.7	8.71	12.32	5	15.7	6.4	1.3

Explanation of symbols used in Table A3.1

1 $hc=4$ m
2 $z_w=(2+4)$ m
3 $z_{om}=0.492$ m
4 $z_{ov}=0.0492$ m
5 $z_p=(2+4)$
6 $d=0.67*4=2.6667$
7 The following terms were calculated according to the following equations:
8 $r_c= 26.36$ (Eq.3.8).
9 r_a (Eq.3.7).
10 $\gamma^*=(1+26.36Uz/47.98).\gamma$
11 γ = psychrometric constant (kPa/°C) = 0.0667 at 20 °C and 1kPa
12 $P_w^{sat}(T_{dp})$ (from Eq.3.10)
13 R_a (Table 3.1)
14 R_s (Eq.3.12)
15 R_b (Eq.3.13)
16 R_n (Eq.3.11)
17 Δ (Eq.3.9)
18 λ = volumetric latent heat of vaporization of water (= 2453 MJ/m^3)
19 The Penman equation: $\lambda.E_w = \phi(Rn - G) + (1-\phi)\lambda.K_e[P_w^{sat}(T) - P_w^{sat}(T_{dp})]$,
20 from which E_w was determined.
21 $K_e.\lambda = 4.82 + 6.38u$ (u = wind velocity, m/s, at 2m above ground level, K_e =
22 water vapour mass transfer coefficient (m/d.kPa)
23 λ.ET (Eq. 3.6)
24 $\lambda.E_w = (R_n - G) - \gamma\lambda K_e \Delta T$, $G \approx 0$ (MJ/m^2/d) = Conductive heat transfer
25 T = water temperature (°C)
26 The Penman estimator, ϕ, given by: $\phi = [\frac{\Delta}{\Delta + \gamma}]$

APPENDIX 3.2 Table A3.2 Summary of meteorological and evapotranspiration data for the Nakivubo swamp

	Apr	May	Jun	July	Aug	Sept	Oct	Nov	Dec	Jan	Feb	Mar	Apr	May	Jun	Jul	Aug
Count	18	30	30	30	30	30	30	30	30	31	29	31	30	31	30	31	31
Depth (m)																	
Mean	0.44	0.51	0.4	0.55	0.42	0.54	0.5	0.42	0.43	0.43	0.43	0.38	0.39	0.43	0.39	0.37	0.35
SEM	0.02	0.02	0	0.03	0.02	0.03	0.02	0.01	0.01	0.01	0.02	0	0	0.02	0.01	0.01	0.01
Minimum	0.35	0.4	0.38	0.39	0.32	0.4	0.43	0.38	0.36	0.38	0.34	0.36	0.37	0.35	0.36	0.33	0.32
Maximum	0.6	0.91	0.45	0.94	0.74	0.94	0.86	0.51	0.58	0.54	0.87	0.44	0.48	0.9	0.47	0.75	0.46
Discharge(m^3/d)																	
Mean	109392	165297	76686	198922	95152	190008	156477	89916	97219	98798	97813	60122	67453	101331	66350	52521	37317
SEM	13893	19355	3054	24964	16637	26583	12519	5689	8973	5526	15526	3290	3743	15350	4437	10625	5746
Minimum	30383	71725	59322	67591	9711	72632	100665	63456	38651	59322	22114	38651	51054	30383	42785	17980	9711
Maximum	237095	497552	117202	522357	356988	522357	456209	166812	224692	191618	464478	108933	137873	489283	133739	361122	121336
Rain (mm/d)																	
Mean	9.1	6.8	2.5	4.5	1.3	6.4	6.2	4.2	1.4	4	2.9	6.9	2.7	5.9	2.2	2.1	4.4
SEM	2.9	1.8	1.3	1.8	0.9	2.5	2.6	1.5	0.6	1.3	0.9	2.1	1.2	2.7	0.8	1.2	1.6
Minimum	0	0	0	0	0	0	0	0	0	0	0	0	0	0	0	0	0
Maximum	64.1	35.1	33.3	41.4	28.4	60.1	76.8	33	11.5	30.8	18	50.6	32	78.1	15.5	34.3	36
E_w (water, mm/d)																	
Mean	4.2	3.9	4	3.3	4.4	4.2	4.1	4	4.6	4.3	4.5	4.4	4.4	3.8	3.6	3.6	4.1
SEM	0.1	0.1	0.1	0.1	0.1	0.1	0.2	0.1	0.1	0.1	0.1	0.2	0.1	0.1	0.1	0.1	0.1
Minimum	2.4	2.2	2.2	2	2.2	2.6	2.5	2.7	2.8	2.4	2.8	2.9	2.8	2.7	2.3	2.3	2.9
Maximum	5.5	5.2	5.1	4.7	5.5	5.7	5.5	5.1	5.8	5.2	5.6	5.7	5.5	5	4.9	4.9	5.5
ET swamp(mm/d)																	
Mean	5.1	4.4	4.8	3.9	5.3	5.2	4.9	4.5	5.7	5.2	5.3	5.3	5.1	4.3	4.0	4.2	4.9
SEM	0.2	0.2	0.2	0.2	0.2	0.3	0.3	0.2	0.3	0.2	0.2	0.3	0.2	0.2	0.2	0.2	0.2
Minimum	3	2	2	2	2	3	2	3	3	2	3	3	3	2	2	2	3
Maximum	7	7	7	7	8	8	8	7	9	7	7	8	7	6	6	6	7

APPENDIX 3.3 Table A3.3 Summary of surface runoff data from Luzira and Bukasa watersheds

month	Luzira m^3/month	Bukasa m^3/month	Both (mm/month)	Rainfall, (mm/month)	% runoff
Jan 1995	13097	17312	21	29	71
Feb	3089	4083	5	8	61
Mar	96411	127438	151	170	89
Apr	157573	208284	247	273	91
May	116556	154066	183	210	87
June	42586	56291	67	75	89
July	76694	101376	120	133	90
Aug	20607	27238	32	39	83
Sept	109017	144101	171	191	90
Oct	112074	148142	176	203	87
Nov	68606	90685	108	124	87
Dec	20711	27376	32	42	77
Jan 1996	65759	86922	103	123	84
Feb	44357	58632	70	84	83
Mar	121995	161256	191	214	90
Apr	67187	88809	105	119	89
May	103776	137173	163	181	90
June	34282	45315	54	65	83
July	36310	47995	57	66	87
Aug	78329	103537	123	137	90
Mean	69451	91802	109	124	85
SEM	9488	12542	15	16	2
Min	3089	4083	5	8	61
Max	157573	208284	247	273	91

APPENDIX 3.4 Calculation of typical seiche movement and effect on flow in the swamp

General:
Time per cycle = 135 minutes, average flow depth = 1.5m, width = 400m.

Influent:
Velocity = 0.03 - 0.07 m/s
Q daily (balanced) = 108,845 m^3/d
Q (1 cycle) = 10,204 m^3
Q (½ cycle) = 5,102 m3

Seiches:
Velocity = 0.03 - 0.07 m/s
ΔH = 0.20 m/cycle
Distance per half cycle = 202.5 m
Q per width = 40.5 m^3/m
ΔQ per half cycle = 16,200 m^3

Time series:
At t=0, equilibrium

At t=1,

Q net (per half cycle)	= Q (waste) + Q (seiche backflow)
	=10,204 + 16,200 = 26,404 m^3
Distance (per half cycle)	= Q/A = 26,404/(1.5*400)
	= 44 m towards lake

At t=2, reversal of flow direction.

At t=3,

Q net (per half cycle)	= Q (waste) + Q (seiche backflow)
	=16200 - 10204 = 5996 m^3
Distance (per half cycle)	= Q/A = 5996/(1.5*400)
	= 10 m towards lake

Net distance travelled = 34 m in 1 cycle. Hence the waste water element travels from the inlet to the bay (1,200 m) in 1,200/34 cycles =35.3 cycles or 79.4 hours or approximately 3 days.

Similar calculations for a storm (assuming no return seiches) and the same travel velocities/areas as above give a one cycle distance of 70.9 *2 m, or a total of 8.46 cycles for the 1,200 m distance = 19 hours travel time.

At t=4, Reversal of flow direction.

APPENDIX 3.5 Modeling the Water Flows

APPENDIX 3.5.1 Model structure

Duflow is based on one-dimensional partial differential equations which describe unsteady flow in open channels. The equations, which are mathematical translations of the laws of conservation of mass and momentum are represented in equations 3.4.1 and 3.4.2, respectively.

$$B.\frac{\partial H}{\partial t} + \frac{\partial Q}{\partial x} = 0 \qquad (3.5.1)$$

$$\frac{\partial Q}{\partial t} + g.A\frac{\partial H}{\partial x} + \frac{\partial(\alpha.Q.v)}{\partial x} + \frac{g.|Q|.Q}{C^2.A.R} = b.\gamma.w^2.\cos(\Phi - \phi) \qquad (3.5.2)$$

t:	time [s]
x:	distance as measured along the channel axis [m]
H(x,t):	water level with respect to reference level [m]
v(x,t):	mean velocity [m/s]
Q(x,t):	discharge at location x and at time t [m^3/s]
R(x,t):	hydraulic radius of cross-section [m]
A(x,H):	cross-sectional flow area [m^2]
b(x,H):	cross-sectional flow width [m]
B(x,H):	cross-sectional storage width [m]
g:	acceleration due to gravity [m/s^2]
C(x,H):	Chezy coefficient [m^2/s]
w(t):	wind velocity [m/s]
Φ(t):	wind direction in degrees [degrees]
ϕ(x):	direction of channel axis, clockwise from N [degrees]
γ(x):	wind stress coefficient [-]
α:	correction factor for non-uniformity of the velocity distribution in the advection term, defined as equation 3.4.3, with A integrated over the cross section, A [m^2].

$$\alpha = \frac{A}{Q^2}\Sigma v(y,z)^2 dy.dz \qquad (3.4.3)$$

Equation 3.5.1 states that a change in water level at a location will be the net result of local inflow minus outflow, whereas equation 3.5.2 depicts that the net change of momentum in the system is the sum of interior and exterior forces like friction, wind and gravity.

Appendix 3.5.2 Table 3.5.1 Network definition (to be read in conjuction with Fig. 3.21).

Section	From node	To node	Section	From node	To node
1	1	2	80	80	81
2	2	3	81	81	82
3	3	4	82	82	83
4	4	5	83	83	84
5	5	6	84	84	85
11	1	41	95	85	45
12	2	42	91	81	41
14	4	43	92	82	42
15	5	44	93	83	43
16	6	45	94	84	44
40	40	41	13	3	43
41	41	42	45	45	100
42	42	43	85	85	101
43	43	44	6	6	102
44	44	45			

Orientation of network: 0.00 ° from N to Y axis

Appendix 3.5.3 Table 3.5.2 General and flow related parameters of sections

Section	Length (m)	Direction °CW fr.N	Bottom Level (m) Begin	End	Resistance(n) Pos. dir.	Neg. dir.	Windconv. (10^-6)
1	267	110.00	134.25	134.25	5.00	5.00	3.600
2	215	114.00	134.25	133.96	1.00	1.00	3.600
3	282	106.00	133.96	132.50	2.00	2.00	3.600
4	73	81.00	132.50	132.25	2.00	2.00	3.600
5	222	109.00	132.25	132.15	2.00	2.00	3.600
11	172	222.00	134.25	133.96	2.00	2.00	3.600
12	232	211.00	134.25	133.84	2.00	2.00	3.600
14	296	233.00	132.50	132.45	2.00	2.00	3.600
15	319	219.00	132.25	132.30	3.00	3.00	3.600
16	448	224.00	132.15	132.23	3.00	3.00	3.600
40	231	140.00	134.22	133.96	5.00	5.00	3.600
41	294	123.00	133.96	133.84	0.20	0.20	3.600
42	381	113.00	133.84	132.45	0.50	0.50	3.600
43	121	118.00	132.45	132.30	0.50	0.50	3.600
44	173	146.00	132.30	132.23	1.00	1.00	3.600
80	463	121.00	134.82	134.00	5.00	5.00	3.600
81	345	126.00	134.00	133.85	1.00	1.00	3.600
82	149	122.00	133.85	133.15	2.00	2.00	3.600
83	271	111.00	133.15	132.94	2.00	2.00	3.600
84	249	163.00	132.94	132.71	2.00	2.00	3.600
95	370	39.00	132.71	132.23	3.00	3.00	3.600
91	247	43.00	134.00	133.96	2.00	2.00	3.600
92	259	31.00	133.85	133.84	2.00	2.00	3.600
93	388	67.00	133.15	132.45	2.00	2.00	3.600
94	284	48.00	132.94	132.22	2.00	2.00	3.600
13	259	172.00	133.96	132.45	1.00	1.00	3.600
45	1728	122.00	132.23	132.10	5.00	5.00	3.600
85	1435	138.00	132.71	132.10	5.00	5.00	3.600
6	1574	108.00	132.15	132.10	5.00	5.00	3.600

APPENDIX 3.5.4 Table 3.5.3 System cross sections

SECTION	Depth to bott.(m)	Flow width (m) at begin	at end	Storage Width (m) at begin	at end
1	0.00	0.00	82.63	0.00	82.63
	0.99	0.00	82.63	0.00	82.63
	1.00	68.93	82.63	68.93	82.63
2	0.00	0.00	60.00	0.00	60.00
	0.28	0.00	60.00	0.00	60.00
	0.29	60.00	60.00	82.63	82.63
	1.29	60.00	60.00	82.63	82.63
3	0.00	0.00	70.50	0.00	70.50
	1.45	0.00	70.50	0.00	70.50
	1.46	60.00	70.50	82.63	82.70
	2.46	60.00	70.50	82.63	82.70
4	0.00	0.00	79.85	0.00	79.85
	0.24	0.00	79.85	0.00	79.85
	0.25	70.50	79.85	70.50	79.85
	1.25	70.50	79.85	70.50	79.85
5	0.00	0.00	141.95	0.00	141.95
	0.09	0.00	141.95	0.00	141.95
	0.10	79.85	141.95	79.85	141.95
	1.10	79.85	141.95	79.85	141.95
6	0.00	0.00	345.70	0.00	345.70
	0.29	0.00	345.70	0.00	345.70
	0.30	141.95	345.70	141.95	345.70
	1.30	141.95	345.70	141.95	345.70
11	0.00	0.00	10.00	0.00	10.00
	0.28	0.00	10.00	0.00	10.00
	0.29	10.00	10.00	10.00	10.00
	1.29	10.00	10.00	10.00	10.00
12	0.00	0.00	10.00	0.00	10.00
	0.40	0.00	10.00	0.00	10.00
	0.41	10.00	10.00	10.00	10.00
	1.41	10.00	10.00	10.00	10.00
13	0.00	0.00	10.00	0.00	10.00
	1.50	0.00	10.00	0.00	10.00
	1.51	10.00	10.00	10.00	10.00
	2.51	10.00	10.00	10.00	10.00
14	0.00	0.00	10.00	0.00	10.00
	0.04	0.00	10.00	0.00	10.00
	0.05	10.00	10.00	10.00	10.00
	1.05	10.00	10.00	10.00	10.00
15	0.00	0.00	10.00	0.00	10.00
	0.94	0.00	10.00	0.00	10.00
	0.95	10.00	10.00	10.00	10.00
16	0.00	0.00	10.00	0.00	10.00
	0.91	0.00	10.00	0.00	10.00
	0.92	10.00	10.00	10.00	10.00
40	0.00	0.00	50.00	100.00	150.00
	0.75	0.00	50.00	100.00	150.00
	0.76	9.00	150.00	230.00	280.00
	1.76	9.00	150.00	230.00	280.00
41	0.00	0.00	200.00	0.00	300.00
	0.11	0.00	200.00	0.00	300.00
	0.12	150.0	200.00	280.00	300.00
	1.12	150.0	200.00	280.00	300.00
42	0.00	0.00	500.00	0.00	617.84
	1.38	0.00	500.00	0.00	617.84
	1.39	200.00	500.00	382.82	617.84
	2.63	200.00	500.00	382.82	617.84
43	0.00	0.00	564.18	0.00	564.18
	0.14	0.00	564.18	0.00	564.18
	0.15	500.00	564.18	617.84	565.00
	1.15	500.00	564.18	617.84	565.00

44	0.00	0.00	671.37	0.00	671.37
	0.06	0.00	671.37	0.00	671.37
	0.07	564.18	671.37	564.18	671.37
	1.07	564.18	671.37	564.18	671.37
45	0.00	0.00	1500.00	0.00	1500.00
	0.12	0.00	1500.00	0.00	1500.00
	0.13	669.82	1500.00	669.82	1500.00
	1.13	669.82	1500.00	669.82	1500.00
80	0.00	0.00	219.58	0.00	219.58
	0.81	0.00	219.58	0.00	219.58
	0.82	9.00	219.58	100.00	220.00
	1.82	9.00	219.58	100.00	220.00
81	0.00	0.00	135.39	0.00	135.39
	0.14	0.00	135.39	0.00	135.39
	0.15	219.58	135.39	219.58	135.39
	1.15	219.58	135.39	219.58	135.39
82	0.00	0.00	125.51	0.00	125.51
	0.69	0.00	125.51	0.00	125.51
	0.70	135.39	125.51	135.39	125.51
	1.70	135.39	125.51	135.39	125.51
83	0.00	0.00	78.40	0.00	78.40
	0.20	0.00	78.40	0.00	78.40
	0.21	100.00	78.40	125.00	125.51
	1.21	100.00	78.40	125.00	125.51
84	0.00	0.00	152.19	0.00	152.19
	0.22	0.00	152.19	0.00	152.19
	0.23	78.40	152.19	78.40	152.19
	1.23	78.40	152.19	78.40	152.19
85	0.00	0.00	345.70	0.00	345.70
	0.60	0.00	345.70	0.00	345.70
	0.61	153.72	345.70	153.72	345.70
	1.61	153.72	345.70	153.72	345.70
91	0.00	0.00	10.00	0.00	10.00
	0.03	0.00	10.00	0.00	10.00
	0.04	10.00	10.00	10.00	10.00
	1.04	10.00	10.00	10.00	10.00
92	0.00	0.00	10.00	0.00	10.00
	0.01	0.00	10.00	0.00	10.00
	1.01	10.00	10.00	10.00	10.00
93	0.00	0.00	10.00	0.00	10.00
	0.69	0.00	10.00	0.00	10.00
	0.70	10.00	10.00	10.00	10.00
	1.70	10.00	10.00	10.00	10.00
94	0.00	0.00	10.00	0.00	10.00
	0.71	0.00	10.00	0.00	10.00
	0.72	10.00	10.00	10.00	10.00
	1.72	10.00	10.00	10.00	10.00
95	0.00	0.00	10.00	0.00	10.00
	0.47	0.00	10.00	0.00	10.00
	0.48	10.00	10.00	10.00	10.00
	1.48	10.00	10.00	10.00	10.00

APPENDIX 3.5.5 Table 3.5.4 Summary of water velocities in the swamp generated from DUFLOW for the different seasons

	MEAN CONDITIONS						*DRY CONDITIONS*						*WET CONDITIONS*					
Velocity, V	sect	sect 41	sect42	sect43	sect 44	sect 45	sect 40	sect 41	sect 42	sect 43	sect 44	sect 45	sect 40	sect 41	sect 42	sect 43	sect 44	sect 45
Mean	0.0172	0.0030	0.0007	0.0003	0.0002	0.0001	0.0106	0.0018	0.0006	0.0003	0.0002	0.0002	0.0502	0.0121	0.0032	0.0015	0.0013	0.0009
Standard Error	0.0009	0.0004	0.0004	0.0004	0.0004	0.0009	0.0009	0.0003	0.0004	0.0004	0.0004	0.0009	0.0005	0.0003	0.0004	0.0004	0.0004	0.0009
Minimum	-0.0291	-0.0194	-0.0147	-0.0207	-0.0203	-0.0235	-0.0281	-0.0115	-0.0146	-0.0208	-0.0203	-0.0235	0.0220	0.0037	-0.0093	-0.0190	-0.0199	-0.0228
Maximum	0.0718	0.0284	0.0196	0.0186	0.0177	0.0321	0.0610	0.0145	0.0186	0.0185	0.0176	0.0318	0.0741	0.0223	0.0201	0.0166	0.0157	0.0326
Count	173	173	173	173	173	173	185	185	185	185	185	185	185	185	185	185	185	185
section length X (m)	231	294	381	121	173	1728	231	294	381	121	173	1728	231	294	381	121	173	1728
*V*X*	3.9694	0.8865	0.2764	0.0343	0.0360		2.4371	0.5307	0.2144	0.0322	0.0370		11.5968	3.5622	1.2060	0.1773	0.2186	
*Sum V*X (minus lake)*	5.2026						3.2514						16.7610					
Sum X	1200						1200						1200					
*V avg = {SumV*X}/{Sum X}*	0.0043						0.0027						0.0140					
Retention time (h) = X/V avg	76.9						123.0						23.9					
Retention time (d) = X/V avg	3.2						5.1						1.0					
Velocity	sect 80	sect 81	sect 82	sect 83	sect 84	sect 85	sect 80	sect 81	sect 82	sect 83	sect 84	sect 85	sect 80	sect 81	sect82	sect 83	sect 84	sect 85
Mean	0.00004	0.00154	0.00197	0.00149	0.00084	0.00034	-0.0001	0.00101	0.0014	0.00113	0.00067	0.00035	0.00425	0.01143	0.01359	0.01106	0.00666	0.00271
Standard Error	0.00038	0.00061	0.00094	0.00105	0.00091	0.00096	0.00035	0.00057	0.0009	0.00100	0.00086	0.00091	0.00020	0.00036	0.00067	0.00085	0.00076	0.00087
Minimum	-0.0158	-0.0193	-0.0295	-0.0263	-0.0235	-0.0250	-0.0174	-0.0191	-0.0240	-0.0252	-0.0236	-0.0248	-0.0063	-0.0038	-0.0118	-0.0168	-0.0177	-0.0229
Maximum	0.01539	0.02326	0.02995	0.03199	0.02827	0.03373	0.01551	0.01762	0.0265	0.02694	0.02794	0.03297	0.01363	0.02422	0.04006	0.05750	0.03775	0.03418
Count	173	173	173	173	173	173	185	185	185	185	185	185	185	185	185	185	185	185
section length X (m)	463	385	149	271	249	1435	463	385	149	271	249	1435	463	385	149	271	249	1435
*V*X*	0.018	0.5943	0.293	0.4038	0.2089		-0.03	0.3895	0.2072	0.3075	0.1668		1.9666	4.399	2.0243	2.996	1.6579	
*Sum V*X (minus lake)*	1.5184						1.0457						13.044					
Sum X	1517						1517						1517					
*V avg = {SumV*X}/{Sum X}*	0.0010						0.0007						0.0086					
Retention time (h) = X/V avg	421.0						611.3						49.0					
Retention time (d) = X/V avg	17.5						25.5						2.0					
Velocity	sect 1	sect 2	sect 3	sect 4	sect 5	sect 6	sect 1	sect 2	sect 3	sect 4	sect 5	sect 6	sect 1	sect 2	sect 3	sect 4	sect 5	sect 6
Mean	0.00566	0.00308	0.00135	0.00057	0.00031	0.00008	0.00348	0.0019	0.0009	0.0004	0.0002	0.0002	0.0368	0.0209	0.0109	0.0052	0.0032	0.0015
Standard Error	0.0008	0.0008	0.0007	0.0006	0.0005	0.0009	0.0008	0.0008	0.0007	0.0006	0.0004	0.0008	0.0004	0.0003	0.0004	0.0004	0.0004	0.0008
Minimum	-0.0245	-0.0301	-0.0311	-0.0261	-0.0173	-0.0227	-0.0314	-0.0552	-0.0475	-0.0254	-0.0165	-0.0229	0.0246	0.0111	-0.0099	-0.0231	-0.0159	-0.0203
Maximum	0.0331	0.0283	0.0313	0.0221	0.0156	0.0304	0.0279	0.0310	0.0281	0.0212	0.0150	0.0302	0.0539	0.0300	0.0241	0.0187	0.0160	0.0311
Count	173	173	173	173	173	173	185	185	185	185	185	185	185	185	185	185	185	185
section length X (m)	267	215	282	73	222	1574	267	215	282	73	222	1574	267	215	282	73	222	1574
*V*X*	1.5119	0.6614	0.3816	0.0417	0.0686		0.9294	0.4115	0.2522	0.0300	0.0512		9.8205	4.4994	3.0700	0.3825	0.7019	
*Sum V*X (minus lake)*	2.6652						1.6744						18.474					
Sum X	1059						1059						1059					
*V avg = {SumV*X}/{Sum X}*	0.0025						0.0016						0.0174					
Retention time (h) = X/V avg	116.9						186.1						16.9					
Retention time (d) = X/V avg	4.9						7.8						0.7					

APPENDIX 8.1 Transport and resuspension

In the swamp, physical transport as effected by horizontal flows is a function of the shear forces on the surface of the sediment (represented by the drag coefficient, C_d), the particle size (d) in metres and density (ρ_s) in kg/m^3, gravitational constant (g) in m/s^2, and the density of water (ρ_w) in kg/m^3.

Assuming quiescent conditions, the terminal settling velocity of particles, v (m/s) in the system was determined using Stoke's law:

$$v = \sqrt{\frac{4}{3}\frac{gd}{C_d}\frac{(\rho_s - \rho)}{\rho}} ,$$

where $C_d = f\ (Re)$ and $Re = d\rho v/\mu$ is the Reynold's number, μ is the viscosity of water ($\approx$ 10E-3 kg/m/s at 20°C).

In the swamp, resuspension mechanisms include flow, wind-driven turbulence, bioturbation and gas ebullition. For a low velocity regime, resuspension is avoided if the water velocity in m/s is: $u \leq 8.33 * 10^{-5} * (\frac{v^{1/3}.H^{1/6}}{n.d^{2/3}})$

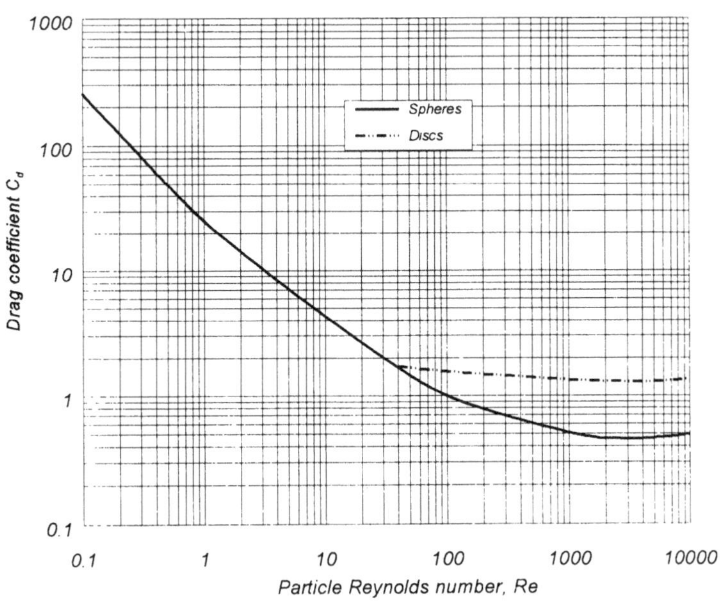

Fig. A8.1. Drag coefficients for discs and spheres as a function of the Reynolds number (From Kadlec and Knight, 1996).

for Re $\leq$ 1 and Re $\leq$ $[2H^{1/6}/(n.g^{1/2})]$ (Kadlec and Knight, 1996) with n =Manning's friction factor (Maidment, 1992) taken as 0.05 $s/m^{1/3}$ (Chapter 3, section 3.3.3) and H as the water depth =1.5m.

Worksheet for determination of particle settling velocities (V_s) in Nakivubo swamp

Using Stoke's law, with C_d, found as above:

$\rho(s)$ = 1830 kg/m^3

ρ (w)= 1000 kg/m^3 (at 20°C)

Assume spherical particles d = 75 μm (since over 90% of particles are generally > 75μm

$g=9.81\ m/s^2$ $\mu=0.001$ kg/m/s at 20 °C

Cd	Re	V_s	Re	m/d
250	0.10	1.80E-03	1.35E-01	156
200	0.14	2.02E-03	1.51E-01	174
170	0.15	2.19E-03	1.64E-01	189
130	0.20	2.50E-03	1.88E-01	216
140	0.19	2.41E-03	1.81E-01	208
150	0.18	2.33E-03	1.75E-01	201
160	0.17	2.26E-03	1.69E-01	195
165	0.17	2.22E-03	1.67E-01	192

Checks for laminar flow:

i) Re $\leq$ 1, satisfied in spreadsheet

ii) $$u \leq 8.33 * 10^{-5} * \frac{(2.22 * 10^{-3})^{1/3} * 1.5^{1/6}}{0.05 * (75 * 10)^{2/3}}$$

yields $u \leq 0.1307$ m/s.

This means that resuspension would not be expected till the water velocity exceeded 0.13 m/s.The resuspension experiment (section 8.7) showed that particles would be readily resuspended at velocities as low as 0.0045 m/s.

APPENDIX 9.1 Water quality model conditions

INITIAL CONCENTRATIONS

Node	ec	fc	n	p
	µs/cm	no/1ml	mgn/l	mgp/l
1	423.120	7578.0	12.437	1.326
2	379.410	6101.0	10.641	1.216
3	400.690	5452.9	10.695	1.325
4	376.440	3255.7	9.028	1.077
5	375.640	2922.9	8.871	1.052
6	357.310	1859.8	7.960	0.93088
40	475.120	11103.0	17.528	1.655
41	411.030	8817.6	13.291	1.287
42	391.270	4920.3	10.122	1.021
43	362.490	1969.5	7.686	0.80647
44	360.580	1470.7	7.278	0.76858
45	357.710	672.710	6.456	0.69030
80	540.820	3781.4	12.552	1.250
81	351.990	2379.8	8.115	0.80892
82	351.200	1954.3	7.793	0.78007
83	355.980	1684.0	7.661	0.76932
84	339.410	957.870	6.656	0.67169
85	306.890	320.910	5.145	0.52254
100	93.972	1.000	1.000	0.08362
101	98.045	1.000	0.81128	0.09086
102	98.620	1.000	0.90084	0.10203

QUALITY BOUNDARY CONDITIONS

Node	Condition	Type	Value
40	ec (µS/cm) N (mg/l) P (mg/L)	Constant Constant Constant	450 17 1.6
3	ec (µS/cm) N (mg/l) P (mg/L)	Constant Constant Constant	900 80 20
80	ec (µS/cm) N (mg/l) P (mg/L)	Constant Constant Constant	350 5 0.5
100	ec (µS/cm) N (mg/l) P (mg/L)	Constant Constant Constant	90 0.7 0.08
102	ec (µS/cm) N (mg/l) P (mg/L)	Constant Constant Constant	90 0.7 0.08
102	ec (µS/cm) N (mg/l) P (mg/L)	Constant Constant Constant	90 0.7 0.08

External Variables Specification

Node	d (m^2/s)	Node	d (m^2/s)
1	5	80	5
2	3	81	4
3	2	82	3
4	2	83	2
5	1	84	0.5
40	3	85	0.5
41	2	6	0.5
42	1	100	0.05
43	0.8	101	0.05
44	0.5	102	0.05
45	0.5		

Orientation of network: 0.00 ° from N to Y axis

Background values at the nodes in the lake:

EC 90 μS/cm

P 0.08 mg/l

N 0.70 mg/l

APPENDIX 9.2 Determination of Swamp Efficiency from Measurement of EC

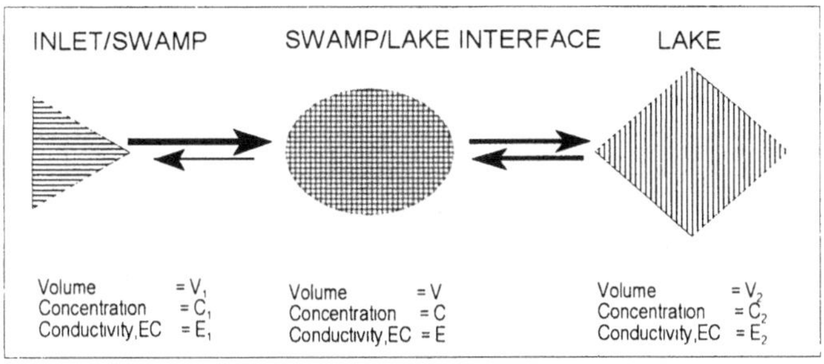

Description of terms. Volume V_1 of the polluted water mixing with volume V_2 of lake water to form a volume $V=V_1+V_2$

Major assumptions: EC is a conservative variable in the system, changes are due to dilution; complete mixing takes place. The following derivation is based on steady-state conditions.

Mass balance for a pollutant

$$dCV = \sum inflow - \sum outflow - sink \quad \text{(A9.1)}$$

Under steady state conditions, $dCV=0$, $dV=0$ and $dC=0$. The mass balance equation therefore becomes:

$$dCV = 0 = C_1V_1 + C_2V_2 - C(V_1 + V_2) - X \quad \text{(A9.2)}$$

where X are the transformations/conversions/decay within the system.

Similarly for the conservative dissolved solids EC (given as E):

$$E_1V_1 + E_2V_2 - E(V_1 + V_2) = 0 \quad \text{(A9.3)}$$

From which

$$V_2 = (E_1 - E)/(E - E_2) \quad \text{(A9.4)}$$

Giving the **dilution factor** $(V_1+V_2)/V_1$ or F_d as:

$$F_d = \frac{V_1 + V_2}{V_1} = \frac{E_1 - E_2}{E - E_2} \quad \text{(A9.5)}$$

The removal/conversion (amount X from equation A9.2) is given by:

$$X = V_1(C_1 - C) + V_2(C_2 - C) \quad \text{(A9.6)}$$

Substituting for V_2 (also $= F_dV_1-V_1$) from equation A9.5 and rearranging yields

$$X = V_1(C_1 - C_2) + V_1 F_d (C_2 - C) \quad \text{(A9.7)}$$

The **removal/purification efficiency** is, after some rearrangement equal to:

$$\frac{X}{C_1V_1} = 1 + \frac{C_2}{C_1}(F_d - 1) - \frac{C}{C_1}F_d \quad \text{(A9.8)}$$

Thus knowing the influent and effluent concentrations and EC plus the background values can give a good indication of the purification rate of water within the swamp.

Appendix 9.3 Data and calculation for the purification efficiencies over different sections of the swamp

Location	Thelake values are subtracted from all						In to lk	NH4	TRP			NH4	TRP	
	NH_4-N	NH_4-Lk	TRP	TRP-Lk	EC	EC-Lk	F_d	*X/CV*	*X/CV*		F_d	*X/CV*	*X/CV*	
Inlet	20.6	19.9	1.93	1.85	453	363	1.08	0.23	0.36	inlet to T4 400	1.11	0.23	0.10	inlet to T2 200
Inlet	13.1	12.4	1.72	1.64	427	337	0.97	0.40	0.00	inlet to T4 400	1.01	0.22	0.02	inlet to T2 200
Inlet	18.5	17.8	1.97	1.89	479	389	1.20	0.28	0.23	inlet to T4 400	1.26	0.01	-0.04	inlet to T2 200
Inlet	17.0	16.3	1.82	1.74	454	364	1.24	0.59	0.42	inlet to T4 400	1.17	0.13	-0.08	inlet to T2 200
T2 200	14.6	13.9	1.58	1.50	418	328	1.10	0.43	0.63	T2 200 to T4 200	1.04	0.28	0.15	inlet to T2 400
T2 200	10.3	9.6	1.66	1.58	423	333	1.24	0.21	0.24	T2 200 to T4 200	0.91	0.30	0.12	inlet to T2 400
T2 200	14.8	14.1	1.64	1.56	398	308	1.20	0.57	0.53	T2 200 to T4 200	1.04	0.17	0.13	inlet to T2 400
T2 200	12.9	12.2	1.68	1.60	400	310	1.05	0.32	0.73	T2 200 to T4 200	0.94	0.30	0.13	inlet to T2 400
T2 400	15.9	15.2	1.38	1.30	440	350	1.04	0.03	0.13	T2 400 to T4 400	1.21	0.56	0.67	inlet to T4 200
T2 400	12.6	11.9	1.48	1.40	462	372	1.07	0.31	-0.27	T2 400 to T4 400	1.25	0.39	0.26	inlet to T4 200
T2 400	8.9	8.2	1.42	1.34	464	374	1.15	-0.50	-0.03	T2 400 to T4 400	1.51	0.58	0.52	inlet to T4 200
T2 400	16.5	15.8	1.36	1.28	479	389	1.33	0.56	0.17	T2 400 to T4 400	1.23	0.41	0.71	inlet to T4 200
T4 200	8.0	7.3	0.54	0.46	389	299								
T4 200	6.9	6.2	1.03	0.95	359	269								Background values:
T4 200	5.9	5.2	0.66	0.58	348	258				Average removals	NH4	TRP	%	EC 90
T4 200	8.6	7.9	0.44	0.36	385	295				inlet to T4 400	37.66	25.19		TRP 0.08 mg/l
T4 400	14.8	14.1	1.15	1.07	425	335				T2 200 to T4 200	38.43	53.31		NH_4-N 0.7mg/l
T4 400	8.4	7.7	1.77	1.69	439	349				T2 400 to T4 400	9.93	0.05		
T4 400	11.5	10.8	1.28	1.20	415	325				inlet to T2 200	14.69	0.07		
T4 400	6.2	5.5	0.87	0.79	383	293				inlet to T2 400	26.15	13.52		
										inlet to T4 200	48.57	54.00		

Curriculum Vitae

Maimuna Nalubega was born on 17 July 1964 in Mulago - Kampala District, Uganda. She finished her advanced level education at Gayaza High School-Uganda, in March 1984. In October 1984, she joined Makerere University in Uganda from where she obtained a Bachelor of Science Degree in Engineering (Civil) in January 1989. Between 1989 and 1990, she worked as a Water Treatment Engineer at the Gaba Water Works for Kampala city. In 1990, she was awarded a NUFFIC scholarship to undertake a Diploma course in Sanitary Engineering at the International Institute for Infrastructural Hydraulic and Environmental Engineering (IHE) which she obtained with distinction September 1991. Thereafter she pursued her MSc. Degree in the same institute (in collaboration with the Technical University of Delft and The Biesbosch Water Reservoirs) and graduated with distinction in 1992. Her thesis was entitled 'Algae and the Direct Filtration of Biesbosch Water: Preoxidation and Scaling-up'. In October 1992, she started to pursue a PhD programme under the sponsorship of the Cooperation in Environmental Ecotechnology with Developing Countries (CEEDC) projects initiated by IHE and Wageningen Agricultural University, in The Netherlands. The research was carried out in Uganda and focused on the role of the Nakivubo swamp, at the Northern shores of Lake Victoria, in treating waste water from Kampala with emphasis on processes and implications. Since January 1993, she has worked as a lecturer in Public Health and Environmental Engineering in the Department of Civil Engineering, at Makerere University.

Curriculum Vitae

Frank Kansiime was born on 28 November 1961, in Buhweju - Busheyi District, Uganda. He finished his advanced level education at Caltec Academy in Kampala-Uganda, in 1984. In 1985, he was awarded a fellowship by the Interuniversity Exchange Council of East Africa to peruse a BSc at the University of Dar es Salaam, Tanzania. In 1988, he obtained a BSc with Education (majoring in Biology, Chemistry and Education). Thereafter he worked in the Applied Microbiology Laboratory at the same university as a research fellow where, for the first time, he demonstrated the presence of the ciliate *Nyctotherus ovalis* in the hind gut of a cockroach, *Americana periplaneta*. In 1989, he was awarded a NUFFIC sponsorship (cooperation between the Applied Microbiology Unit of the University of Dar es Salaam) to undertake an MSc. In 1991, he was awarded an MSc (by thesis alone) in Applied / Environmental Microbiology and the thesis was entitled `Growth and Maintenance of Bacteria During High-Rate Anaerobic Wastewater Treatment'. Thereafter, he joined Makerere University Institute of Environment & Natural Resources, Kampala - Uganda, as a lecturer. In September 1992, received a fellowship from the International Institute for Infrastructural Hydraulic and Environmental Engineering (IHE), to follow an MSc programme (by thesis alone) in Sanitary Engineering and his research focused on the `Role of floating *Cyperus papyrus* purifying waste water in a segmented wetland under green house conditions. After obtaining an MSc in June 1993, he started on a PhD programme under the sponsorship of the Cooperation in Environmental Ecotechnology with Developing Countries (CEEDC) projects initiated by IHE and Wageningen Agricultural University, in The Netherlands. The research was carried out in Uganda and focused on the role of the Nakivubo swamp, located at the Norther shores of Lake Victoria, in treating waste water from the City Kampala with emphasis on processes and implications.

For Product Safety Concerns and Information please contact our EU representative GPSR@taylorandfrancis.com
Taylor & Francis Verlag GmbH, Kaufingerstraße 24, 80331 München, Germany

www.ingramcontent.com/pod-product-compliance
Lightning Source LLC
LaVergne TN
LVHW080642011026
849160LV00014B/1289
* 9 7 8 9 0 5 4 1 0 4 2 0 9 *